Livestock Production and Management

NIPA® GENX ELECTRONIC RESOURCES & SOLUTIONS P. LTD.
New Delhi-110 034

www.nipaers.com
Browse, Search, Read & Buy...
Print Books
eChapters
eBooks
Publishers
Forthcoming
Subject Catalogues
New Titles
Online Resources on:
✓ Current Affairs
✓ Reasoning, Logic & Aptitude
✓ Competitive Examinations
✓ English Language Lab
✓ Writing & Pronunciation Tools
✓ Personality Development
✓ Online Programmes for Professionals
✓ Interview Preparation
Pay Using
PhonePe
Google Pay
paytm
PayPal
UPI
CC Avenue

Livestock Production and Management:
Recent Trends and Future Prospects

Editors
Sunil Kumar
B.K. Mishra

NIPA® GENX ELECTRONIC RESOURCES & SOLUTIONS P. LTD.
New Delhi-110 034

NIPA® GENX ELECTRONIC
RESOURCES & SOLUTIONS P. LTD.

101,103, Vikas Surya Plaza, CU Block
L.S.C. Market, Pitam Pura, New Delhi-110 034
Ph : +91 11 27341616, 27341717, 27341718
E-mail: newindiapublishingagency@gmail.com
web: www.nipabooks.com

For customer assistance, please contact
Phone: + 91-11-27 34 17 17
Fax: + 91-11- 27 34 16 16
E-Mail: feedbacks@nipabooks.com

ISBN: 978-81-19215-86-7

Composed and Designed by NIPA®.

KHANDELWAL COLLEGE
of Management Science & Technology
(Approved by: A.I.C.T.E., New Delhi
Affilated to G.B.T.U. Lko. & M.J.P. Rohilkhand University, Bly.)

Dr. Amresh Kumar
B.V.Sc. & A.H., M.V.Sc., Ph.D. (Ill.U.S.A.)
FNAAS, FISVS; FNAVSC.
(Ex-Dean, V.A.Sc., Pantnagar)

Foreword

Its gives me an immense pleasure to express my views about the book entitled "*Livestock Production and Management: Recent Trends and Future Prospects*" edited by Dr. Sunil Kumar and Dr. B.K. Mishra. Livestock sector is highly dynamic and livestock production is growing and evolving rapidly as a result of increasing demand for animal products. The demand for livestock products is largely driven by human population, income growth, urbanization and production responsible in different livestock. Livestock production has been growing faster than any other agricultural sub-sector and it is predicted that by 2020, it will account for more than half of the total global agricultural output in financial terms. Therefore, due emphasis needs to be given in developing strategies to improve livestock productivity. This compilation is an attempt in this direction.

Though India has world's largest dairy herds comparising of cows and buffaloes and Stands 1st in milk production, 3rd largest in egg production and 6th largest producer of the poultry meat, yet milk and meat production from our animals are of very poor quality which needs to be greatly augmented. Developments in breeding, nutrition and animal health will continue to contribute to increasing substantial production and further efficiency and genetic gains. In this compilation, the editor have tried to include the latest information on livestock improvement, be it embryo transfer technology, trends on nutrition, and alternative source of feed, bioinformatics in livestock production, diagnosis and management of diseases etc. The various chapter have been written in simple language including presentation in tabular form, which will be helpful for the students and faculty members

I would like to compliment the editors for bringing out a useful scientific publication not only for the benefit of the students but persons engaged in livestock production.

Amresh Kumar

Introduction

Animal-rearing has its origins in the transition of cultures to settled farming communities rather than hunter-gatherer lifestyles. Animals are 'domesticated' when their breeding and living conditions are controlled by humans. Over time, the collective behaviour, lifecycle, and physiology of livestock have changed radically. Many modern farm animals are unsuited to life in the wild. Dogs were domesticated in East Asia about 15,000 years ago, Goats and Sheep were domesticated around 8000 BCE in Asia. Swine or pigs were domesticated by 7000 BCE in the Middle East and China. The earliest evidence of horse domestication dates to around 4000 BCE.

Older English sources, such as the King James Version of The Bible, refer to livestock in general as "*cattle*", as opposed to the word "*deer*", which then was used for wild animals which were not owned. The word *cattle* is derived from Middle English *chatel*, which meant all kinds of movable personal property, including ofcourse livestock, which was differentiated from non-movable real-estate (real property). In later English, sometimes smaller livestock was called "*small cattle*" in that sense of movable property on land, which was not automatically bought or sold with the land. Today, the modern meaning of "*cattle*", without a qualifier, usually refers to domesticated bovines. Other species of the genus *Bos* sometimes are called wild cattle.

Livestock sector combines with animal husbandry, dairy and fisheries in three major sectors. They play an important role in the national economy as well as in the socio-economic development of our country. These three major sectors play significant role in the rural economy as supplementing family incomes and generating employment in the rural sector. After India's independence, it has realized substantial economic growth and structural change; these changes are also reflected in trends in the livestock sector which has shown considerable growth in recent decades especially since the late 1990s. There have also been major changes in the production of the Indian livestock sector.

According to 17th Census of the livestock population was 485002. It is increasing day by day due to its importance as an alternative source of income and food also. The importance of livestock in India goes beyond the function of food production. It is an important source of draught power, manure for crop production and fuel for domestic use. Thus, by minimizing use of nonrenewable energy, livestock make a positive contribution to the economic development. Livestock sector is an important source of income for the farmers and rural poor people. The growth in the livestock subsector is expected to contribute to poverty alleviation, as the livestock elements are largely concentrated among the marginal and small farmers in rural areas. Near about 70% of livestock market in India is owned by 67% of small and marginal farmers and by the land less. Livestock are an important source of income for the rural poor also. This sector contributed important share in export of Indian international trade. Recently the livestock subsectors playing very important role in poverty alleviation in rural India. The livestock sector contributed over 5.26 percent to the total GDP during 2006-07. According to estimates of the Central Statistical Organization (CSO), the value of output from livestock and fisheries sectors together at current prices was about Rs.2,82,779 crore during 2007-08 which is about 31.6 per cent of the value of the output of Rs.8,94,420 crore from agriculture & allied Sector.

India has huge livestock population with 185 million cattle, 98 million buffaloes, 124 million goats, 61 million sheep, 14 million pigs and 489 million poultry birds. Out of the total livestock in the country, around 38.2 percent are cattle, 20.2 percent are buffaloes, 12.7 percent are sheep, 25.6 percent are goats and only 2.8 percent are pigs. All other animals are less than 0.50 percent of the total livestock population. The composition of livestock population in broad groups like bovine (cattle and buffaloes), ovine (sheep and goats), pigs and poultry, however, has changed over the last two decades. Cattle population that had been increasing until 1992 has started declining and between 1992 and 2003, it declined by 9 percent. The decline in the cattle population is confined to indigenous stock that comprised 87 percent of the total cattle population in 2003. The number of indigenous cattle declined by 15 percent, while that of the crossbred increased by 62 percent. Within the indigenous stock, decline was drastic for males (22%). The main reasons for decline in indigenous cattle population are: increasing substitution of draught animals with mechanical power and low milk yield. The buffalo population has increased from 70 million in 1982 to 98 million in 2003. There has been a small decrease in total bovines in the country by 1.9% between 1997 and 2003.

We are the largest producer of milk in the world which 13.1 percent of the total milk produced in the world. Hence, India has attained the first rank in milk production in the world. At present the first five countries in the world producing maximum milk are India, USA, Russia, Germany and France. At the beginning production of milk was only 17 million tonnes (MT) in 1950-51 in India. Now it is increased to 108.5 million tonnes in 2008-09. World milk production is estimated at 693 million tonnes during 2007-08 and Indian milk production stands at 104.8

million tonnes. Despite a higher growth rate, the per capita availability of milk in India is 252 grams per day is lower than the world average 265 grams per day. This has not only placed the country on top in the world, but also represents sustained growth in the availability of milk and milk products for the growing population of the country. Concentrated dairy products such as skimmed milk continues to be the largest item of export, which together accounts for nearly 78% of net milk and milk product exports during 2007-08.

Poultry is one of the fastest growing segments of the agricultural sector in India today. Their growth rate has been rising at 8 to 10 percent per annum. As a result, India is now the world's fifth largest egg producer and the eighteenth largest producer of broilers. Eggs and broiler meat are the major end products of the poultry sector in India. The organized sector of poultry industry is contributing nearly 70% of the total output and the rest 30% in the unorganized sector in India. The Andhra Pradesh, Tamil Nadu and Maharashtra are producing nearly 70% of the country's egg production. Presently production of eggs is estimated to a number about 37 billion, that of broilers 895 million, and that of poultry meat 735,000 tonnes. Egg production has increased from 21 billion in 1990-91 to 51 billion numbers in 2006-07 and 53.5 billion numbers in 2008. India ranks fifth in the world with annual egg production of 1.61 million tonnes. Poultry exports are mostly to Maldives and Oman. Indian poultry meat products have good markets in Japan, Malaysia, Indonesia and Singapore.

India has an about 8041 Km of coastline, and about 5.70 million ha of fresh water area suitable for fisheries production. Fishing, aquaculture and allied activities are reported to have provided livelihood to over 14 million persons in 2006-07. In 2001 overall fish production was 5666 thousand tonnes, it gradually increasing continuously. At present, India's total fish production is about 76.21 thousand tonnes. This sector has great potential to export fish and fish product also. Since 1991, overall export of fish is raised at considerable rate. According to the data provided by Ministry of Agriculture and Commerce of India, the export of fish and fish product increased in last ten years. In 1991 India has exported 140 thousand tonnes of fish and fish products but till it reached to 541 thousands tonnes, its money value is Rs. 7621 crore in 2008.

The meat products industry in India is largely in the un-organized sector. With rapid urbanization, higher income levels and changes in lifestyle, market for scientifically produced and hygienically packed meat and meat products are expanding rapidly. Today, the increasing demands of the meat and meat products for in domestic and foreign market, particularly to the Gulf and West Asia and neighbouring countries. In 2003 India had a livestock population of 470 million that included 205 million cattle and 90 million buffaloes. Cattle, buffaloes, sheep and goat, pigs and poultry are the types of animals, which are generally used for production of meat.

There is a huge scope for expanding exports, especially in buffalo and poultry meat, eggs and dairy products. Slaughter rate for cattle as a whole is 20%, for buffaloes it is 41%, pigs 99%, sheep 30% and 40% for goats. The country has 3,600 slaughter houses, 9 modern abattoirs and 171 meat-processing units licensed under the meat products order. The production of meat has increased 1.9 million tonnes to 23 million tonnes from 2001 to 2007.

In meat and meat processing sector, poultry meat is the fastest growing animal protein in India. The estimated production of meat was 6.5 million tonnes during 2007-08. Per capita consumption increased from 870 grams in 2000 and expected to reach 2 Kg during 2009. According to APEDA, the export of buffalo meat was increased from 343817.08 tonnes (value Rs 1536.77crore) in 2003-04, to 483478 tonnes (Rs.3549.70 crore) in 2007-08. The export of sheep/goat meat is increased from 16820.53 tonnes (Rs 110.39 crore) in 2003-04 to 8908 tonnes (Rs.134.09 crore) in 2007-08. The processed meat export was 986.13 tonnes (Rs 7.63 crore) in 2003-04 and now it is 1245 tonnes (value Rs 12.96 crore). The export of poultry products was 415228.17 tonnes (Rs 202.40 crore) in 2003-04 it is also increased near about 1355246 tonnes (Rs 401.08 crore) in 2007-08.

Animal Husbandry sector provides large self-employment opportunities. Presuming that one family member is employed in looking after the livestock, 25 million people are estimated to be employed with the livestock rearing activity. This sector is playing very important role in the rural economy as support sector of the economy. Especially 70 million rural households primarily, small and marginal farmers and landless labourers in the country are getting employment opportunities in dairy. Dairying has become an important secondary source of income for millions of rural families.

Poultry is also another way of getting food and food security in India. Apart from food security it has provides employment to about 1.5 million people. Livestock sector not only provides essential protein and nutritious human diet through milk, eggs, meat etc but also plays an important role in utilization of non-edible agricultural by-products. Livestock also provides raw material/by products such as hides and skins, blood, bone, fat etc. This provides subsidiary occupation to a large section of the society particularly to the people living in the drought prone, hilly, tribal and other remote areas where crop production on its own may not be capable of engaging them fully. In the adverse climatic conditions and national calamities like drought, flood etc., animal husbandry practices shall be proved to be boon for sustaining the livelihood of the landless and marginal farmers in the state.

Livestock sector contributes about 6 percent to the Gross Domestic Product and 25 percent to the Agricultural Gross Domestic Product. Over the last two decades, this sector has grown at an annual rate of 5.6 percent, which is higher than the growth of agricultural sector (3.3 percent). This suggests that livestock is likely to emerge as an engine of agricultural growth in the coming decades. It is also considered as one of the potential sector for export earnings. It provides draught power and organic manure to crop sector and hides, skin, bones, blood and fibers to the industrial sector. Livestock sector also makes significant contributions towards conservation of environment. Livestock sector supplements income from crop production and other sources and absorbs income shocks due to crop failure. It generates a continuous stream of income and employment and reduces seasonality in livelihood patterns particularly of the rural poor.

Rural poverty is largely concentrated among the landless and the marginal households comprising about 70 percent of rural population. Several empirical studies indicate that livestock rearing has significant positive impact on equity in terms of income and employment and poverty reduction in rural areas as distribution of livestock is more egalitarian compared to. In our country more than 70 percent of the rural households have own livestock and majority of livestock owning households are small, marginal and landless households. Small animals like sheep, goats, pigs and poultry are largely kept by the land scarce poor households for commercial purposes because of their low initial investment and operational costs. This paper analyses the development of livestock sector in terms of population, production, trade and employment on one hand and the role of livestock sector in reducing rural poverty on the other hand.

There has been a tremendous increase in the production of livestock products and this is expected to continue in the coming future. This is especially in developing countries. The greatest increase is in the production of poultry and pigs, as well as eggs and milk. However, this overall increase obscures the fact that the increased supply is restricted to certain countries and regions, and is not for most of the countries. Although there is over consumption of meat in developed regions, there is clearly a need to promote the development of livestock production and the consumption of livestock products in many developing countries, to improve physical and mental health and fitness. Livestock production can make good use of resources, some of which may otherwise not be used, and contributes high quality protein and important micronutrients to the human diet.

The present edited volume consisting of twenty papers each has been supported with necessary tables, figures and references at the end of the chapter. We hope this book will be valued for teachers, students, researchers, extension specialist, policy makers and for those who are working in Krishi Vigyan Kendras and other related Non-Government-Organizations.

Sunil Kumar
B.K. Mishra

Acknowledgements

We owe our sincere gratitude to Vice-chancellor, GB Pant University of Agriculture and Technology, Pantnagar who has kindly consented to allow us to edit this book.

With great respect, immense pleasure and esteemed reverence, we express our thanks to Dr. PS Bisht, Dean, College of Forestry and Hill Agriculture, Hill Campus, Ranichauri for his critical and continuous guidance throughout shaping the manuscript and deserve sincere thanks for his blessing and to write foreword of the book.

The moral and academic appreciation also received from Dr R.B. Prasad, Ex Professor, Animal Genetics and Breeding, College of Veterinary and Animal Science, GBPUA&T, Pantnagar. We would never have been able to finish the assignment without the guidance of Dr RB Verma, Dr VK Yadav, Dr SK Sharma.

We owe our thanks to all the contributors of the book who have submitted the valuable information on various aspects of Livestock Production and encourage to publish this book.

Editors wish to record a word of appreciation to New India Publishing Agency, New Delhi for accepting the manuscript and shaping this book in a beautiful manner. All the words in the lexicon will be futile and meaningless if we fail to express our reverence to family members and Mr Abhishek Bahuguna for their blessing, affection, sacrifice and cheerful cooperation to overcome the hurdle.

Sunil Kumar
B.K. Mishra

List of Contributors

Abed M Al-Bial
Department of A.H. & Dairying
Institute of Agricultural Sciences
Banaras Hindu University
Varanasi – 221005

Ajay Kumar Upadhyay
Veterinary Public Health and Epidemiology
College of Veterinary and Animal Sciences
G. B. Pant University of Agri. & Tech.
Pantnagar – 263 145 (Uttarakhand)

B.K. Mishra
Department of RDAP,
North-Eastern Hill University
Tura Campus,Tura – 794 002 (Meghalaya)

Bharat Bhushan
(Animal Genetics & Breeding) & Incharge
Livestock Production and Management Section
Indian Veterinary Research Institute
Izatnagar – 243 122 (Bareilly)

Brijesh Singh
Department of Genetics and Animal Breeding
College of Veterinary and Animal Sciences
G.B. Pant University of Agriculture and
Technology Pantnagar – 263 145
(Uttarakhand)

D.P. Singh
Department of A.H. & Dairying
Institute of Agricultural Sciences
Banaras Hindu University
Varanasi – 221 005

Dhirendra Kumar
Division of Animal Genetics & Breeding
FVSc & AH, SKUAST-J
(J&K) – 181 102

G K Gaur
(Animal Genetics & Breeding) & Incharge
Cattle and Buffalo Farm
Indian Veterinary Research Institute
Izatnagar – 243 122, Bareilly
Uttar Pradesh

Gyanendra Singh
Division of Physiology and Climatology
Indian Veterinary Research Institute
Izatnagar, Bareilly – 243 122
Uttar Pradesh

Jai Singh
Department of A.H. and Dairying
Institute of Agriculture Science
Banaras Hindu University
Varanasi – 221 005
Uttar Pradesh

Lalit Tiwari
Homeopathic Pharmacopoeia Laboratory
Kamla Nehru Nagar
Ghaziabad, Uttar Pradesh

Manish Kumar
Veterinary Shrgem
Department of Animal Husbandry & Dairying
Haryana

Manjit Panigrahi
Scientist (Animal Genetics & Breeding)
Cattle and Buffalo Farm (IVRI)
Izatnagar – 243 122, Bareilly
Uttar Pradesh

Mihir Sarkar
Division of Physiology and Climatology
Indian Veterinary Research Institute
Izatnagar, Bareilly – 243 122
Uttar Pradesh

Nitin Rai
Department of Botany
M.M.H. College
Ghaziabad – 201 001 (Uttar Pradesh)

P C Pande
Department of Botany
S.S.J. Campus, Kumaun University
Almora, Uttarakhand

Prabhat Tripathi
Division of Nutrition
Feed Resources and Product Technology
Central Institute for Research on Goats
Makhdoom, Mathura – 281 122 (U.P.)

Pranav Kumar
Division of AHE, Faculty of Veterinary Science & Animal Husbandry
Sher-e-Kashmir University of Agricultural Sciences and Technology of Jammu
RS Pura, Jammu-181102 (J&K)

Rajeev Kumar Sharma
Pharmacopoeial Laboratory for Indian Medicine Ghaziabad – 201 002
Uttar Pradesh

Rajendra Kumar Pandey
Ex-Head, Department of Animal Husbandry & Dairying
Director ICAR Winter and Summer School
Banaras Hindu University
Varanasi –221005 (Uttar Pradesh)

Rajesh Kumar Sharma
Department of Instructional Livestock Farm
College of Veterinary and Animal Sciences
CSK Himachal Pradesh Agricultural University Palampur, Himachal Pradesh

Rajinder Singh Kishtwaria
Department of Instructional Livestock Farm
College of Veterinary and Animal Sciences
CSK Himachal Pradesh Agricultural University Palampur, Himachal Pradesh

Ram Niwas
Department of A.H. & Dairying
Institute of Agricultural Sciences
Banaras Hindu University
Varanasi – 221 005 (Uttar Pradeh)

Ran Vir Singh
National Project on Organic Farming
Animal Genetics Division, IVRI
Izatnagar – 243 122
Uttar Pradesh

Ravindra Kumar Tiwari
SMS, KVK (ICAR-VPKAS)
Chinyalisaur, Uttarkashi
Uttarakhand

Ravindra Kumar
Division of Nutrition
Feed Resources and Product Technology
Central Institute for Research on Goats
Makhdoom, Mathura-281122
Uttar Pradesh

Shalik Gram Shukla
Department of A.H. & Dairying
R.M.P. (P.G.) College, Gurukul Narsan
Haridwar (Uttarakhand)

Sunil Kumar
Department of Animal Science
GBPUA&T Hill Campus
Ranichauri, Tehri Garhwal
Uttarakhand – 249 199

Teena Rani
Department of Molecular Biology and Biotechnology CCS HAU, Hisar
Haryana

V P Maurya
Division of Physiology and Climatology
Indian Veterinary Research Institute
Izatnagar, Bareilly – 304 5501
Uttar Pradesh

V P Singh
Department of Livestock Products Technology
College of Veterinary Sciences and Animal Husbandry
U.P. Pt. Deen Dayal Upadhyay Veterinary University and Go Anusandhan Sansthan
Mathura – 281 001
Uttar Pradesh

V. Sejian
Division of Physiology and Biochemistry
Central Sheep and Wool Research Institute
Avikanagar, Rajasthan – 304501

Vibha Raj Shanti
Scholar- Master of Intellectual Property Law
SUKAST-J
J&K – 181 102

Vikas Mahajan
Division of Animal Breeding and Genetics
F.V.Sc.&A.H., SKUAST-J
(J&K) – 181102

Vikas Pathak
Department of Livestock Products Technology
College of Veterinary Sciences and Animal Husbandry U.P. Pt. Deen Dayal Upadhyay Veterinary University and Go Anusandhan Sansthan
Mathura – 281 001
Uttar Pradesh

Contents

1

Recent Advances and Future Prospects of Embryo Transfer Technology

Rajesh Kumar Sharma and Rajinder Singh Kishtwaria

Preamble

The recent advances in embryo transfer protocols have placed these techniques as widely adopted means of developing experimental models to understand complex biological processes such as feto-maternal dialogue, fetal programming and maternal effects on early development. This has noticeably widened the scope of embryo transfer programs which are now not only limited to transferring the genetically superior embryos from donors to recipients for faster genetic improvement, or as only the therapeutic tools in human medicine, but has embarked these techniques as basic research tools in biological research. It is however critical to achieve effective synchronization between donors and recipients along with optimal superovulation in donors to get more number of transferable embryos and subsequently greater conception rates of recipients. For transfer, the embryos are collected Day 6 or 7 post insemination of donors. An overall discussion of sequential steps to achieve successful embryo transfer along with recent advances in the subject area and its possible future scope is discussed in this chapter.

Introduction

Since the first successful report of embryo transfer in rabbits by Walter Heape in 1890 and later on in sheep by Warwick *et al.* in 1934, much advancement has been reported in this subject. Traditionally speaking, embryo transfer is a technique in which embryos are collected from genetically superior female donor(s) and transferred to the recipient females. This is because these procedures have a widely accepted attestation for genetic and non-genetic improvement of farm animals, commercial application in animal production, conservation of indigenous germplasm and endangered species, animal birth control, and for a relatively low-cost means of international livestock trade (Schneider *et al.* 1980, Nicholas and Smith 1983, Smith 1988, Gearheart *et al.* 1989). Over the years, these techniques are also gaining significance as a therapeutic tool in human and veterinary medicine. With recent developments and standardization in embryo transfer protocols, it is possible to develop animal models for understanding various biological activities in the humans and animals that have complex underlying mechanisms such as feto-maternal interaction, fetal programming and epigenetics effects on early development (Allen *et al.* 2002a, Allen *et al.* 2002b, Giussani *et al.* 2002, Sharma *et al.* 2011). This usage of embryo transfer protocols provide with approaches to gain experimental knowledge in genetics, cytology, animal breeding, immunology, evolution and other physiological and biochemical processes (Hafez and Hafez 2000). Due to the critical role of these protocols for developing various cutting edge experiments in developmental biology including evolutionary development; embryo transfer protocols can also be regarded as one of the basic tools of research. These tools have enabled the transfer of embryos between breeds or even species irrespective of the superiority of germplamsm in order to exclusively understand the effects of maternal environments upon embryonic and fetal development, and this recent significance of embryo transfer has lead its role beyond the limit of merely achieving faster genetic improvement in livestock; which no doubt still however is the largest commercial application of embryo transfer.

Synchronization of oestrus

This is the first critical step in transferring embryos from donors to recipients. Achieving effective synchrony between the oestrous cycle of donors and recipients is essential to ensure congenial environment corresponding to the stage of embryo for its subsequent development in the uterus of recipient. An asynchrony in the reproductive tract of donor and recipient females can lead to failure of an embryo to implant and or survival post transfer. Synchronization can be achieved either by shortening life span of corpus luteum or prolonging luteal phase. The period of progesterone dominance can be shortened by using luteolytic agents such as prostaglandin (PGF_2 alpha). This has been widely used to cause premature luteolysis in sheep (Pereira *et al.* 1998, Uphale *et al.* 1998, Naqvi and Gulyani 1999, Naqvi *et al.* 2006), goats and cattle (Gallo *et al.* 1992, Pivko and Majerciak 1981). However, the functional corpus luteum is essential for its action which limits its use to a specific period during oestrous cycle or warrants administering two injections at ten days interval. Moreover, this protocol is of little use during non-breeding season in seasonal breeders. In cattle, two injections of PGF_2 alpha are intramuscularly administered at an interval of 11 to 12 days. In general the animals are inseminated 72 to 96 hours after second PGF_2 alpha (Jainudeen *et al.* 2008).

The period of progesterone dominance can be prolonged using exogenous progesterone preparations. These protocols are frequently used as they are effective in both breeding and non-breeding seasons. The common progesterone protocols for synchronization of oestrus in livestock include norgestomet ear implants (Sharma *et al.* 1994, Sharma *et al.* 2004, Ozturtlu *et al.* 2009), intravaginal progesterone pessaries (Wright *et al.* 1981, Bretzlaff and Madrid 1985), controlled internal drug release (CIDR) (Thompson *et al.* 1990, Macmillan and Peterson 1993, Wheaton *et al.* 1993, Sharma *et al.* 2011), progesterone sponges (Carrick and Shelton 1967, Scanlon *et al.* 1971, Thompson *et al.* 1990, Bari *et al.* 2001, Lima-Verde *et al.* 2003) or oral progestogens (Huelt and Foote 1967, Kinser *et al.* 1983, Naqvi *et al.* 1996). In cattle, a combination of using a progesterone vaginal pessary for 7 days and administration of PGF_2 alpha on the sixth day is widely reported to achieve effective oestrus synchronization in large cattle herds. Such cows can be inseminated around 84 hours after the PGF_2 alpha injection (Jainudeen *et al.* 2008).

In sheep, following completion of these protocols, the oestrus generally commences within 2 to 5 days and around 80% to 89% of ewes are generally observed in oestrus (Baril *et al.* 1989, Lima-Verde *et al.* 2003). In general, the controlled internal drug release has been found more reliable and effective means of synchronizing oestrus in ewes compared to oral progesterones or progesterone sponges (Rhodes and Nathanielsz 1988). In some geographical regions of the world where the cyclicity of sheep is dependent upon photoperiod such as those at poles, the synchronization and superovulatory outcome could be markedly variable. This might be due to the fact that with increase in the day length, there is a gradual decline in the pulstality of GnRH to the extent that a preovulatory surge of LH might not embark resulting in cessation of oestrous cyclicity (Gonzalez-Bulnes *et al.* 2004). Therefore in general FSH-p is not effective in embarking superovulatory treatment during non-breeding season in Gaddi ewes (Sharma 2002).

A novel protocol of synchronization of oestrus in cows comprises of administration of GnRH on Day 0, followed by injection of PGF_2 alpha on Day 6. This procedure has been found effective in synchronization of oestrous cycle in 70 to 80 % of cyclic cows within a range of 4 days (Twagiramungu *et al.* 1995). An advantage of this procedure is that it can be used at any time of the oestrous cycle because GnRH causes luteinization of mature follicles and allows development of new follicles, whereas PGF_2 alpha causes lysis of corpora lutea present on the ovary and thus animal exhibit oestrus (Britt and Gaska 1998).

In mares synchronization can be achieved by using progestogens or PGF_2 alpha. The progesterone preparation is are generally kept in situ for 15 days and the mares exhibit oestrus around 3 days after withdrawal of progesterone. Mares in dioestrus exhibit oestrus within 3 to 5 days after a single intramuscular injection of PGF_2 alpha (Jainudeen *et al.* 2008). Oestrus in cycling gilts and sows commences 4th to 5th day following termination of progesterone feeding for a period of 15 days. However, PGF_2 alpha is not effective in synchronizing oestrus in cyclic sows because CL is non-responsive to PGF_2 alpha for 12 to 13 days of the oestrous cycle (Jainudeen *et al.* 2008).

In an embryo transfer program it is essential to achieve enhanced conception rates. These can be obtained by achieving effective synchronization between donors and recipients. In general, a difference of two days in the oestrus of donor or recipient generally has no adverse effects in the pregnancy rates of recipients. However, greater conception rates can be achieved in the recipients who manifest oestrus 12 h before the donor contrary to those that exhibit oestrus 36 h after the donor (Wright *et al.* 1981, Lindner and Raymond 1983). The other major factor that should be considered for achieving higher conception rates is the type of drug used for synchronization. The usage of Control Internal Drug Release (CIDR) generally results in enhanced pregnancy rates in recipients (Hambra *et al.* 1986). Recent studies have also attempted combination of PGF_2 alpha and GnRH protocols to achieve greater synchronization and consequently conception rates. In these protocols, prostaglandins causes regression of CL or any luteinized follicles that might be formed as a result of GnRH administration.

For a successful fertilization, it is critical to achieve optimal timing between the AI and ovulation. For example if AI is carried out much before the ovulation, the sperm might lose the fertilization potential by the time of actual fertilization. Similarly a delay in insemination post-ovulation results in ageing of oocyte and enhances reduced fertility (Wilmut and Hunter 1984). Therefore, over recent years the concept of timed insemination following a hormonal treatment is increasingly recognized. Use of fixed time AI in single or double ovsynch protocols is rapidly gaining acceptability over recent years as these procedures have been shown to enhance the conception rates following embryo transfer. Furthermore the administration of GnRH 16 hours before fixed time AI in these protocols results in enhanced conception rate in dairy cows. This might be due to optimal timing between AI and ovulation. Similarly, using a double-Ovsynch protocol followed 7 days later by a normal Ovsynch 56-fixed time artificial insemination protocol results in greater conception rates in bovine. These protocols have been found particularly more effective in primiparous cows than the pleuriparous cows (Brusveen *et al.* 2008, Souza *et al.* 2008). In general, the timed insemination performed at 17-24 hours after GnRH results in greater conception rates. However the timing of second injection of GnRH is critical as it affects size of ovulatory follicle and consequently the luteal function and fertilization rates (Peters and Pursley 2003).

Superovulation

Superovulation involves development and ovulation of more than normal number of follicles. An effective superovulatory protocol should not only ensure supply of greater number of embryos from the donors without causing excessive ovarian edema and development of abnormally large follicles, but should also ensure conducive uterine environment including endocrine levels required for normal embryonic development (Boland *et al.* 1991). The use of non-pituitary gonadotropins such as equine chorionic gonadotrophin (eCG) (Gonzalez *et al.* 1994, Uphale *et al.* 1998, Lozano *et al.* 2003, Sharma *et al.* 2004, Naqvi *et al.* 2006) pituitary gonadotropins viz. follicle stimulating hormone (FSH) (Sharma *et al.* 1994, Boscos *et al.* 1997, Kelly *et al.* 1997, Pereira *et al.* 1998, Gonzalez-Bulnes *et al.* 2002, Lima-Verde *et al.* 2003, Windorski *et al.* 2007, Zambrano-Varón and BonDurant 2007) or a combination of eCG and FSH (Jabbour and Evans 1991, Ryan *et al.* 1991, Naqvi and Gulyani 1999) has been reported for multiple ovulation. FSH has short biological half life; therefore this warrants administration of FSH for around 3 to 4 days at 12 h intervals to achieve effective superovulation. Conversely, eCG has long half life and thus only a single injection is adequate for superstimulation of ovaries.

The efficacy of a superovulatory protocol in both cattle and sheep depends upon various extrinsic and intrinsic factors (Kafi and McGowan 1997, Bari *et al.* 2001, Ramon-Ugalde *et al.* 2008). The critical external factors include type of hormone used, batch of hormone, diluents used, frequency of administration of drug, and the administration protocol such as fixed dosage or in decreasing dosage (Donaldson 1990, Driancourt and Fry 1992, Gonzalez-Bulnes *et al.* 2004, D'Alessandro *et al.* 2005). The intrinsic factors are of animal origin e.g. the ovarian status at the time of initiation of superovulatory protocol. The presence of a dominant follicle on the ovary can reduce the efficacy of superovulatory drug. Under-nutrition and lactation also adversely affect the number of ova produced during superovulatory treatment. In general FSH is widely used in inducing superovualtion in sheep compared with eCG. In Merino ewes Nottle *et al.* (1984) reported greater ovulation rates (14.9 ± 1.1) by using 12 hourly multiple injections of a total of 22.5 mg porcine FSH compared with a single intramuscular injection of 1200 IU of eCG (8.0 ± 0.8). The efficacy of a superovulatory drug in inducing superovualtion also depends upon the breed of the animal. For example FSH is highly effective in superovulating Rambouillet ewes compared with a combination of eCG plus FSH or eCG plus gonadotrophin releasing hormone (GnRH) (Naqvi and Gulyani 1999). However in Merino ewes, both FSH and eCG are equally effective in superstimulating ovaries with an average ovulation rate of 8.4 ± 0.81 and 7.3 ± 1.21, respectively (Walker *et al.*

1986). Similarly, a combination eCG plus a single injection of FSH or six tapering injections of FSH induces comparable superovulatory response in Ojalada ewes, except that the ewes treated with eCG plus FSH combination protocol manifested oestrus earlier than those treated with FSH (Forcada *et al.* 2010). Conversely, a protocol of using eCG plus FSH in combination or FSH alone did not result in any differences in the onset of oestrus in Sarda ewes (Leoni *et al.* 2001). This indicates that not only the superovulatory protocol, but also the breed of ewes influences the response to superovulatory treatments. Administration of PGF_2 alpha on the day of removal of progesterone devices positively affects the number of total and transferable embryos recovered (Scudamore *et al.* 1994). In fact a higher priming of progesterone is essential to achieve greater superovulatory response.

The superovultory response is generally same for first three consecutive treatments; however subsequent responses might be lower due to production of antibodies against the gonadotropins. The age of the donors markedly affect the superovulatory response such that younger animals result in production of lower number of ova, however with increase in parity at least up to 6 to 7 year of age, the ovultory response increases, although this again decreases in older animals (Bari *et al.* 2001). Thus in sheep, it is advisable to have donors between the ages of 3 to 5 years, particularly for higher number of ova production during superovulation.

The combination of GnRH and FSH has also been advocated in inducing oestrus in cattle and sheep (Evans *et al.* 1994). Using this protocol, the preovulatory LH surge in donor heifers can be blocked with GnRH agonist such as deslorelin kept as implant along with CIDR. FSH is used to achieve superovulation. Subsequently, the donors are injected with LH and simultaneously inseminated. Thus the observation of oestrus is not required (D'Occhio *et al.* 1997).

Nutritional levels of dams influence both ovulation rates and subsequent conception rates. Nutritional alteration, particularly *ad libitum* feeding during superovualtion has adverse effects upon superovulatory response. A low calorie or restricted diet during embryonic period adversely affects early embryonic development (Lozano *et al.* 2003). Supplementing ewes with 0.5 times or 1.5 times more than the maintenance ration 16 days before the ovulatory protocol and till 6 days after the protocols significantly increases the ovulation rates and as well as the total number of cells in the blastocyst for a given day of development. This indicates enhanced embryonic development which would mean lesser chances of early embryonic mortality thereby enhanced conception rates and subsequent fetal growth (Kakar *et al.* 2005).

Insemination of Donors and Collection of Embryos

Insemination of the donors can be achieved by non-surgically or using laparoscopy. In large animal such as cows and mare non-surgical method of insemination is used, whereas both non-surgical insemination and laparoscopic insemination can be used in small animals such as sheep and goats. The factors that can affect the success rate of insemination include the proper restraining of animals, time of insemination after completion of progesterone treatment. Insemination of donors carried out 48 hours after the progesterone pessary withdrawal yield greater conception rates compared to the insemination performed 60 hours post progesterone pessary withdrawal (Scudamore *et al.* 1991a). The pregnancy rates are critically dependent on the method of superovulation, embryo collection, embryo handling and even the geographical location (Wright *et al.* 1981).

Collection of day 6 or 7 embryos

Embryos can be recovered from the donors by surgical or non-surgical methods, depending upon the type of animals. In general, the embryos can be recovered at day 6 or 7 from donors for subsequent immediate transfer to recipients.

i) Surgical method

This involves collection of embryos generally at Day 6 or 7 post mating or insemination. Animal is maintained under dorsal recumbency and genitalia is exteriorised by mid ventral surgical incision after preparation of surgical site. A puncture wound is then made into one uterine horn at the uterine bifurcation towards the uterotubular junction. A two way female urethral catheter is then inserted about 4 cm in to the horn. The cuff is then gently inflated with air to keep the catheter in place and avoid any spillage of flushing. A disposable 20 mL syringe is then fitted with a blunt 20-gauge needle and is inserted through uterine tube through ampulla, which is then flushed 3 to 4 times with 20 ml of flushing medium. Normograde flushing of uterine horn and tube is then performed via a catheter inserted in to the uterine horn, just anterior to the bifurcation. Alternatively, retrograde flushing can also be achieved by canulating through infundibulum (Fig 1) by using a catheter of appropriate diameter (Tervit and Havik 1976, Sharma 2002). The disposable 20 mL syringe is fitted with blunt 20-gauge needle

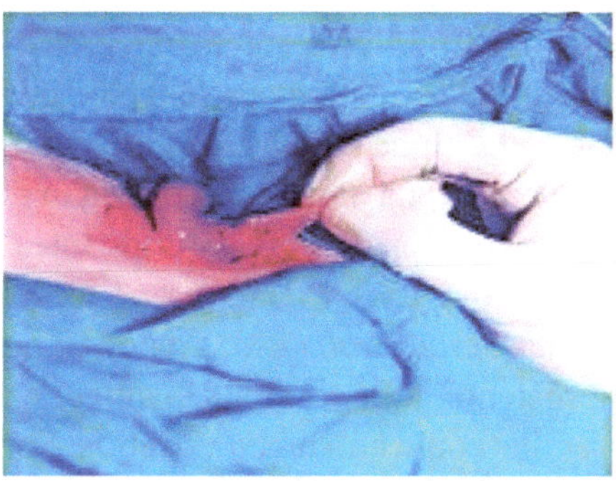

Fig. 1: Intubation of fallopian tube

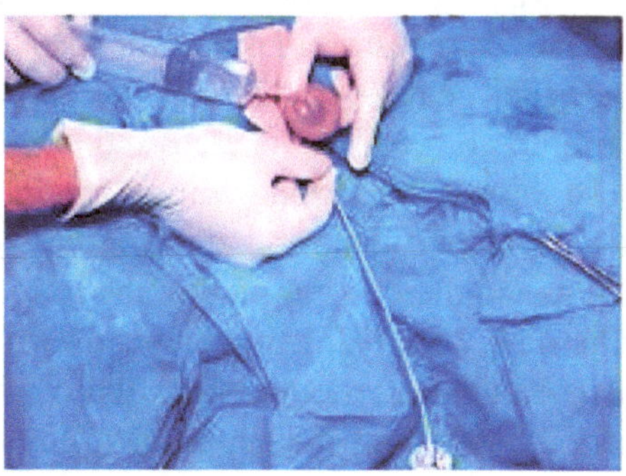

Fig. 2: Retrograde flushing (Sharma 2002)

and is inserted in to the uterine horn, just anterior to the bifurcation and flushings are collected in the petri dish (Fig 2). The other oviduct is also flushed in the same way. Skin sutures are applied to close the wound. Complete asepsis is very essential during the entire procedure. Tervit and Havik (1976) reported 83% of recovery rate following this method.

ii) Non- surgical method

Surgical method sometimes poses a potential risk for the health of donor due to formation of adhesions (Baril *et al.* 1989; Flores-Foxworth *et al.* 1992). Therefore, non-surgical embryo collection is preferred as not only it avoids risk of adhesions; but also enables repeated superovulation of the donors for collection of embryos and thus (Melican and Gavin 2008). The common methods of non-surgical embryo collection are:

a) Transcervical method

This method is used in cattle, buffalo and mares by using flexible plastic catheters and has also been attempted in sheep and goats with variable success rates (Tischner and Bielanski 1980, Mylne *et al.* 1992, Pereira *et al.* 1998). The flexible plastic catheter, along with the stylet is guided through the cervix by rectal manipulation and is placed in the uterine horn ipsilateral to the corpus luteum and the balloon is then inflated to secure the catheter in place (Pereira *et al.* 1998, Hafez and Hafez 2000). The stylet is then removed and the uterine horn is filled with 60 - 125 mL of PBS medium, which is then allowed to flow into the collection vessel. The uterus is gently massaged while collecting the fluid. The process is repeated until about 300 ml to 800 mL of flushing medium is collected. Similarly, the other uterine horn is also flushed. The technique is same in the mare, but in mare both the horns are flushed together and the balloon is inflated in the cervix. In goats, injections of PGF_2 alpha and oxytocin has been recommended to facilitate introduction of catheter (Pereira *et al.* 1998). This method is reported safer in mares as it did not had any adverse effect upon the subsequent fertility. Moreover the infusion of flushing medium and the necessary manipulation of uterine horn might trigger luteolysis and consequently onset of next oestrus (Tischner and Bielanski 1980). In small ruminants such as sheep and goat the structure of the cervix limits the use of non-surgical methods for embryo recovery. Therefore the use of PGF_2 alpha along with oestradiol, oestradiol 17 beta along with oxytocin, or PGF_2 alpha along with oxytocin is advocated prior to introduction of the catheter to soften the cervix during luteal phase and thereby facilitating the introduction of catheter (Barry *et al.* 1990, Khalifa *et al.* 1992, Pereira *et al.* 1998). Lima-Verde *et al.* (2003) reported no beneficial effect of administration of cloprostenol, a PGF_2 alpha analogue in cervical softening of Sannen goats; however this could be a breed variation. For easy introduction of the catheter, the brass tubing is also used inside the catheters to support entry into the cervix and prevent occlusion of the flexible catheter tubing with the pressure of cervix and these techniques have been reliable in embryo collection from multiparous ewes but not from nulliparous ewes (Flohr *et al.* 1999). Trancervical method of embryo collection has advantages over surgical method in goats as the recovery rate of embryos is generally more (53%) compared with laparatomy (37%) (Lima-Verde *et al.* 2003). Therefore transcervical method can be used as a simpler alternate of surgical method for recovery of embryos in goats. Even in miniature breeds of goats, this method has been successfully used for collection of embryos (Nagashima *et al.* 1987).

b) Laparoscopy

The use of laparoscopy for collection of embryos was first reported by McKelvey *et al.* in 1986. This method is rapid and effectively safer as it results lesser trauma and pain compared with surgical methods. Therefore, it is advocated as a preferred method of embryo collection in small ruminants (Kramer 1989, Scudamore *et al.* 1991b, Ramon-Ugalde *et al.* 2008). This is also referred as keyhole surgery or band-aid surgery. The major difference between surgical technique and laparoscopy is that it involves only stab wounds to insert instruments for flushing rather than a mid-ventral incision. Thus allows the repeated recovery of embryos from donor ewes without causing trauma or adhesions or affecting embryo recovery rates (McKelvey *et al.* 1986). It is now successfully used in goats and pigs as well. The animal is placed in the dorsal recumbency; site is prepared for surgery and incised to allow placement of 3 trocars through the ventral abdominal musculature (Fig 3). The trocars are placed in a triangle about 8 to 10 cm apart, one on midline and two parallel to midline, 8-10 cm anterior to the udder. Thus the cranial trocar is used for endoscope and the other two trocars are used for the caudal and lateral forceps.

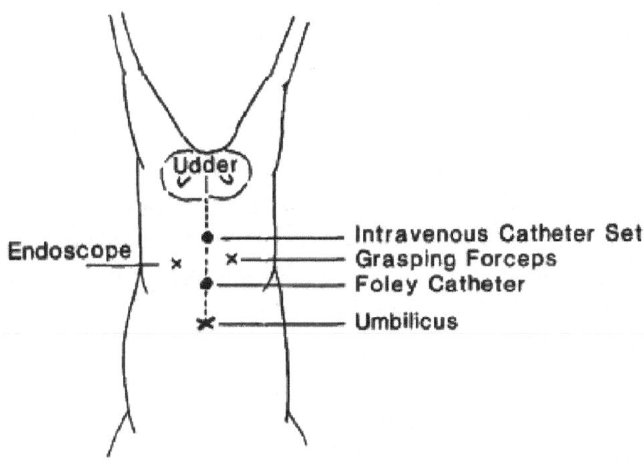

Fig. 3: Illustration of sites of introduction of laparoscopic instrument (*McKelvey et. al.,* 1986)

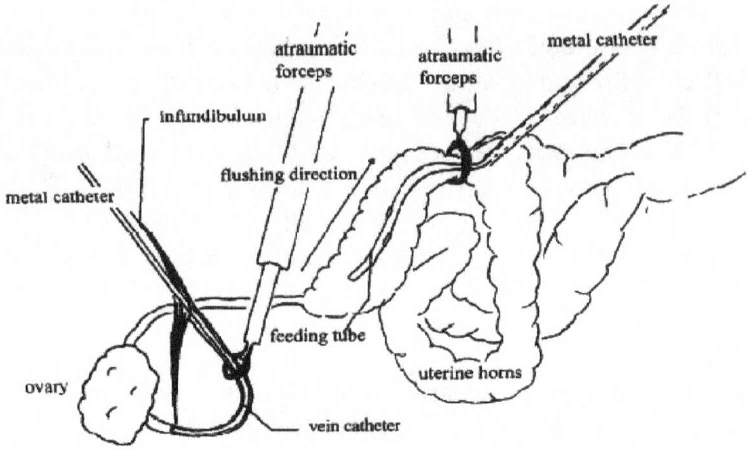

Fig. 4: Illustration of laparoscopic collection of embryos (Besenfelder *et al.* 1997)

Evaluation of pre-transfer embryos

The infundibulum of the ipsilateral ovary is fixed by the lateral forceps and a catheter is inserted through it to the ampulla. The uterine horn is fixed by the caudal forceps and the catheter is inserted in to the uterine horn. The laparoscope is inserted through one stab wound. A 2 way Foley catheter is then inserted through another stab wound and guided into one uterine horn and the balloon is then inflated. Then an intravenous catheter is inserted into the uterine lumen in proximity to the utero-tubular junction. About 40 to 50 mL flushing medium is injected through the catheter and is collected via Foley catheter (Fig 4). The other horn is also flushed in the same way and the skin sutures are applied to close the wound (Besenfelder *et al.* 1997). This whole procedure is performed by visualization through laparoscope. Laparoscopy has been shown equally effective as the cervical method of insemination to achieve conception rates (Scudamore *et al.* 1991b).

A careful embryo evaluation is an essential prior to transfer to as this helps achieving greater pregnancy rates. The evaluation of embryos includes gross morphological evaluation, dye exclusion tests, measuring of enzyme activity, glucose uptake and live dead stains. Most of these methods are time consuming and require complex equipments. Therefore, when embryo transfer has to be immediately performed after collection; gross morphological evaluation is widely used. The parameters commonly used to evaluate embryo quality include shape, colour, number and compactness of cells, size of the perivitelline space, number of extruded and degenerated cells and number and size of vesicles (Lindner and Raymond 1983). After collection, the flushing medium is examined under the stereomicroscope. Paraffin oil is generally used to cover the medium containing embryos to prevent evaporation and avoid any contamination. The stages of embryonic development can be variable. Usually transfer of very early or very late stage embryos results in low pregnancy rates, whereas, embryos transferred between eight-cell to blastocyst stage yields greater pregnancy rates. Generally only good quality embryos are selected for transfer (Hafez and Hafez 2000). Overall, the recovered embryos are usually considered viable if their development correspondes to the physiological age without any signs of cellular degeneration (Winterberger-Torres and Sevellec 1987).

Wright (1981) classified the embryos to the stage of development using the following definitions:

Morula: An embryo containing at least 16 cells that are distinct individuals and have not yet coalesced together to form a compact mass.

Advanced morula: An embryo containing more than 16 cells that have coalesced together to form a compact mass but has not developed a blastocoele.

Early blastocyst: An embryo that has formed a blastocoele and has the appearance of a signet ring.

Blastocyst: An embryo which has a blastocoele that has expanded so that it has lost the signet ring appearance but has not expanded enough to completely fill the zona pellucida.

Advanced blastocyst: A blastocyst that has expanded to fill the entire zona pelucida. Lindner and Raymond (1983) classified the embryos as below (Fig 5):

Morula: This is commonly called as “ball of cells” because the individual blastomeres are tightly attached with one another. The cellular mass of the embryo occupies majority of the space.

Compact morula: Individual blastomeres are coalesced to form a compact mass. The embryo mass occupies 60 to 70% of the perivitelline space.

Early blastocyst: An embryo that has a fluid-filled cavity called blastocoele and looks like a signet-ring. The embryo occupies 70 to 80% of the perivitelline space. In this embryo, the visual differentiation between trophoblast and the inner cell mass is however possible.

Blastocyst: The differentiation of the outer trophoblast layer and the darker, compact inner cell mass is evident. The blastocoele is well developed and the embryo occupies most of the perivitelline space.

Expanded blastocyst: Overall diameter of the embryo increases to the extent of about 1.2 to 1.5X, along with a thinning of the zona pellucida to approximately $1/3^{rd}$ of its original thickness. Embryos at the expanded blastocyst stage generally appear collapsed with a complete or partial loss of the blastocoele; however, the zona pellucida sometimes regains its original thickness.

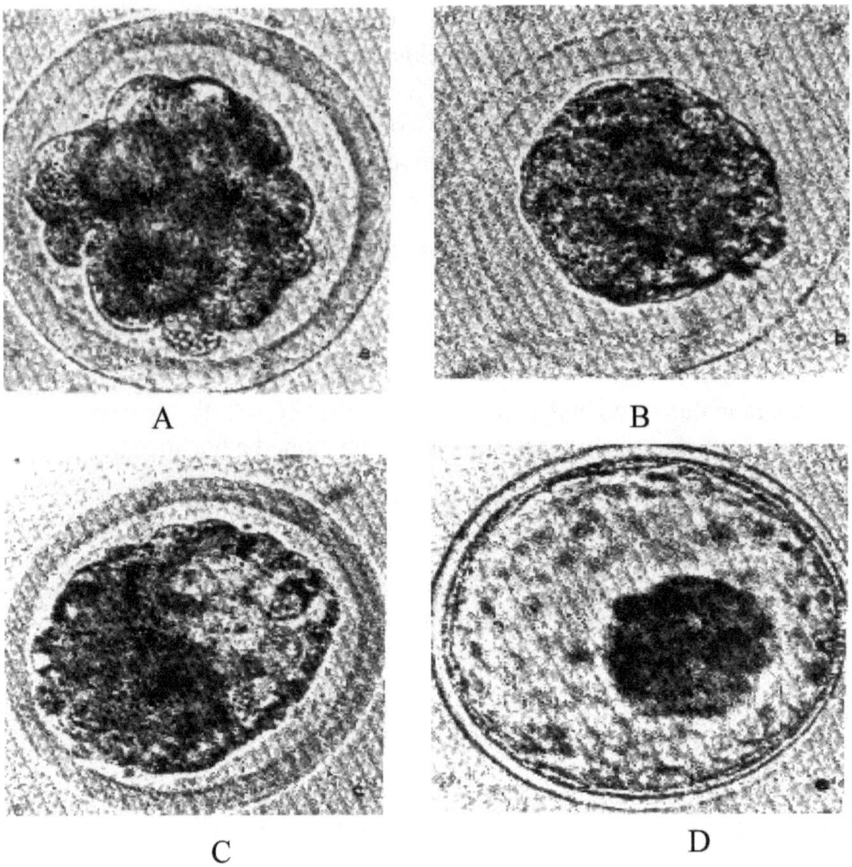

A. Morula, B. Compact morula, C. Early blastocyst, D. Expanded blastocyst

Fig. 5: Stages of Bovine embryo development (Lindner and Raymond 1983)

Hatched blastocyst: These embryos rmight be at the stage of hatching or may have completely shed the zona pellucida. These hatched blastocysts may be spherical with a well-defined blastocoele.

These stages of embryo development significantly affect the pregnancy rates. In cattle, following transcervical method of embryo transfer to the uterine horn, Wright (1981) reported pregnancy rates of 44% with morulae and 53% with advanced morulae as comparison to 65%, 66% and 64% with early blastocyst, blastocyst and advanced blastocyst. Whereas, Schneider *et al.* (1980) reported pregnancy rates of 61% with morulae, 67% with late morulae, 67% with early blastocysts and 71% with late blastocysts. However using laparoscopy method of embryo transfer in sheep, no significant differences in pregnancy rates are observed following transfer of embryos at morulae or blastocyst (Ishida *et al.*

1999). It appears that not only the stage of embryonic development but also the quality of embryo affects conception rates. The various qualities of morulae in bovine are depicted in Fig 6 (Lindner and Raymond 1983).

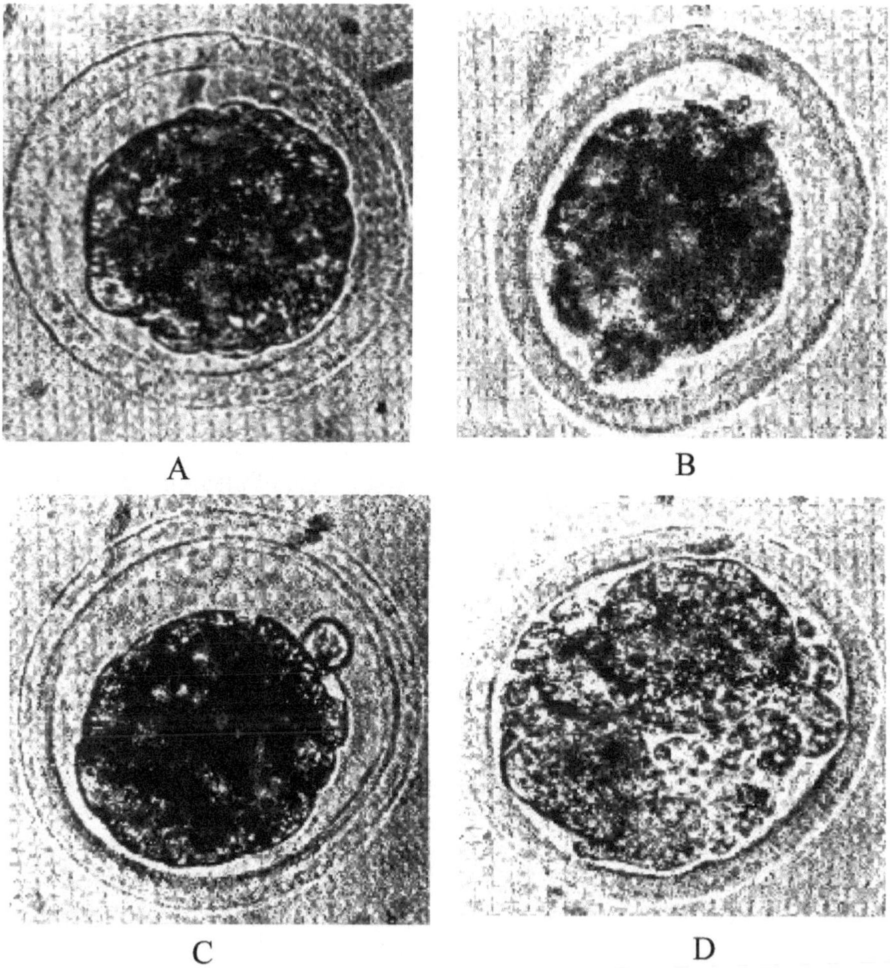

A. Excellent quality, **B.** Good quality (one extruded blastomere), **C.** Fair quality (extruded blastomere and degenerated cells), **D.** Poor quality (irregular shape and cellular degeneration)

Fig. 6: Bovine morulae of different qualities (Lindner and Raymond 1983)

Based on the morphological evaluation, the embryos can be categorized as below (McKinnon and Squires 1988).

Grade1 ++++	Excellent	An ideal embryo, symmetrical and spherical embryo mass with cells of uniform size, colour, and density. Embryo mass has clear edges without indentation. No evidence of cellular fragmentation or loose cytoplasmic granules in the perivitelline space.
Grade 2 +++	Good	Moderate irregularities in the overall shape of the embryo, for example, a minor degree of cytoplasmic fragmentation evident as slight convolution of the edges of the embryo and small cytoplasmic granules in the perivitelline space.
Grade 3 ++	Fair	Major irregularities in the shape of the embryonic mass or in the size, color, or density of the constituent cells. High degree of cytoplasmic fragmentation with extrusion or degeneration of cells of variable size reflected by highly lobulated, irregular edges of the embryonic mass and a large perivitelline space.
Grade 4 +	Poor	High degree of cytoplasmic fragmentation. Embryonic cells of irregular size and color. Absence of cellular compaction and failure to form a clear embryonic mass. This group includes embryos with irregularly sized cells of a low number (< 4) inconsistent with the expected stage of development
Grade 5	Unfertilized or dead	Unfertilized oocytes or total degenerate embryos

The embryos showing non-uniform and variable size of blastomeres, cellular debris in morula, degenerated and collapsed blastocyst with in an oblong shaped zona pellucida, disintegrating mitotic figures, indistinct foamy blastomeres, fragmentation of cytoplasmic and nuclear material and abnormal shape of morula or blastocyst are rendered unfit for transfer. International Embryo Transfer Society has outlined criteria for evaluation of embryos. This include evaluating the degree of compactness, size and appearance of the perivitelline space, color, presence of extruded cells, and the degree of cell granulation and cytoplasmic fragmentation (Tremoleda *et al.* 2003). Based on these criteria, the embryos can be categorized as below (Fig 7).

Embryo Transfer

After embryos has been evaluated and categorized, they can be either cryopreserved or freshly transferred to the recipient animals. The cryopreserved embryos can be easily and safely transported and their thawing allows the transfer of embryos as and when required. The use of vitrification technology is rapidly gaining wide acceptation and thus replacing the older method of slow cooling of embryos for cryopreservation. A careful and optimal cryopreservation not only ensures the use of embryos for later on transfer with good survival rates, but also has enables the safer international trade of livestock. This is because at this

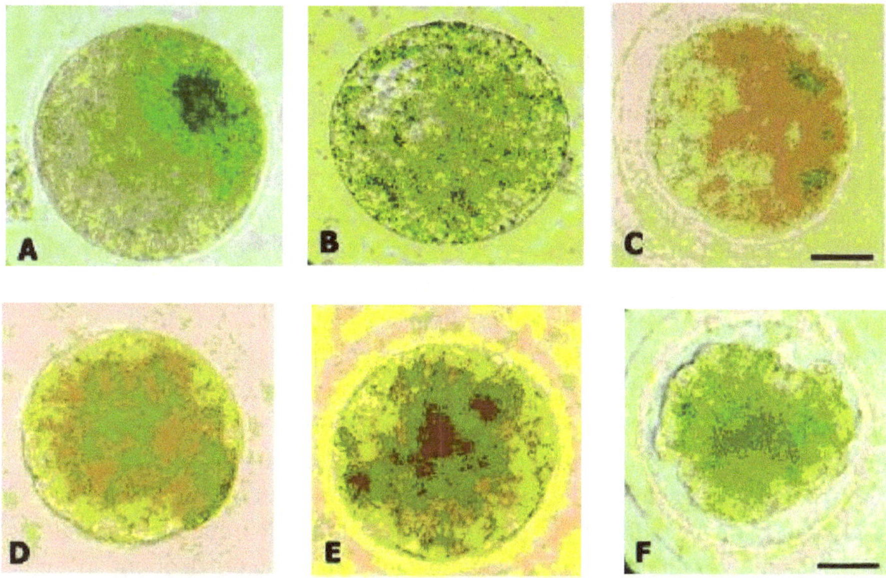

A.and **B.** Grade 1, **C.** and **D.** Grade 2, **E.** Grade 3, **F.** Grade IV
Fig. 7: Day 7 evaluation of *in vitro* produced equine embryos (Tremoleda et al. 2003)

stage the embryos are enclosed by Zona pellucida which does not allow entry of viruses and other pathogens to the embryo and thus enables transport of embryos without a threat of disease spread compared to transport of live animal or semen.

The transfer of embryos to the recipients can be done either surgically or non-surgically. The pregnancy rates using embryo transfer technology can be variable depending up on the method of superovulation, embryo collection, embryo handling, transfer and the geographical location (Wright 1981).

Surgical method of embryo transfer

This is used in sheep and goats, but its use is relatively decreasing day by day as most farmers now prefer transfer by laparoscopy. The animal is restrained and mid ventral laparotomy is performed under general or local anaesthesia. Through the midventral incision, embryos are transferred with one or two drops of the flushing medium into the oviduct or the uterine horn. To transfer the embryos into the oviduct, the tip of the capillary pipette containing the embryos is inserted through the infundibulum to deposit embryos into the ampulla. When transfer is made to the uterus, the wall of the uterine horn is punctured with a blunt needle, and the embryos are expelled from the tip of the capillary pipette inserted into the uterine lumen. Surgical method of embryo transfer has many disadvantages as it poses the risk of trauma and adhesions, and can adversely affect subsequent reproductive performance of the animal.

Non surgical embryo transfer

i) Transfer by laparoscopy *(Transpic technique)*

The key instrument in this technique is a laparoscope and it can be of two types (1) a telescopic rod lens system, which is usually connected to a video camera, (2) a digital laparoscope in which a charge-coupled device is attached at the end of the laparoscope and thus eliminating the rod lens system (Eubanks *et al.* 2004). Recent advances in fiber optics has enabled laparoscopy as a convenient and rapid technique to visualize the uterine horns and ovary with much clarity, thus making the embryo transfer readily feasible. The operative field is visualized through a fiber optic cable system attached with cold light source such as halogen or xenon and inserted through a 5 mm or 10 mm cannula or trocar. Due to its minimum invasiveness, small incision, easy post-operative management, and shortened recovery time, this technique has advantages over laparatomy. Moreover the images observed through laparoscope can be visualized on TV monitor. This further adds to the ease of the operative technique. In this method, the recipient ewe is held in dorsal recumbency, with head down about an angle of 45^0 in a laparoscopy cradle. The abdominal area anterior to the udder is clipped and sprayed with 70% alcohol solution to achieve asepsis, and 2% lignocaine hydrochloride is injected at the two puncture sites to achieve adequate local anaesthesia. Two abdominal puncture are then made using a 7 mm diameter trocar about 5 to 7 cm anterior to the udder, one of them at the midline and other 5 to 7 cm right lateral to the midline. The abdomen is generally inflated with carbon dioxide gas; however more inert nitrous oxide gas is more common in use, which further minimizes the discomfort to the animal. The holding forceps are then introduced through the midline puncture and slightly opened to stretch the surrounding tissue. Ovaries are visualized to confirm as well as count the corpora lutea through the laparoscope. After completion of the ovarian examination, the uterine horn ipsilateral to the corpus luteum is gently lifted to exteriorize a small loop (2 to 3 cm) to allow the embryologist to transfer the embryo. The embryo is then directly sucked from the observation dish under the microscope with a small volume of the fluid followed by the same volume of air. It is loaded in modified sheep AI aspic by cutting the sheath to 10 to 12 cm in length and attached to a 1mL syringe through 2 cm silastic tubing. A small volume of the flushing medium is sucked on to the aspic followed by similar volume of air. This liquid-air-liquid-air profile is essential to prevent any loss of fluid from the aspic (McMillan and Hall 1994). These steps ensure a high pregnancy rates following transfer by this technique. Using this technique, McMillan and Hall (1994) reported conception rates of 73% following transferring single embryo, 84% by transferring two embryos and 100% following transfer of three embryos. This technique is now commercially used for embryo transfer particularly in larger flocks. The authors experience also suggest that in sheep around 85 to 90% conception rates can be achieved by following embryo transfer by laproscopic method.

The complication of this method includes sudden movement by the animal resulting in spillage from the aspic thereby causing embryo loss and incorrect site of transfer like transfer of embryo in to the uterine wall. However the authors experience suggests that this rarely occurs if adequate anaesthesia and retrain has been achieved.

ii) Transcervical method

This is performed in large animals such as cattle. In this method, the recipient is rectally palpated to determine the ovary containing the corpus luteum. The posterior epidural anaesthesia is administered to prevent straining by animal, but the dose of local anaesthetic should be such that animal should remain in the standing position during the entire procedure. The embryo is loaded in a 0.25 mL straw and is fitted into the embryo transfer gun (Cassou transfer gun) and covered by a sterile sheath. The gun is then inserted into the vagina and passed through the cervix by rectal manipulation and guided into the uterine horn ipsilateral to the corpus luteum. The contents of the straw are deposited in the uterine horn.

The skill of the person performing the transfer is an important factor affecting the pregnancies rates in cows (Schneider *et al.* 1980). Generally by transferring embryos at a relatively faster speed (straw method) results in greater pregnancy rates compared to the slower rate of transfer (catheter method) in cows (Tervit *et al.* 1980).

Conclusion

Reproductive biotechnology techniques such as embryo transfer are critically important for faster genetic gain and increasing used to develop animal models for developmental biology research. It is pivotal to achieve effective synchronization of oestrus between the donors and recipients and superovualtion of donors. The synchronization of oestrus between donors and recipients ensures that post-transfer the embryonic development stage corresponds to the right stage of recipient uterus to supports implantation and further embryonic and fetal growth. As a thumb rule, Day 6 or 7 embryos are generally used for transfer to recipients either by surgically, non-surgically or laparoscopically depending upon the animal species, as this corresponds to the embryonic stage at which higher conception rates could be obtained. The limitation of the embryo transfer technology is that the genetic gain by this technique is usually slower than artificial insemination and is not always cost effective. It is also critical to carefully evaluate the production of embryos as infective gametes can result in introduction of pathogens in the embryo. This can result in widespread transfer of diseases during subsequent livestock trade. Hence, it is pivotal that embryo production should be achieved free from pathogens by regularly monitoring the quality and health of embryos.

Future perspectives

In addition to the commercial application of embryo transfer protocols in animal production and trade; the use of such techniques is gaining wider adaptability because of its use in basic biological research. Recent development in establishment of animal models by using embryo transfer techniques has made it possible to underpin complex biological research in developmental biology. Therefore, in addition to commercial programs, the future of such techniques also appears well in developing various animal models in cutting edge novel biological research such towards understanding of embryo-maternal interaction, early embryonic development and epignetics and in various other fields of developmental biology.

References

Allen WR, Wilsher S, Stewart F, Ousey J, Fowden A 2002a. The influence of maternal size on placental, fetal and postnatal growth in the horse. II. Endocrinology of pregnancy. Journal of Endocrinology 172: 237-46.

Allen WR, Wilsher S, Turnbull C, Stewart F, Ousey J, Rossdale PD, Fowden AL 2002b. Influence of maternal size on placental, fetal and postnatal growth in the horse. I Development in utero. Reproduction 123: 445-53.

Bari FM, Khalid M, Wolf B, Haresign W, Murray A, Merrell B 2001. The repeatability of superovulatory response and embryo recovery in sheep. Theriogenology 56: 147-55.

Baril G, Casamitjana P, Perrin J, Vallet JC 1989. Embryo production, freezing and transfer in Angora, Alpine and Saanen goats. Zuchthygiene 24: 101-15.

Barry DM, van Niekerk CH, Rust J, van der Walt T 1990. Cervical embryo collection in sheep after ripening of the cervix with prostaglandin E2 and estradiol. Theriogenology 33: 190.

Besenfelder U, Modl J, Muller M, Brem G 1997. Endoscopic embryo collection and embryo transfer into the oviduct and the uterus of pigs. Theriogenology 47: 1051-60.

Boland MP, Goulding D, Roche JF 1991. Alternative Gonadotropins for Superovulation in Cattle. Theriogenology 35: 5-17.

Boscos C, Vainas E, Kouskoura T, Samartzi F, Vafiadis D, Dellis S 1997. Superovulatory response of Chios and Friesian ewes to two FSH-P dose levels. Reproduction in Domestic Animals 32: 195-8.

Bretzlaff KN, Madrid N 1985. Synchronization of estrus and fertility in goats with norgestomet ear implants. Theriogenology 24: 351-7.

Britt JS, Gaska J Comparision of two estrus synchronization programs in a large confinement-housed dairy herd 1998. Journal of American Veterinary Medical Association 212: 210-6.

Brusveen DJ, Cunha AP, Silva CD, Cunha PM, Sterry RA, Silva EPB, Guenther JN, Wiltbank MC 2008. Altering the time of the second gonadotropin-releasing hormone injection and artificial insemination (AI) during Ovsynch affects pregnancies per AI in lactating dairy cows. Journal of Dairy Science 91: 1044-52.

Carrick MJ, Shelton JN 1967. The synchronozation of oestrus in cattle with progesterone-impregnanted intravaginal sponges. Journal of Reproduction and Fertility 14: 21-32.

D'Occhio MJ, Sudha G, Jillella D, Whyte D, Maclellan LJ, Walsh J, Trigg TE, Miller D. Use of a GnRH agonist to prevent endogenpus LH surge and injection of exgenous LH to induce ovulation in heifers stimulated with FSH: A new model for superovualtion. 1997. Theriogenolgy 47: 601 – 7.

D'Alessandro AG, Martemucci G, Taibi L. How the FSH/LH ratio and dose numbers in the p-FSH administration treatment regimen and insemination schedule affect superovulatory response in ewes 2005. Theriogenology 6: 1764-74.

Donaldson LE. FSHp batch variation. 1990. Theriogenology 33: 215 p.

Driancourt MA, Fry RC. Effect of superovulation with pFSH or PMSG on growth and maturation of the ovulatory follicles in sheep. 1992. Animal Reproduction Science 27: 279-92.

Eubanks SW, Eubanks S, Swanstrom LL, Soper NJ 2004. Mastery of Endoscopic and Laparoscopic Surgery 2nd Edition, Lippincott Williams & Wilkins.

Evans G, Brooks J, Struthers W, McNeilly AS 1994. Superovulation and embryo recovery in ewes treated with gonadotropin releasing hormone agonist and purified stimulating hormone. Reproduction Fertility and Development 6: 247-50.

Flohr SF, Wulster-Radcliffe MC, Lewis GS 1999. Development of a Transcervical Oocyte Recovery Procedure for Sheep. Journal of Animal Science 77: 2583-6.

Flores-Foxworth G, McBride BM, Nuti LC 1992. A comparison between laparoscopic and transcervical embryo collection and transfer in goats. Theriogenology 37: 213.

Forcada F, Ait Amer-Meziane M, Abecia JA, Maurel MC, Cebrián-Pérez JA, Muiño-Blanco T, Asenjo B, Vázquez MI, Casao A 2011. Repeated superovulation using a simplified FSH/eCG treatment for in vivo embryo production in sheep. Theriogenology 15: 769-776.

Gallo GF, Algire JE, Srikandakumar A, Downey BR 1992. Effects of a prostaglandin $F_2\alpha$ analogue on the ovulatory response of superovulated heifers. Animal Reproduction Science 27: 83-90.

Gearheart WW, Smith C, Teepker G 1989. Multiple ovulation and embryo manipulation in the improvement of beef cattle: relative theoretical rates of genetic change Synchronization of oestrus. Journal of Animal Science 67: 2863-71.

Giussani DA, Forehead AJ, Gardner DS, Fletcher AJW, Allen WR 2002. Postnatal cardiovascular function after manipulation of fetal growth by embryo transfer in the horse. Journal of Physiology 547: 67-76.

Gonzalez-Bulnes A, Garcia-Garcia RM, CJH Souza CJH, Santiago-Moreno J, Lopez-Sebastian A, Cocero MJ, Baird DT 2002. Patterns of Follicular Growth in Superovulated Sheep and Influence on Endocrine and Ovarian Response. Reproduction in Domestic Animals 37: 357-61.

Gonzalez-Bulnes A, Baird DT, Campbell BK, Cocero MJ, Garcia-Garcia RM, Inskeep EK, Lopez-Sebastian A, McNeilly AS, Santiago-Moreno J, Souza CJ, Veiga-Lopez A. Multiple factors affecting the efficiency of multiple ovulation and embryo transfer in sheep and goats. 2004. Reproduction Fertility and Development 16: 421-35.

Gonzalez A, Wang H, Carruthers TD, Murphy BD, Mapletoft RJ 1994. Superovulation in the cow with pregnant mare serum gonadotrophin: effects of dose and antipregnant mare serum gonadotrophin serum. Canadian Veterinary Journal 35: 158-62.

Hafez B, Hafez ESE 2000. Reproduction in Farm Animals, Lippincott Williams & Wilkins, Philadelphia.

Hambra AH, Massri YG, Marcek JM, Wheaton JE 1986. Plasma progesterone levels in ewes treated with progesterone-controlled internal drug-release dispensers, implants and sponges. Animal Reproduction Science 11: 187-94.

Huelt CV, Foote WC. Induction of fertile estrus in lactating and dry anestrous ewes using oral progestogens and repeated PMSG treatment. 1967, Journal of Animal Science 26: 545-8.

Ishida N, Jung YG, Itagaki R, Okada M, Ogiso T, Ishikawa D, Fukui Y. Non - surgical transfer of fresh or frozen - thawed ovine embryos by laparoscopy. 1999. Journal of Reproduction and Development 45: 289-293

Jabbour HN, Evans G. Ovarian and Endocrine Responses of Merino Ewes to Treatment with Pmsg and or Fsh-P. 1991. Animal Reproduction Science 26: 93-106.

Jainudeen MR, Wahid H, Hafez ESE. (eds) (2008) 'Ovulation induction, embryo production and transfer. Reproduction in Farm Animals Lippincott Williams & Wilkins, Philadelphia.

Kafi M, McGowan MR. Factors associated with variation in the superovulatory response of cattle. 1997. Animal Reproduction Science 48: 137-57.

Kakar MA, Maddocks S, Lorimer MF, Kleeman DO, Rudiger SR, Hartwich KM, Walker SK. The effect of peri-conception on embryo quality in the superovulated ewe. Theriogenology 64: 1090-103.

Kelly P, Duffy P, Roche JF, Boland MP 1997. Superovulation in cattle: effect of FSH type and method of administration on follicular growth, ovulatory response and endocrine patterns. 46: 1-14.

Khalifa RME, Sayre BL, Lewis GS 1992. Exogenous oxytocin dilates the cervix in ewes. Journal of Animal Science 70: 38-42.

Kinser AR, Gibson MF, Vincent DL, Scheffrahn NS, Kesler DJ 1983. Ovarian responses of seasonally anestrous ewes administered progesterone, PMS, HCG and (or) GnRH. Theriogenology 19: 449-64.

Kramer DC. Embryo collection and transfer in small ruminants 1989. Theriogenology 31: 141-8.

Leoni G, Bogliolo L, Pintus P, Ledda S, Naitana S 2001. Sheep embryos derived from FSH/eCG treatment have a lower in vitro viability after vitrification than those derived from FSH treatment. Reproduction Nutrition Development 41: 239-46.

Lima-Verde JB, Lopes Júnior ES, Teixeira DIA, Paula NRO, Medeiros AA, Rondina D, Freitas VJF 2003. Transcervical embryo recovery in Saanen goats South African Journal of Animal Science 33: 127-31.

Lindner GM, Raymond WWJ 1983. Bovine embryo morphology and evaluation. Theriogenology 20: 407-16.

Lozano JM, Lonergan P, Boland MP, Callaghan DO 2003. Influence of nutrition on the effectiveness of superovulation programmes in ewes: effect on oocyte quality and post-fertilization development. Reproduction 125: 543-53.

Macmillan KL, Peterson AJ 1993. A new intravaginal progesterone releasing device for cattle (CIDR-B) for oestrous synchronisation, increasing pregnancy rates and the treatment of post-partum anoestrus. Animal Production Science 33: 1-25.

McKelvey WAC, Robinson JJ, Aitken RP, Robertson IS 1986. Repeated recoveries of embryos from ewes by laparoscopy. Theriogenology 25: 855-65.

McKinnon AO, Squires EL 1988. Morphologic assessment of the equine embryo. Journal of the American Veterinary Medical Association 192: 401-6.

McMillan WH, Hall DRH 1994. Laparoscopic transfer of ovine and cervine embryos using the transpic technique. Theriogenology 42: 137-46.

Melican D, Gavin W 2008. Repeat superovulation, non-surgical embryo recovery, and surgical embryo transfer in transgenic dairy goats. Theriogenology 69: 187-203.

Mylne MJA, McKelvev WAC, Fernie K, Matthews K 1992. Use of transcervical technique for embryo recovery in sheep. Veterinary Record 130: 450-1.

Nagashima H, Matsui K, Sawasaki T, Kanyo Y 1987. Nonsurgical collection of embryos in Shiba goats. Jikken Dobutsu 36: 51-60.

Naqvi SMK, Mathur AK, Kalra DB 1996. Estrus synchronization using PGF 2 alpha with or without progesterone in ewes treated for superovulation. International Journal of Animal Sciences. 11: 139- 140.

Naqvi SMK, Gulyani R 1999. Ovarian response and embryo recovery to different superovulatory regimens in Rambouillet ewes under semiarid conditions. Small Ruminant Research 34: 127-31.

Naqvi SMK, Joshi A, Gulyani R, Kumar D, Kolte AP, Kumar S, Maurya VP, Saha S, Mittal JP,

Singh VK 2006. Production of prolific microsheep by embryo transfer into large non-prolific sheep. Veterinary Record 159: 522-6.

Nicholas FW, Smith C 1983. Increased rates of genetic change in dairy cattle by embryo transfer and splitting. Animal Science 36: 341-53.

Nottle MB, Armstrong DT, Holland MK, Sharpe PH 1984. Superovulation and embryo recovery in the Merino. In: Sixteenth Animal Conference of the Australian Society for Reproduction Biology. (Melbourne, Australia).

Ozturtlu N, Cetin Y, Kucukaslan I, Kocamuftuoglu M 2009. Induction of Oestrus with Norgestomet Ear Implant and PRID in Acyclic Holstein Heifers. Journal of Animal and Veterinary Advances 8: 1035-9.

Pereira RJTA, Sohnrey B, Holtz W 1998. Nonsurgical embryo collection in goats treated with prostaglandin F2 alpha and oxytocin. Journal of Animal Science 76: 360-3.

Peters MW, Pursley JR 2003. Timing of final GnRH of the Ovsynch protocol affects ovulatory follicle size, subsequent luteal function, and fertility in dairy cows. Theriogenology 60: 1197-204.

Pivko J, Majerciak P 1981. The effect of gonadotropins on superovulation and of prostaglandin F2 alpha on estrus synchronization in advance management of reproduction. Vedecke – Prace – Vyskumneho – Ustavu – Zivocrisney – Vyroby – V – Nitre 19, 159 – 71.

Ramon-Ugalde RP, Folch J, Cocero MJ, Piña-Aguilar RE, Alabart JL 2008. Embryo recovery from the oviduct in superovulated ewes: a method to improve MOET systems. Czech Journal of Animal Science 53: 145-51.

Rhodes L, Nathanielsz PW 1988. Comparison or a controlled internal drug release device containing progesterone with intravaginal medroxyprogesterone sponges for estrus synchronization in ewes. Theriogenology 30: 831-6.

Ryan JP, Hunton JR, Maxwell WM 1991. Increased production of sheep embryos following superovulation of Merino ewes with a combination of pregnant mare serum gonadotrophin and follicle stimulating hormone. Reproduction Fertility and Development 3: 551-60.

Scanlon PF, Neville WJ, Burgess TD, Macpherson JW 1971. Synchronization of estrus in cattle by intravaginal application of progesterone with estrogen administration. Canadian Journal of Animal Science 51: 250-1.

Schneider HJ, Jr. Castleberry RS, Griffin JL 1980. Commercial aspects of bovine embryo transfer. Theriogenology 13: 73-85.

Scudamore CL, Robbinson JJ, Aiken RP 1991a. The effect of timing of laparoscopic insemination in superovulated ewes, with or without sedation, on the recovery of embryos, their stage of development and subsequent viability. Theriogenology 35: 907-914.

Scudamore CL, Robinson JJ, Aitken RP, Kennedy DJ, Ireland S, Robertson IS 1991b. Laparoscopy for intrauterine insemination and embryo recovery in superovulated ewes at a commercial embryo transfer unit. Theriogenology 35: 329-37.

Scudamore CL, Robinson JJ, Aitken RP, Robertson IS 1994. The effect of method of oestrous synchronisation on the response of ewes to superovulation with porcine follicle stimulating hormone. Animal Reproduction Science 34: 127-33.

Sharma R 2002. Study on superovulatory response and embryo recovery rates using different drugs in Gaddi sheep. Traditional thesis, CSK Himachal Pradesh Agricultural University, Palampur.

Sharma R, Vasishta N, Singh M, Sood P 2004. Efficacy of equine chorionic gonadotropin for superovulation of Gaddi ewes during non breeding season. Indian Veterinary Journal 81: 838-839.

Sharma RK, Blair HT, Jenkinson CMC, Kenyon PR, Cockrem JF, Parkinson TJ 2011. Uterine environment as a regulator of birth weight and body dimensions of newborn lambs. Journal of Animal Science PMID: 22079991.

Sharma VK, Gupta RC, Khurana NK, Khar SK 1994. Oestrus synchronisation and superovulation using equine FSH in crossbred ewes. Veterinary Record 135: 164-5.

Smith C 1988. Applications of embryo transfer in animal breeding. Theriogenology 29: 203-12.

Souza AH, Ayres H, Ferreira RM, Wiltbank MC 2008. A new presynchronization system (Double-Ovsynch) increases fertility at first postpartum timed AI in lactating dairy cows. Theriogenology 70: 208-15.

Tervit HR, Havik PG 1976. A modified technique for flushing ova from the sheep uterus. New Zealand Veterinary Journal 24: 138-40.

Tervit HR, Cooper MW, Goold PG, Haszard GM 1980. Non-surgical embryo transfer in cattle. Theriogenology 13: 63-71.

Thompson JGE, Simpson AC, James RW, Tervit HR. 1990. The application of progesterone-containing CIDRTM devices to superovulated ewes. Theriogenology 33: 1297-304.

Tischner M, Bielanski A 1980. Non-surgical embryo collection in the mare and subsequent fertility of donor animals. 58: 357-61.

Tremoleda JL, Stout TAE, Lagutina I, Lazzari G, Bevers MM, Colenbrander B, Galli C 2003. Effects of in vitro production on horse embryo morphology, cytoskeltal characteristics and blastocyst capsule formation. Biology of Reproduction 69: 1895-906.

Twagiramungu H, Guilbault LA, Dufour JJ 1995. Synchronization of ovarian follicular waves with a gonadotropin- releasing hormone agonist to increase the precision of estrus in cattle: a review. Journal of Animal Science 73: 3143 – 52.

Uphale DS, Takarkhede RC, Kolte AY, Barmase BS, Dhore RN 1998. Oestrous synchronization and superovulation by PGF2 alpha in the presence of different level of PMSG in Malpura ewes. Indian Journal of Animal Reproduction 19: 13-14.

Walker SK, Smith DH, Seamark RF 1986. Timing of multiple ovulations in the ewe after treatment with FSH or PMSG with and without GnRH. Journal of Reproduction and Fertility 77: 135-42.

Wheaton JE, Carlson KM, Windels HF, Johnston LJ 1993. CIDR: a new progesterone-releasing intravaginal device for induction of estrus and cycle control in sheep and goats. Animal Reproduction Science 33: 127-41.

Wilmut I, Hunter RH 1984. Sperm transport into the oviducts of heifers mated early in oestrus. Reproduction Nutrition Development 24: 461-8.

Windorski EJ, Redmer DA, Luther JS, Bilski JJ, Kirsch JD, Kraft KC, Borowczyk E, Vonnahme KA, Reynolds LP, Grazul-Bilska AT 2007. Superovulation in sheep: number and weight of the corpora lutea and serum progesterone. Proceedings, Western Section, American Society of American Science 58. Sheep and Goat Research Journal, 22: 26-31

Winterberger-Torres S, Sevellec C 1987. Atlas du développement embryonaire precoce chez les ovins. Institute National de la Recherche Agronomique. Paris, France.

Wright JM 1981. Nonsurgical embryo transfer in cattle: Embryo-recipient interactions. Theriogenology 15: 43-56.

Wright RW Jr, Bondioli K, Grammer J, Kuzan F, Menino A Jr, 1981. FSH or FSH plus LH superovulation in ewes following estrus synchronization with medroxyprogesterone acetate pessaries. Journal of Animal Science 52: 115-8.

Zambrano-Varón JL, BonDurant RH 2007. Superovulatory response of water buffalo (Bubalus bubalis) to gonadotrophins out of the breeding season. Italian Journal of Animal Science 6: 626-8.

2

Application of Molecular Markers in Livestock Improvement

Vikas Mahajan and Dhirendra Kumar

Brief History

1953: discovery of the molecular structure of DNA

1961-1965: decoding of the genetic code

1977: first method developed for DNA sequencing

1980: the principle of genetic mapping by anonymous molecular markers

1987: first "automatic" DNA sequencing machine developed

1990: publication of the BLAST algorithm for the comparison of DNA sequences

Sequencing of the human genome

- First map of genetic markers of the human genome in 1987
- Birth of the sequencing project in 1988 and the first version in 2001
- The genome sequence was considered as complete in 2004

Sequencing of farm animal genomes

- The construction of maps of genetic markers for cattle, pigs and chicken began in the 1990s
- Transcriptomic analysis began in the year 2000

- First version of the sequence of the chicken genome was obtained in 2004 (Sequencing began in 2002), publication of the genome sequence of the dog in 2005, the cattle in 2006, the horse in 2007 and the pig in 2009.

Introduction

During the past few decades, several molecular marker technologies have been developed and applied for genome analysis, predominantly assessing the differences between individuals within a species and for the association of genomic regions with heritable traits. Biological level of differentiation between all organisms is DNA and among DNA sequences; there is an enormous proportion of genetic variation. Molecular techniques allow detecting variation or polymorphisms exists among individuals in the population for specific regions of the DNA. These polymorphisms can be used to build up genetic maps and to evaluate differences between markers in the expression of particular traits in a family that might indicate a direct effect of these differences in terms of genetic determination on the trait. Molecular markers are used for determining such genetic variation and biodiversity with high levels of accuracy and reproducibility.

It is now possible to uncover a large number of genetic polymorphisms at the DNA sequence level, and to use them as markers for evaluation of the genetic basis for the observed phenotypic variability. Though theoretically DNA sequencing is the direct approach to reveal such DNA polymorphism, it has two practical limitations:

i. Sequencing needs initial cloning of the gene or DNA fragment at which allelic variation of interest exists.

ii. It requires suitable and cost-effective method for scoring DNA sequence variation.

However, the indirect approach for uncovering of genetic variation at the DNA level using molecular or DNA markers obviates the above limitations.

Molecular Markers and Their Different Types

The markers are the genetic basis for assessing of the observed phenotypic variability. A marker is usually considered as a constituent that determines the function of a construction (Webster's Dictionary). Genetic marker can be defined as any stable and inherited variation that can be measured or detected by a suitable method, and can be used subsequently to detect the presence of a specific genotype or phenotype other than itself, which otherwise is non-measurable or very difficult to detect.

Properties that define the genetic markers are	An ideal marker should be:
(i) Locus specificity	(i) Multi-allelic
(ii) Polymorphism	(ii) Highly polymorphic
(iii) Ease of genotyping	(iii) Widely available
	(iv) Inexpensive to genotype
	(v) Evenly distributed

Molecular markers are distinguished on four principle classes

i. Based on visually evaluated traits (morphological and productivetraits)

ii. Based on gene product (biochemical markers)

iii. Immunological markers (blood group and major histo-compatibility complex)

iv. Based on DNA analysis (molecular/genetic markers)

The markers revealing variations at the DNA level are referred to as the molecular markers. The idea of using the genetic markers came forward very early in the scientific studies (Sax 1923) but the development of electrophoretic separation of isozymes (Market and Moller, 1959) and molecular markers (Botstein *et al.* 1980, Nakamura *et al.* 1987, Jaccoud *et al.*, 2001) made better the comprehensions in biological based sciences. Molecular markers are specific fragments of DNA that can be identified within the whole genome and transmitted by the Mendel's laws of inheritance in the generations. They are not considered as normal genes and as a rule do not have phenotypic effect. The existence of different molecular techniques requires attentive consideration in choosing the relevant marker.

The quality of genetic marker is measured by

i. The level of heterozygosity (H) of a locus is defined as the probability that an individual is heterozygous for that locus for a population. When P is at least 1% it is described as polymorphic. A genetic marker with H higher than 70% is considered as highly polymorphic.

$$\text{PCI} = 1 - \sum_{i=1}^{I} \text{p}^{i^2}$$

Where, pi is the frequency of the ith allele among a total of I alleles

ii. The polymorphism information content (PIC) value is commonly used in genetics as a measure of polymorphism for a marker locus used in linkage analysis.PIC is described as the probability that for a given progeny one could identify which homologous of a given parent was transmitted to the progeny, the other parent being genotyped as well. PIC is the product of H of the parents and informativeness of the offspring.

$$PCI = 1-\sum_{n=1}^{n} p^{i^2} - 2\left[\sum_{n=1}^{n-1}\sum_{j=i+1}^{n} p^{i^2} p^{j^2}\right]$$

Where pi is the frequency of the ith allele, and n is the number of alleles (Botstein *et al.* 1980)

The modern developments in molecular biological research connected with livestock breeding include by Teneva and Petrovic (2010).

a. Assignment of the entire genome sequence of the most important livestock species.

b. Development of technology to measure DNA polymorphisms at loci widespread in the genome (microsatellites and SNPs)

c. Development of microarray technology to measure gene transcription. The molecular markers are mainly classified into following types

On the basis of techniques used for their detection, these have been classified into two major categories: Hybridization-based markers and PCR-based markers.

The hybridization-based markers

These include the traditional Restriction Fragment Length Polymorphism (RFLP) analysis (*Botstein et al. 1980*) as well where in appropriately labeled probes for the genes of importance (e.g., cDNA or genomic DNA sequences) are hybridized on to filter membranes containing Restriction Enzyme (RE)-digested DNA separated by gel electrophoresis and subsequently transferred onto these filters by Southern blotting. The polymorphisms are then visualized as hybridization bands. The individuals carrying different allelic variants for a locus will show different banding patterns. Hybridization can also be carried out with the probes (e.g., genomic or synthetic oligonucleotide) for the different families of hypervariable repetitive DNA sequences namely; minisatellite simple repeats (All *et al.*, 1986) Variable Number of Tandem Repeats (VNTR) (Jeffreys *et al.*, 1985) (Nakamura *et al.*, 1987) Litt and Luty, 1989) and microsatellite to reveal highly polymorphic DNA Fingerprinting Patterns (DFP).

The PCR-based markers

These have however, removed the necessity of probe-hybridization step and have led to the discovery of several useful and easy-to-screen methods. Depending on the type of primers (i.e., primers of specific sequences targeted to a particular region of a genome or primers of arbitrary sequences) used for PCR, these markers can be further sub-divided into the following two groups:

The sequence-targeted PCR assays

In this assay system, a particular fragment of interest is amplified using a pair of sequence-specific primers. In this category, PCR-RFLP or Cleaved Amplified Polymorphic Sequence (CAPS) analysis is a useful technique for screening of sequence variations that give rise to the polymorphic RE sites. Such analysis involves amplification of a specific region of DNA encompassing the polymorphic RE site and digestion of the amplified DNA fragment with the respective RE. However for the screening of the sequence variations that do not lead to creation or abolition of restriction sites while other approaches namely allele specific PCR (Nichols *et al.* 1989). PCR Amplification of Specific Alleles (PASA) (Sarkar *et al.*, 1990). Allele Specific Oligonucleotide hybridization assay (*Saiki et al. 1986*), Amplification Refractory Mutation System (ARMS) (*Newton et al. 1989*) and Oligonucleotide Ligation Assay (OLA) are used. These assays are based on the principle of high specificity of PCR to selectively amplify specific alleles using primers that match the nucleotide sequence of one but mismatch the sequence of other allele. The sequence-targeted PCR approach is also employed to reveal Simple Sequence Length Polymorphism (SSLP), using a pair of primers that flank the Simple Sequence Repeat (SSR) motifs. If cloned and sequenced microsatellite loci can be subjected to PCR amplification and such microsatellite loci can be recovered by PCR such loci are termed as Sequence Tagged Microsatellite Site (STMS) markers (Beckmann and Soller, 1990).

The arbitrary PCR assays

In this assay system, however unlike the standard PCR protocol, randomly designed single primer is used to amplify a set of anonymous polymorphic DNA fragments. It is based on the principle that when the primer is short (usually 8-10 mm) there is a high probability that priming may take place at several sites in the genome that are located within amplifiable distance and are in inverted orientation. Polymorphism detected using this method is called Randomly Amplified Polymorphic DNA (RAPD) Based on this principle, several techniques which do not require any prior sequence knowledge have been developed. However, they differ in number and length of primers used, stringency of PCR conditions and the method of fragment separation and detection. In Arbitrary Primed PCR (AP-PCR) (*Welsh and McClelland 1990*), slightly longer primer is used (e.g., universal M13 primer) and amplification products are detected by radiactive or nonradioactive method following polyacrylamide gel electrophoresis. In DNA Amplification Fingerprinting (DAF) analysis, shorter primer is used (5-8 mm) which reveals relatively greater number of amplification fragments by polyacrylamide gel electrophoresis and silver staining. All these techniques having similar features can be described by a common term Multiple Arbitrary Amplicon Profiling (MAAP) Besides these, a number of modifications of the basic MAAP assays (namely, template endonuclease cleavage MAAP and RAPD-RFLP) have been developed as well.

In addition to arbitrary primers, semi-arbitrary primers designed on the basis of RE sites or sequences that are interspersed in the genome such as repetitive sequence elements (Alu repeats or SINEs), microsatellites and transposable elements are also used. In the Amplified Fragment Length Polymorphism (AFLP) assay ,template DNA is digested with two REs and the resulting restriction fragments are then ligated with adapters and subsequently, PCR amplification is carried out using specially designed primers which comprise; a unique part corresponding to selective bases and a common part corresponding to the adapters and the RE site. Microsatellite-Primed PCR (MAP-PCR) assay is carried out using microsatellite as the primer.

On the basis of markers: mitochondrial (mt) DNA and nuclear markers

Mitochondrial (mt) DNA and nuclear markers

The mitochondrial genome comprises a circular 'chromosome' of DNA. Animal mt DNA ordinarily contain 36 or 37 genes; two for ribosomal RNAs, 22 for tRNAs and 12 or 13 for subunits of multimeric proteins of the inner mitochondrial membrane. In addition, there is a noncoding sequence termed the control region (CR) due to its role in replication and transcription of mtDNA molecules. Exons in the mtDNA circle are tightly packed, with no spacing introns. Mitochondrial DNA is histone-free, has limited repair ability, and therefore has a relatively high mutationfixation rate (5–10 times that of single-copy nuclear DNA). mtDNA is a circular DNA molecule of the mitochondrion, a cellular organelle found in hundreds to thousands of copies in the cytoplasm of each cell. mtDNA is inherited uniparentally from mother to offspring (Brown *et al.* 1979). mtDNA markers can provide only information concerning female evolutionary history, gene flow, and genetic diversity; and Y chromosome markers can only provide information about paternal evolutionary history, gene flow, and genetic diversity. The widely used mitochondrial DNA markers with decreasing order of conserved sequences are 12S rDNA > 16S rDNA > cytochrome b > control region (CR); thus the 12S rDNA is highly conserved and the CR is highly variable. Nuclear DNA resides in the nucleus of cells and is inherited biparentally. Thus, cells have 2 copies of each nuclear chromosome - one from the mother and one from the father. Nuclear DNA markers can provide information about both maternal and paternal evolutionary history, gene flow, genetic diversity and relatedness.

Table 1: Characteristics of various mtDNA markers

Molecular marker	Characteristics
mtDNA	Inherited from the mother (maternal lineage); rare exceptions do exist*
mtDNA	Degrades slower than nuclear DNA. It can be used in degraded or old samples
mtDNA	Evolves about 10–fold faster than nuclear DNA
12s rDNA	Highly conserved; used for high–category levels: phyla and subphyla
16S rDNA	Usually used in mid–category differentiation such as families
Protein–coding genes	Conserved; used in low–categories such as families, genera and species
Control region	Hypervariate; used for identification of species and sub–species

*Not all mitochondria in animals are inherited maternally. Rare instances of paternal leakage have been found in some species including mice, birds and humans. However, the extent of paternal leakage is thought to be low in most animals, with the exception of certain mussel species which follow double biparental inheritance (Freeland 2005).

Table 2: Classification of the DNA markers on the bases of locus (Crawford et al. 2000)

DNA markers	
Multilocus markers	Single-locus markers
Minisatellite or Variable number tandem repeat (VNTR)	Restriction fragment length polymorphisms (RFLPs)
Random amplified polymorphic	Microsatellites
	DNA fragments (RAPD)
Amplified fragment length polymorphisms (AFLP)	Single nucleotide polymorphisms (SNPs)

Multilocus markers

Minisatellite or Variable number tandem repeat (VNTR) markers

The minisatellites were the first tandem repeat markers with multiple alleles to be developed. Discovered by Jeffreys *et al.* (1985), they were the first markers to be sufficiently informative to reveal a unique genotype in each individual. With the advent of microsatellites they have fallen out of favour for linkage studies because they are technically demanding and require a large quantity of DNA for analysis. They were mostly dominant markers with only one identifiable allele. Minisatellites were also located mainly in telomeric regions so that good coverage of the genome using these markers could not be achieved.

A few highly informative single-loci minisatellites were identified in livestock (Georges *et al.* 1990) and these have been found useful. The first marker linked to the callipyge gene of sheep was a single-locus minisatellite (Cockett *et al.* 1994).

Silent points

a. Multi-allelic marker

b. Hyper-variability due to change in number of repeats due to unequal recombination and replication slippage of replication forks.

c. 11-60 bp repeats

Random amplified polymorphic DNA fragments (RAPD) markers

These were one of the first PCR-based markers to be used (William *et al.*, 1990). Small primers (8—10 bp) were used to amplify a mixture of random fragments of DNA from the genome. The size of the primers was set such that about 20 bands were amplified by each PCR reaction. Some of these bands would be polymorphic and could be used as genetic markers. These markers have the great advantage of being very easily generated and requiring only small quantities of DNA. For this reason many linkage maps, especially in plants, have used RAPD markers. Since heterozygous and homozygous individuals cannot be differentiated, the markers are dominant. The appearance and disappearance of bands is very sensitive to slight changes in PCR conditions, so that RAPD markers are not easily reproducible. The most serious disadvantage, however, is that a new map must be generated for each new pedigree being examined, as there is no locus specificity in the primers used. The bands generated by a particular primer in one pedigree may not bear any relation to bands generated by the same primer in a second pedigree.

Amplified fragment length polymorphisms (AFLP) markers

These are now the multilocus markers of choice (Vos *et al.* 1995). Genomic DNA is cut with restriction endonucleases and linkers are ligated to each end of the restriction fragments. Selective PCR primers are then used to amplify subsets of fragments from the mixture of genomic restriction fragments. The selective primers encompass the linkers added on to the end of the restriction fragments and include additional bases to the 3^1 end of the primer to give additional specificity. The amplified fragments are then separated according to size. Those bands that are present in some individuals but absent in others can be used as genetic markers.

These markers have the same advantage as RAPD, being easily generated, but they are much less reliant on PCR conditions staying exactly the same in order to obtain reproducible amplification products. To obtain a new set of markers, small changes in the 3^1 bases of the amplifying primers are all that is required, so this technology can provide a limitless supply of new markers. The markers have the disadvantages of being dominant, requiring the generation of a new map for each new set of pedigrees being studied.

Single-locus markers

Restriction fragment length polymorphisms (RFLPs)

These, the first DNA markers developed, predate the development of PCR methods. The RFLP method utilizes restriction endonuclease digestion of genomic DNA, its separation by size using agarose gel electrophoresis, and detection and analysis of the DNA sequence by Southern blotting. The RFLP marker detects the presence or absence of a restriction site. RFLP markers are codominant, biallelic and can detect heterozygosity upto 50%. The two major disadvantages of these markers are that they are technically demanding and require large quantities of DNA.

Microsatellites

Microsatellites are the markers responsible for the recent expansion in genetic linkage maps of farm animals. Like the minisatellites, they are multiallelic tandem repeats. However, they are single-locus, codominant, spread throughout the genome, require only small amounts of template DNA and are relatively easy to find and characterize.

At the heart of any microsatellite is a simple sequence, either a mono-, di-, tri- or tetranucleotide, that is repeated between 10 and 50 times. Virtually all of the microsatellites that have been found for livestock have the sequence (AC/GT) as the repeat unit. The reason for this is not that the other types do not exist but that this type is the most abundant within the livestock genomes and hence is much easier to find and characterize. The variation between alleles of the microsatellites is due to variation in the number of simple sequence repeats. Microsatellites are typed by designing primers to the unique DNA sequences on either side of the repeat and, using PCR, amplifying the region containing the repeat. The size of the PCR product is then measured, usually by electrophoresis on a DNA sequencing gel. Alleles differ in size according to the difference in number of simple-sequence repeats.

Silent points:-

a. More frequent in eukaryotes and less in prokaryotes.

b. 1-6 bp repeats

c. Multialleic and codominant

Single nucleotide polymorphisms (SNPs)

The remaining variation found within the genome of animals occurs as changes in single nucleotides. The RFLP the first DNA marker developed for, eukaryotes detects the variations that occur within a restriction endonuclease cut site. This, however, is only a small subset of all the polymorphic nucleotides in the genome. Very few of the nucleotide changes have any deleterious effect, so the preferred name is single nucleotide polymorphisms (SNPs) rather than point mutations. It has been estimated that an SNP occurs about once every kilobase of unique sequence in humans (Cooper *et al.* 1985). In other mammals, such as livestock, the figure is likely to be similar, so at least two to three million SNPs remain to be identified and characterized in any livestock species. SNPs represent by far the richest source of genetic variation available for research purposes.

With the advent of PCR any unique region of DNA could be amplified and scanned for variation in its DNA sequence. These SNPs can be detected in numerous ways. Once detected and characterized a wide variety of systems is now available for typing SNPs.

If SNPs occur at sufficiently high density, such as the hypervariable exon 2 region of DRB and DQB genes, direct sequencing of the DNA can be a cost-effective method of typing. More often, however, SNPs are hundreds of base pairs from their nearest neighbour, which means that other methods of detection are employed.

A base change in a short (approximately 100 bp) amplified fragment of DNA will often change the secondary structure of the DNA when it is melted and allowed to reanneal as single strands (Orita *et al.* 1989). The differences in secondary structure can be detected by electrophoresis on non-denaturing polyacrylamide gels, where one structure travels at a different speed through the gel. Ion gradients, gel density gradients and strictly controlled gel temperature have all been found useful in resolving these differences in secondary structure. The different methods of band resolution have led to many different acronyms, but the most commonly used name for this methodology is single-stranded conformational polymorphism (SSCP).

If the DNA sequence around the SNP is known, a number of typing methods become available. The most simple is that of Zhu and Clarke (1996). Four primers are designed. Two primers amplify the region containing the point mutation. Two further primers are designed such that their 32 ends correspond to one of the bases at the SNP Depending on which primer was present in the reaction and which mutation(s) were in the template DNA; a smaller band would or would not be amplified. In this way, two PCR reactions can be used to type the SNP.

The invention of the 'Taqman' system of typing (Holland *et al.* 1991) allows typing of SNP markers without needing an electrophoresis step. Instead, the different alleles are detected by a change in the colour of the PCR reaction. Using a fluorogenic probe complementary for the target DNA sequences being amplified, the system detects and quantifies cycle by cycle increases in the level of PCR products. The probe consists of an oligonucleotide with a reporter and quencher dye attached. Uncoupling of the two dyes occurs when the probe, bound to the internal sequence of the PCR product, is cleaved by the nucleolytic activity of the T polymerase, which results in an increase inaq the fluorescence intensity of the reporter dye.

For SNP analysis, competition between oligonucleotides differing only at the point mutation is used. The two allele-specific oligonucleotides, which differ only by one base-pair change, are labelled with different dyes. Binding of the 'correct' oligonucleotide is detected by increased fluorescence dye used to label the 'correct' oligonucleotide. With a heterozygote, a mixture of the two dyes is released and detected. This method should reduce both the cost and time required to type SNPs.

The recent development of 'DNA chips' containing high-density arrays of DNA (Chee *et al.* 1996) may speed up the analysis process even further. Perhaps now is the time to begin searching for more SNPs in livestock, so that a large battery of SNPs is available when these new technologies become available.

Genetic Mapping

Genetic maps are sets of loci arranged in order and separated by distances (in units depending on the type of map), with each set corresponding to a chromosome pair. Locating genes of interest by scanning the genome in a linkage study is best conducted using a set of markers evenly spaced throughout the genome. These may be chosen from a marker map. Once linkage has been established, these and other types of maps help refine the location. Maps of genes or expressed sequences may provide clues to an effect's cause. Maps are also useful in selecting markers for other uses, such as population studies, tracking the inheritance of DNA segments using flanking markers, and parentage identification.

Table 3: Showing list of molecular marker with most suitable use

Molecular marker	Suitable for
mtDNA markers	Useful to study evolutionary relationship and biodiversity
RAPD and AFLP	Do not require prior molecular information
AFLP	Has greater differentiation power than RAPD
RAPD	Is a simple method and is the least–expensive
Microsatellite and AFLP	Are highly powerful in determining genetic diversity
Microsatellite	Are best for parentage and strain analysis

Properties of molecular markers

i. Morphological (e.g. pigmentation or other features) and chromosomal (e.g. structural or numerical variations) markers usually show low degree of polymorphism and, hence, are not very useful for genetic markers.

ii. Biochemical markers have been tried out extensively, but have not been found encouraging as they are often sex limited, age-dependent, and are significantly influenced by the environment. Sometimes, the various genotypic classes are indistinguishable at the phenotypic level owing to dominance effect. Furthermore, these markers reflect variability in their coding sequences that constitute less than 10 per cent of the total genome.

iii. The molecular markers, capable of detecting the genetic variation at the DNA sequence level, they are numerous and distributed ubiquitously throughout the genome. These follow a typical Mendelian inheritance which usually expresses in a co-dominant fashion, and are often multiallelic giving mean heterozygosity of more than 70 per cent. They remain unaffected by the environmental factors, and generally do not have pleiotropic effects on quantitative trait loci (QTL) Gelderman, 1975. Since gene expression is not a prerequisite, virtually the entire genome including the non-coding regions can be visualized.

For genetic analysis, molecular markers offer several methodological advantages that are both attractive as the DNA samples can be

a Isolated from blood, tissues like sperm, hair follicle, and even from archival preparations.

b. Stored for longer periods and can readily be exchanged between the laboratories

c. Analysis at an early age or even at the embryonic stage, irrespective of the sex

d. Once the DNA is transferred on to a solid support, such as filter membranes, it can be repeatedly hybridized with the different probes, and moreover, heterologous probe and in vitro-synthesized oligonucleotide probes can also be used

e. The PCR-based methods can be subjected to automation.

Table 4: Technical requirements and characteristics

Marker name	Technical requirements				Technical characteristics			
	Restric enzyme	PCR	Specific primers	Gel	Development efforts	Genp typing efforts	Repro ducibility[1]	Accuracy[2]
RFLP	+	-	-[3]	+	High	High	High	Very high
PCR-RFLP	+	+	+	+	High	Medium	High	Veryhigh
RAPD	-	+	-	+	Very low	Very low	Low	Verylow
AFLP	+	+	-	+	Low	Very low	High	Medium
SSCP		+	+	+	Medium	Medium	Medium	Medium
Microsatellite	-	+	+	+	High	Low	High	High
SNP	-	+	+	+/-	High	Variable	High	Very High

1. Refers to the genotyping error rate of the method: results may vary from one experiment to another.
2. Refers to the precision at which true allele recognition can be performed.
3. However, the RFLP technique relies on the use of a specific probe for the Southern-blot technique. Nowadays, RFLPs are usually genotyped by PCR-RFLPs, requiring specific primers.

Table 5: The main categories of molecular markers (Vignal *et al.* 2002)

Marker name	Variation type			Information content		
	SNP1	Indel 2	VNTR3	2 dominant alleles	2 co-dominant alleles	Multi allelic co-dominant
RFLP	+	(+)[4]	(+)	-	+	(+)[5]
PCR-RFLP	+	(+)[4]	(+)	-	+	-
RAPD	+	(+)[4]	(+)	+	-	-
AFLP	+	(+)[4]	(+)	+	(+)[6]	-
SSCP	+	(+)[4]	(+)	-	+	(+)[5]
Microsatellite	-	(+)[7]	(+)	-	-	+
SNP	+	(+)[8]	-	-	+	-

1. Single nucleotide polymorphism: any kind of base substitution. The fact that SNPs appear both as a variation type and a marker name, is due to the fact that in reality, many genotyping techniques used for genotyping SNPs are grouped under this generic marker name.
2. Insertions and deletions.
3. Variable number of tandem repeats.
4. Although the RAPD, AFLP, RFLP, PCR-RFLP and SSCP techniques will detect base substitutions in the vast majority of cases, the two other types of DNA variation can also be analysed.
5. In some instances, more than two alleles can be analysed.
6. With an automatic sequencer, some markers can be scored as co-dominant.
7. Variations in PCR product length can be due to a deletion in the sequence flanking the microsatellite.
8. Many SNP detection techniques can also be used for scoring small insertions or deletions (indels).

Table 6: Comparison of commonly used genetic markers by *Mburu and Hanotte (2005)* *

Feature	RFLP	SSR	RAPD	AFLP	Isozymes	STS/EST
Origin	Anonymous/ Genic	Anonymous	Anonymous	Anonymous	Genic	Genic
Maximum theoretical number of possible loci inanalysis	Limited by the restriction site (nucleotide) polymorphism (tens ofthousands)	Limited by the size of genome and number ofSimple repeats in a genome(tens of thousands)	Limited by the size ofgenome, and by nucleotidepolymorphism (tens ofthousands)	Limited by the restriction site (nucleotide) polymorphism (tens ofthousands)	Limited by the number ofenzyme genes & histochemical enzymeassays available (30-50)	Limited by the number of expressed genes (10,000- 30,000)
Dominance	Codominant	Codominant	Dominant	Dominant	Codominant	Codominant
Null alleles	Rare to extremely rare	Occasional to common	Not applicable (presence/ absence type of detection)	Not applicable (presence/ absence type of detection)	Rare	Rare
Transferability Across Related species	Across genera	Within genus or species	Within species	Within species	Across families and genera	Across related species
Reproducibility	High to very high	Medium to high	Low to medium	Medium to high	Very high	High
Amount of sample requiredper sample	2-10 mg DNA	10-20 ng DNA	2-10 ng DNA	0.2-1 μg DNA	Several mg of tissue	10-20 ng DNA
Ease of development	Difficult	Difficult	Easy	Moderate	Moderate	Moderate
Ease of assay	Difficult	Easy to moderate	Easy to moderate	Moderate to difficult	Easy to moderate	Easy to moderate
Automation / multiplexing	Difficult	Possible	Possible	Possible	Difficult	Possible
Genome and QTL mapping Potential	Good	Good	Very good	Very good	Limited	Good
Comparative Mapping potential	Good	Limited	Very limited	Very limited	Excellent	Good tovery good

Contd.

Feature	RFLP	SSR	RAPD	AFLP	Isozymes	STS/EST
Candidate gene mapping Potential	Limited	Useless	Useless	Useless	Limited	Excellent
Potential for studying adaptive genetic variation	Limited	Limited	Limited	Limited	Good	Limited
Development	Moderate	Expensive	Inexpensive	Moderate	Inexpensive	Expensive
Assay	Moderate	Moderate	Inexpensive	Moderate to expensive	Inexpensive	Moderate
Equipment	Moderate	Moderate to expensive	Moderate	Moderate to expensive	Inexpensive	Moderate to expensive

*MBURU D., HANOTTE O. (2005): A practical approach to microsatellite genotyping with special reference to livestock population genetics. ILRI Biodiversity project A manual prepared for the IAEA/ILRI training course on molecular characterisation of small ruminant genetic resources of Asia, October- December 2005, ILRI, Nairobi, Kenya.

Table 7: Different marker systems and their comparison (Ibitoye and Akin-Idowu 2010)

Marker	Advantages	Disadvantages
RFLP (Restriction Fragment Length Polymorphism)	-unlimited number of loci -co-dominant -many detection systems -can be converted to SCARs -robust in usage -good use of probes from other species -detects in related genomes -no sequence information required	-labour intensive -fairly expensive -large quantity of DNA needed -often very low level of polymorphism -can be slow (often long exposure times) -needs considerable degree of skills
RAPD(Random Amplified Polymorphic DNA)	-results obtained quickly -fairly cheap -no sequence information required -relatively small DNA required -high genomic abundance -good polymorphism-can be automated	-highly sensitive to laboratory changes -low reproducibility within and betweenlaboratories -cannot be used across populations noracross species-often see multiple loci
SSR(Simple Sequence Repeat)	-highly polymorphic -fast-robust -can be automated -small quantity of DNA -multi-allelic -does not require radioactive labellling -co-dominant	-high developmental and startup cost -usually single loci even in polyploids -species-specific -difficult interpretation because of stuttering
ISSR (Inter- Simple Sequence Repeat amplification)	-robust in usage -can be automated -highly polymorphic	-usually dominant -species specific
AFLP (Amplified Fragment Length Polymorphism)	-can be automated -no sequence information required -small DNA quantities -can be adapted for different uses e.g. cDNA-AFLP	-marker clustering -dominant -technique is patented -evaluation of up to 100 loci
IRAP/REMAP (Inter-Retrotransposon Amplified Polymorphism/ Retrotransposon-Microsatellite Amplified Polymorphism)	-highly polymorphic depends on the transposon -robust in usage -can be automated -species -specific	-alleles cannot be detected -can be technically challenging
SNP(Single Nucleotide Polymorphism)	-robust -suitable for high throughput -different detection methods available -can be automated -polymorphism are identifiable	-very high development costs -require sequence information -can be technically challenging
Morphological	-usually fast -usually cheap	-few in number -often not compatible with breeding aims -need to know the genetics
Protein and Isozyme	-fairly cheap -protocol for any species -fairly fast	-often rare -often different protocol for each locus -labour intensive

Contd.

Marker	Advantages	Disadvantages
	-co-dominant -require no sequence information	-sometimes difficult to interpret.
Microarray	-highly abundant -co-dominant -single base changes -no gel system -suitable for high throughput -highly polymorphic -highly reliable -small DNA quantity required	-very high development and start -up costs -portability unknown
SCAR/CAPS(Sequence Characterized Amplified Region/ Cleaved Amplified Polymorphic Sequences)	-highly reliable -small DNA required -co-dominant -usually single locus -species-specific	-very labour intensive
STS/EST(Sequence Tagged Site/ Expressed Sequence Tag)	-fast -cDNA sequences -non-radioactive	-substantially decreased levels of polymorphism -sequence information required.

Table 8: Properties of DNA-based markers (Jerry *et al.* 1997)

Variable	Clone sequence-based (CSB)			Fingerprint (FP) type		
	RFLP	Micro-satellite	STS (and EST)	RAPD	Mini-satellite	CR1
No. on EL Map, 7/96 [1]	87	340	30	67	36	47
Genome distribution	Ubiquitous	Ubiquitous	Ubiquitous	Ubiquitous	Hetero chromatin	Ubi quitous
Approx. no./genome	>100,000	>10,000	>>100,000	>100,000	Unknown	100,000
Genome surveyed[2]	sc & mr	sc	sc	sc & mr	mr to hr	sc & mr
Typical PIC3	Low	High	Low	Moderate	High	Moderate
Typical allele no.	2	2–10	2	2	2	2
Type of sc & mr (O'Brien)	I and II	II > I	I > II	II	II	II
Inheritance mode	Co-dominant	Co-dominant	Co-dominant	Dominant	Dominant	-dominant
Reliability	High	High	Variable[4]	Low	High	Moderate
Speed of assay	Low	High	Variable[4]	High	Low	High
Initial investment	Moderate	High	Variable[4]	Low	Low to moderate	Low

[1]Of 617 total markers, including some classical marker loci and some loci not yet publicly available.
[2]Genome surveyed is the portion of the chicken genome that most markers of a class survey; sc = single copy, mr = moderately repetitive, and hr = highly repetitive.
[3]Typical PIC is the common polymorphic information content for a single locus of the class of marker indicated.
[4]Depends on which of the many possible assay methods is used, generally moderate to high reliability, speed, and initial investment.

Applications of molecular markers

Polymorphisms observed at the DNA sequence level have been playing a major role in the field of animal genetics and breeding for gene mapping, pre- and post-natal diagnosis of genetic diseases, and anthropological and molecular evolution studies has opened many vistas in livestock improvement programmes (Beckmann and Soller 1983, Smith and Simpson 1986). Consequently, enormous interest has been generated in determining genetic variability at the DNA sequence level of different livestock species, and in their assessment whether these variations can be exploited efficiently in conventional as well as in transgenic breeding strategies.

Conventional breeding strategies

Molecular markers can play an important role for livestock improvement through conventional breeding strategies. The various possible applications of molecular markers are:

Short-range or immediate applications

Molecular markers have several immediate applications like parentage determination, genetic distance estimation, determination of twin zygosity and freemartinism, sexing of pre-implantation embryos and identification of disease carrier. In the following subsections, each of these applications has been discussed briefly.

(i) Parentage determination

Since the breeding value of an animal is generally estimated using the information available from its relatives, the knowledge of correct parentage is therefore a prerequisite. Parentage testing using molecular markers yields much higher exclusion probability (>90%) than the testing with blood groups (70–90%) or other biochemical markers (40–60%) (Geldermann 1990). Highly polymorphic DNA fingerprinting markers (Jeffreys 1985) are quite useful for this purpose. Recently, DNA fingerprinting with oligoprobes (OAT18 and ONS1) has been successfully used for determining the parentage of IVF buffalo calf (Mattapallil 1988). With the advent of PCR-based microsatellite assays, a large number of microsatellite panels have been reported that are useful for parentage testing in different livestock species. For example in cattle, Glowatzki-Mullis *et al.* 1995 demonstrated that using two triplex microsatellite co-amplification systems, wrong parentage can be excluded with almost 99% accuracy. In addition, molecular markers also serve as a useful tool for animal identification, particularly for verification of the semen used for artificial insemination Geldermann 1990.

(ii) Genetic distance estimation

Genetic distance, a measure of overall evolutionary divergence, i.e. genetic similarities and dissimilarities between two populations (such as between species, breeds, strains), serves as an useful tool for authentication of the pedigree, for characterization of different breeds or strains within a species, and for evaluation of the change in variation in species over time. In principle, genetic distance can be measured on the basis of polymorphic characters occurring at the different levels, viz. morphological, biochemical, cellular and DNA level. Allelic frequencies of blood groups (Kidd 1974) as well as those of other biochemical loci, e.g. serum and milk proteins (Baker 1982), have been used extensively for the estimation of genetic divergence of different livestock species. However, a great amount of genetic variations at protein loci remain undetected, since changes in the underlying nucleotide sequences may not necessarily lead to corresponding change in the amino acid sequences owing to degeneracy in the genetic code. Molecular markers capable of generating individual specific DFP patterns, useful for establishing familial relationships Jeffreys *et al.* 1986, Jeffreys and Morton 1987, can serve as an alternative. The similarities between the DFP patterns that are expressed by band-sharing values, provide a reliable method for evaluating genetic distance amongst populations Kuhnlein *et al.* 1989, Dunnington *et al.* 1991. Presently, the PCR-based RAPD fingerprinting assays are being used for characterization of zebu cattle breeds Gwakisa *et al.* 1994, for detection of genetic variations in cattle and sheep Kantanen et al. 1995, and characterization of highly inbred chicken lines in poultry Plotsky *et al.* 1995.

(iii) Determination of twin zygosity and freemartinism

Correct knowledge of zygosity of twins, particularly in monotocus animal, is very important. Monozygotic twins provide means for epidemiological as well as for genetical studies, and also help in transplant matching. Individual specific DNA fingerprinting techniques have potential applications in determination of twin zygosity Hill *et al.* 1985 and demonstration of spontaneous XX/XY chimaerism Farber *et al.* 1989. Demonstration of XX/XY chimerism in heterosexual bovine twins, by PCR-RFLP assay using sex-chromosome-specific primers, has enabled the identification of freemartin animal Lipkin *et al.* 1993, Yadav and Mitra 1999.

(iv) Sex determination

Sexing of pre-implantation embryos can serve as an important tool for improving herd for a desired purpose. A large number of invasive and noninvasive methods for sexing embryos are available. However, ideally the technique to be applied should not have any adverse effect on embryo survivability, its conception rate and subsequent development. Besides, the technique should be simple and easy to carry out, repeatable, and accurate and time saving. Though embryos can be sexed by cytogenetical method, this method is quite accurate but invasive and needs a large piece of embryo. The molecular markers on the other hand, have potential application in determination of sex of pre-implantation embryos, since the embryos can be sexed using male-specific or Y-chromosome-specific DNA sequence as probes Vaiman *et al.* 1988. However, this method is time consuming as well as tedious.

The PCR-based method of sex determination offers several advantages over all the other methods:

i. It can be carried out in less than five hours with almost 100 per cent accuracy Peura *et al.* 1991.

ii. It is less invasive and requires very small quantities (in nanograms) of DNA for PCR assay, which can be isolated from two to eight cells biopsied from the embryo Thibier *et al.* 1995.

iii. It can be done at an early stage of embryo e.g. blastocyst stage (6 to 8 days) or even earlier at the 16–32 cell stage Machaty *et al.* 1993.

iv. The use of multiplex PCR allows simultaneous genotyping for important loci like milk proteins, diseases carrier, etc.

v. Identification of disease carrier

Many of the most serious incurable diseases result not from infections with bacteria or viruses but defects in genomes of the hosts. Certain allelic variations in the host genome leads to susceptibility or resistance to a particular disease. Kingsbury 1990 reported that a particular RFLP in the Prion protein (Prn P) gene was responsible for the variation in host's response to the causative agent, and the incubation time of bovine spongiform encephalopathy (BSE). DNA polymorphism occurring within a gene helps to understand the molecular mechanism and genetic control of several genetic and metabolic disorders, and allows the identification of heterozygous carrier animals which are otherwise phenotypically indistinguishable from normal individuals.

Long-range applications

The foremost long-range application of molecular markers in conventional breeding includes mapping of the QTL by linkage. Such mapping information, if available, particularly, for those loci which affect the performance traits or disease resistance/susceptibility, can be used in breeding programmes by either within-breed manipulations, like marker-assisted selection of young sires, or between-breeds introgression programmes.

(i) Gene mapping

Molecular markers have three-fold applications in gene mapping:

a. A marker allows the direct identification of the gene of interest instead of the gene product and consequently, it serve as a useful tool for screening somatic cell hybrids.

b. Use of several DNA probes and easy-to-screen techniques, a marker also helps in physical mapping of the genes using *in situ* hybridization.

c. The molecular markers provide sufficient markers for construction of genetic maps using linkage analysis. Genetic maps are constructed on the basis of two classes of molecular markers O'Brien 1991: Type I markers that represent the evolutionary conserved coding sequences (e.g. classical RFLPs and SSLPs), are useful in comparative mapping strategies where polymorphism is not an essential prerequisite. However, these are mostly single locus and di-allelic (SLDA) and thus are not useful for linkage analysis. On the other hand, the type II markers (like microsatellites markers) have higher polymorphism information content conventional RFLPs and can be generated very easily and rapidly. Therefore, major efforts are being made to produce gene maps based on the type II markers. Further utilization of molecular markers developed from DNA sequences information, namely ASO and STMS polymorphic markers are also helpful in rapid progress of gene mapping.

(ii) Marker assisted selection (MAS)

Most of the traits considered in animal breeding programmes are quantitative. They are controlled by many genes as well as are influenced by environmental factors. The underlying genes have small effects on the phenotype observed. In conventional breeding programmes, selection is carried out based on phenotypic traits. There is no knowledge, which genes are being selected. The concept of marker assisted selection (MAS), utilizing the information of polymorphic loci as an aid to selection, was introduced as early as in 1900s (Sax 1923). With the advent of DNA-based genetic markers in the late 1970s, and their subsequent use to detect associations with traits of interest, thus allowing MAS to become a reality. However, its application in genetic improvement of livestock species has been limited due to lack of suitable genetic markers. The discovery of DNA-level polymorphism in eighties and their subsequent use as molecular markers has renewed interest in the use of genetic markers in selection of breeding stocks. Implementation of MAS essentially involves two steps: Identification of the marker loci that is linked to QTL of economic importance, followed by the utilization of linkage association in genetic improvement programme. Once linkage between a QTL and a marker locus is established, it is possible to recognize the alternative QTL allele inherited by the individual. Such information can then be used for the selection of the breeding stock. MAS is a form of indirect selection and their accomplishment includes two steps: identification of the marker loci that is linked to QTL of economic importance and utilization of linkage association in genetic improvement programme. The association between the markers and the QTL is dependent on the distance between the markers and bull traits, type of linkage phase, and degree of linkage disequilibrium. MAS is leading to increased rate of genetic improvement through higher selection intensity, reduction of generation interval, increase in the accuracy of prediction and selection in early age for sex-limited traits.

Salient requirements for MAS

The success of a marker-based breeding depend mainly on following factors.

1. A genetic map with an adequate number of uniformly spaced polymorphic markers to accurately locate desired quantitative trait loci (QTLs) or major genes.
2. Close linkage between the QTL or a major gene of interest and adjacent markers.
    a. The marker is in linkage disequilibrium (LD) with the gene of interest throughout the population. LD is the tendency of certain combination of alleles to be inherited together. Selection using these markers can be called LD-MAS.
    b. The marker is in linkage equilibrium (LE) with the gene of interest throughout the population, which is a most difficult and challenging situation for applying MAS.
3. Adequate recombination between the markers and the rest of the genome.
4. The molecular maker is located within the gene of interest, which is most favourable and preferred situation for MAS but it is difficult to find. It is referred to as gene assisted selection.

MAS is likely to complement rather than replace the conventional breeding systems leading to increased rate of genetic improvement through higher selection intensity, reduction of generation interval and increase in the accuracy of prediction. Furthermore, selection based on markers is possible in early life or in individuals of both sexes for sex-limited traits. However, there is a risk of reduced genetic response if the marker association information is inaccurate, as MAS is a form of indirect selection. The association between the markers and the QTL is a function of distance between the markers and target traits, type of linkage phase, and degree of linkage disequilibrium. Therefore, a high-density gene map with closer linkage is a prerequisite for successful implementation of MAS Smith and Smith 1993. It is estimated, that an average marker density of 10 cm (5–20 cm), with about 200–250 makers, should be sufficient for the detection of marker–QTL association Beckman and Soller, 1983. Till recently, gene maps with average marker interval exceeding 5 cM in pigs Archibald *et al.* 1995 and 10 cM in cattle Ma *et al.* 1996, sheep Crawford *et al.* 1995, goat Vaiman *et al.* 1996 were available. However, currently high resolution maps with 2.5 cM or even less marker density have been published Kappes *et al.* 1997.

MAS can be used efficiently across the population (between breeds) to incorporate the most desirable alleles into a group of selected individuals/strain/ breeds, termed as introgression. Introgression is less common in domestic animals than in plant species because of limited fertility, longer generation interval and the greater expense of each individual Piper *et al.* 1982, nevertheless, introgression can be applied in domestic animals for genes with major effects Beckmann and Soller 1987. Notable examples are the Booroola gene for increased fecundity Piper *et al.* 1982, and genes for trypanosoma tolerance for N'Dama cattle Soller M, Beckmann 1987.

Molecular markers are capable of unravelling genetic variations in both the coding and noncoding sequence regions. Based on this characteristic, Geldermann 1990 suggested two approaches for the identification of markers that influence QTL, as given below.

(iia) Polymorphisms in the coding sequences

DNA polymorphisms that occur in and around the structural and/or regulatory sequences of a gene of physiological significance (e.g. hormone genes, milk protein genes, MHC) may directly affect gene expression (by changing the splicing of mRNA, stability of mRNA, rate of gene transcription or the sequence of gene product) and thereby contribute to the phenotypic variations among the individuals in terms of productivity and health (disease resistance/susceptibility). Consequently, such DNA polymorphisms, occurring in the genes which already have *a priori* possibility to be associated or closely linked with the performance trait of importance, can be selected as markers Beckmann, Soller 1987.

(iib) Polymorphisms in the noncoding sequences

In this approach, the variations occurring in non-coding sequences (e.g. flanking regions or intergenic regions) are utilized indirectly as markers for linkage analysis Beckmann and Soller 1983, Smith and Simpson 1986. Microsatellite markers, which are often highly polymorphic, are presently being exploited to identify QTL for economically important traits. Using 159 microsatellite markers in 14

US Holstein half-sib families, Georges *et al.* 1995 demonstrated the presence of QTL for milk production on five chromosomes (namely chromosome no. 1, 6, 9, 10 and 20). In another study (Weller *et al.* 1995), significant association of microsatellite markers with somatic cell score (SCS, an indicator for susceptibility to mastitis), productive herd life and milk production traits were established. Characterization of QTL for economically important traits using microsatellite markers will help in formulating more efficient breeding programmes using MAS. Meuwissen and Goddard (1996) concluded the following from their study on the use of the MAS:

1. MAS only can increase the rate of genetic gain in the long term, when there is a continuous advantage of new identified QTL (obviously, this has a biological limit determined by the maximum proportion of the genetic variance that can be explained by the segregation of QTL).
2. The extra genetic gain due to the MAS decreases very quickly with the number of generations of selection for a same QTL. The rate of identification of new QTL is difficult to predict.
3. The gain due to MAS for a certain QTL is higher when the characteristic is measured after the selection, as it happens with the fertility and carcass characteristics.

2. Transgenic breeding strategies

The current livestock breeding strategies largely rely on the principle of selective breeding. In this method genetic improvement is brought about by increasing the frequency of advantageous alleles of many loci, though the actual loci are rarely identified. Moreover, in these methods genes cannot be moved from distant sources like different species or genera due to reproductive barrier. The recent developments in molecular biology have given rise to a new technology called transgenesis, which has removed the breeding barrier between different species or genera. Of the many applications of transgenesis, the most convincing one is the development of transgenic dairy animals for the production of pharmaceutical proteins in milk, and animals with altered milk composition (Wall *et al.* 1997).

The starting point for this technology is the identification of the genes of interest. In this context, molecular markers can serve as reference points for mapping the relevant genes that would be the first step towards their identification, isolation, cloning (positional cloning), and their manipulation. After successful production of transgenic animals, appropriate breeding methods could be followed for multiplication of transgenic herd/flock. Molecular markers can also be used for identification of the animals carrying the transgenes. Though most of the QTL are polygenic in nature and in transgenesis presently single gene traits are being manipulated Wall *et al.* 1997, the technology nevertheless holds future promises in moving polygenic QTL across the breeding barriers of animals.

Main problems related to the use of molecular genetics in the improvement of livestock populations (Dekkers and Hospital 2002, Dekkers 2004, Pollak 2005) are.

1. Direct use of a discovered QTL effect for selection across families is not possible.
2. By the time the information about the inferred genotypes is known, frequently the animals involved in the study are not available as candidates for selection, because they will be too old.
3. Advantage from within-family selection for a QTL bracketed by markers over BLUP or phenotypic selection alone is frequently low and the methodology to exploit this information for selection is complex and relatively inefficient.
4. There are statistical estimation errors, causing both false positive and false negative effects, particularly when the effect of the QTL is small.
5. There is a lack of consistency of the effect of the same QTL between studies, caused by QTL by genetic background (epistasis) of QTL by environment interactions.
6. The net economic effect of the QTL may be lower than the effect on single traits, because unfavourable effects on other traits.
7. Selection using QTL is more complex than phenotypic selection alone. QTL information (whether the information on the QTL is direct or indirect), adds to the list of traits used as selection criteria. Issues such as reduction of selection intensities and relative emphasis given to each trait, make optimal selection more difficult, with a need for adequate relative weights for the QTL, and the polygenic portions of the genetic variation for each trait at each generation (year).
8. Short-term gains due to MAS may be at the expense of medium to long-term polygenic responses for important traits.

Conclusions

The genetic improvement of animals is a continuous and complex process. Ever since the domestication of animals by man, he has always remained busy in improving his animals. In this pursuit many methods have been developed and tested. In recent years, the demonstration of genetic polymorphism at the DNA sequence level has provided a large number of marker techniques with variety of applications. This has, in turn, prompted further consideration for the potential utility of these markers in animal breeding. However, utilization of marker-based information for genetic improvement depends on the choice of an appropriate marker system for a given application. Selection of markers for different applications are influenced by several factors, viz. the degree of polymorphism skill or expertise available, possibility of automation, radioisotopes used, reproducibility of the technique, and finally the cost involved. Presently, the pace of development of molecular markers is tremendous, and the trend suggests that explosion in marker development will continue in the near future. It is expected that molecular markers will serve as a potential tool to geneticists and breeders to evaluate the existing germplasm, and to manipulate it to create animals as desired and needed by the society.

Term to understand

1. **Marker**: Any constituent, any characterized the character of individual.
2. **Genetic marker**: Genetic marker can be defined as any stable and inherited variation that can be measured or detected by a suitable method, and can be used subsequently to detect the presence of a specific genotype or phenotype other than itself, which otherwise is non-measurable or very difficult to detect
3. **Molecular markers**: The markers revealing variations at DNA level are referred to as the molecular markers. Molecular markers are found at specific locations of the genome. They are used to flag, the position of a particular gene or the inheritance of a particular characteristic.
4. **Naked eye polymorphism (NEP)**: NEP consists of those genetic markers whose inheritance can be followed with the naked eye and without any specialized biochemical or molecular technique.
5. **Marker assisted selection (MAS)**: MAS is indirect selection process, where a trait of interest is selected not based on the trait itself but on a marker linked to it. MAS can be useful for traits that are difficult to measure exhibit low h^2 and are expressed late in life.
6. **Mapping population**: Any population that is studied for the preparation of genetic maps.
7. **Isozymes**: Isozymes is a variant of an enzyme or protein that shows a different mobility during electrophoresis and can be used as genetic marker.

References

Ali S, Muller CR, Epplen JT 1986. DNA finger printing by oligonucleotide probes specific for simple repeats. Hum Genet 74: 239-243.

Alim MA, Sun DX, Zhang Yi, Faruque MO 2011. Genetic Markers and Their Application in

Buffalo Production . Journal of Animal and Veterinary Advances. 10(14):1789-1800 DOI: 10.3923/javaa.2011.1789.1800 (Urgent)

Archibald AL, Haley CS, Brown JF, Couperwhite S, Mamm Genome, 1995, 6: 157–175.

Baker CMA 1982. Anim. Blood Groups Biochem Genet 4: 141–146.

Beckmann JS, Soller M 1987. Bio/Technology, 5:573–576.

Beckmann JS, Soller M 1983. Restriction Fragment Length Polymorphisms in Genetic Improvement: Methodologies, Mapping and Costs. Theor. Appl. Genet. 67:35-43.

Beckmann JS, Soller M 1990. Toward a unified approach to genetic mapping of eukaryotes based on sequence tagged microsatellite sites. Nat. Biotechnol., 8: 930-932.

Botstein D, White RL, Skolnick M, Davis RW 1980. Construction of a genetic linkage map in man using restriction fragment length polymorphisms. Am. J. Hum. Genet., 32: 314-331.

Brown WM, Georgae MJ, Wilson AC 1979. Rapid evolution of animal mitochondrial DNA. Proceedings of the National Academy of Sciences, USA 76:1967-1971.

Caetano-Anolles G, Bassam BJ, Gressoff PM 1991. DNA amplification fingerprinting using very short arbitrary oligonucleotide primers. Bio Technology, 9: 553-557.

Caetano-Anolles G, Bassaml BJ, Gresshoff PM 1992. DNA fingerprinting: MAAPing out a RAPD redefinition. Bio/Technology, 10: 937-937.

Chee M, Yang R, Hubbell E, Berno A, Huang XC, Stern D, Winkler J, Lockhart DJ, Morris MS, Fodor SP 1996. Accessing genetic information with high-density DNA arrays. Science 274: 610-614.

Cockett NE, Jackson SP, Shay TL, Nielsen D, Moore SS, Steele MR, Barendse W, Green RD, Georges M 1994. Chromosomal localization of the callipyge gene in sheep (Ovis aries) using bovine DNA markers. Proceedings of the National Academy of Sciences USA 91, 3019-3023.

Cooper DN, Smith BA, Cooke HJ, Niemann S, Schmidtke J 1985. An estimate of unique DNA sequence heterozygosity in the human genome. Human Genetics 69: 201-205.

Crawford AM, Dodds KG, Ede AJ, Pierson GW 1995. Genetics, 140: 703–724.

Crawford AM, Dodds KG, McEwan JC. DNA Markers, Genetic Maps and the Identification of QTL: General Principles. Breeding for Disease Resistance in Farm Animals (eds Axford RFE, Bishop SC, Nicholas FW, Owen JB) © CAB International 2000.

Dekkers JCM 2004. Commercial application of marker- and gene-assisted selection in livestock: strategies and lessons. Journal of Animal Science, 82(13): E313-E318.

Dekkers Jack CM, Hospital Frédéric (2002). The use of molecular genetics in the improvement of agricultural populations. Nature Reviews Genetics, 3(1): 22-32.

Dunnington EA, Gal O, Siegel PB, Haberfeld A, Cahaner A, Lavi U, Plotsky Y, Hillel J 1991 Deoxyribonucleic Acid Fingerprint Comparisons Between Selected Populations of Chickens . Poult Sci 70:463-467.

Farber CM, Georges M, Debock G, Verhes A, Simon P, Verschraegenspae M, Human Genet 1989. 82: 1–3.

Freeland JR 2005. Molecular Ecology. John Wiley and Sons, West Essex, UK.

Gelderman H 1975. Investigations on inheritance of quantitative characters in animals by gene markers. I. Methods. Theor. Appl. Genet. 46:319-330.

Geldermann H in Genome Analysis in Domestic Animals (eds Geldermann H, Ellendorff F), Weinheim VCH 1990, pp. 291–323.

Georges M, Lathrop M, Hilbert P, Marcotte A, Schwers A, Swillens S, Vassart G, Hanset R 1990. On the use of DNA fingerprints for linkage studies in cattle. Genomics 6: 461-474.

Georges M, Nielsen D, Mackinnon M 1995. Genetics, 139:907-920.

Glowatzki-Mullis ML, Gaillard C, Wigger G, Fries R 1995. Microsatellite-based parentage control in cattle. Anim Genet 26:7-12.

Gwakisa PS, Kemp SJ, Teale AJ 1994. Characterization of Zebu cattle breeds in Tanzania using random amplified polymorphic DNA markers. Anim. Genet. 25: 89-94.

Hill AVS, Jeffreys AJ 1985. Lancet, 1:1394–1395.

Holland PM, Abrahamson RD, Watson R, Gelfand DH 1991. Detection of specific polymerase chain reaction products by utilising the 52 to 32 exonuclease activity of Thermus aquaticus DNA polymerase. Proceedings of the National Academy of Sciences USA 88: 7276-7280.

Ibitoye DO, Akin-Idowu PE 2010. Marker-assisted-selection (MAS): A fast track to increase genetic gain in horticultural crop breeding. African Journal of Biotechnology. 9 (52): 8889-8895.

Jaccoud D, Peng K, Feinstein D, Kilian A 2001. Diversity arrays: a solid state technology for sequence information independent genotyping. Nucleic Acids Res., 29: 25.

Jeffreys AJ, Morton DB 1987. DNA fingerprints of dogs and cats. Anim. Genet., 18:1–15.

Jeffreys AJ, Wilson V, Thein SL, Weatherall DJ, Ponder BAJ 1986. DNA "fingerprints" and segregation analysis of multiple markers in human pedigrees. Am. J. Hum. Genet.,39:11-24.

Jeffreys AJ, Wilson V, Thien SL 1985. Hypervariable minisatellite regions in human DNA. Nature 314: 67-73.

Jerry B Dodgson, Hans H Cheng, Ronald Okimoto 1997. DNA Marker Technology: A Revolution in Animal Genetics, Poultry Science 76:1108–1114.

Kantanen J, Vikki J, Elo K, Makitanila A 1995. Random amplified polymorphic DNA in cattle and sheep: application for detecting genetic variation. Anim Genet 26: 315-320.

Kappes SM, Keele JW, Stone RT, McGraw RA, Sonstegard TD, Smith TPL, Lopez-Corrales, NL, Beattie CW, Genome Res, 1997. 7: 235–249.

Kerstin Lindblad-Toh, Claire M Wade, Tarjei S Mikkelsen, Elinor K Karlsson, David B Jaffe, Michael Kamal, Michele Clamp, Jean L Chang, Edward J Kulbokas, Michael C Zody, Evan Mauceli, Xiaohui Xie, Matthew Breen, Robert K Wayne, Elaine A Ostrander, Chris P Ponting, Francis Galibert, Douglas R Smith, Pieter J deJong, Ewen Kirkness, Pablo Alvarez, Tara Biagi, William Brockman, Jonathan Butler, Chee-Wye Chin, April Cook, James Cuff, Mark J Daly, David DeCaprio, Sante Gnerre, Manfred Grabherr, Manolis Kellis, Michael Kleber, Carolyne Bardeleben, Leo Goodstadt, Andreas Heger, Christophe Hitte, Lisa Kim, Klaus-Peter Koepfli, Heidi G Parker, John P Pollinger, Stephen MJ Searle, Nathan B Sutter, Rachael Thomas, Caleb Webber, Broad Institute Genome Sequencing Platform & Eric S Lander 2005. Genome sequence, comparative analysis and haplotype structure of the domestic dog. Nature. 438: 803-819. doi:10.1038/nature04338

Kingsbury DT 1990. Annu Rev Genet, 24: 115–132.

Kuhnlein U, Dawe Y, Zadworny D, Gavora JS 1989. DNA fingerprinting: a tool for determining genetic distances between strains of poultry. Theor Appl Genet, 77: 669–672.

Landegren U, Kaiser R, Caskey CT, Hood L 1988. DNA diagnostics-molecular techniques and automation. Science, 242: 229-237.

Lipkin E, Tikoschinsky Y, Arbel R, Sharoni D, Soller M, Friedmann A, Anim Biotech 1993. 4: 195–201.

Litt M, Luty JA 1989. A hypervariable microsatellite revealed by *in vitro* amplification of a dinucleotide repeat within the cardiac muscle actin gene. Am. J. Hum. Genet., 44: 397-401.

Ma RZ, Beever JE, Da Y, Green CA, Russ I, Park C, Heyen DW, Everts RE, Fischer SR, Overton KM, Teale AJ, Kemp SJ, Hines HC, Guerin G, Lewin HA 1996. J Hered, 87: 261–267.

Machaty Z, Paldi A, Caski T, Varga Z, Kiss I, Barndi Z, Vajta GJ Reprod Fertil, 1993. 98: 467–470.

Markert CL, Moller F 1959. Multiple forms of enzymes: Tissue, ontogenetic and species specific patterns. Proc Natl Acad Sci USA, 45: 753-763.

Martin-Burriel I, Garcia-Muro E, Zaragoza P 1999. Genetic diversity analysis of six Spanish native cattle breeds using microsatellites, Anim Genet, 30: 177-182.

Mattapallil MJ, Ali S 1988. Gene, 206:209–214.

Mburu d, Hanotte O 2005. A practical approach to microsatellite genotyping with special reference to livestock population genetics. ILRI Biodiversity project. A manual prepared for the IAEA/ILRI training course on molecular characterisation of small ruminant genetic resources of Asia, October- December 2005, ILRI, Nairobi, Kenya.

Meuwissen THE, Goddard ME 1996. The use of marker-haplotypes in animal breeding schemes. Genetics, Selection, Evolution. 28:161-176.

Nakamura Y, Leppert M, O'Connell P, Wolff R, Holm T, Culver M, Martin C, Fujimoto E, Hoff M, Kumlin E, White R 1987. Variable number tandem repeat (VNTR) markers for human gene mapping. Science, 235: 1616-1622.

Newton CR, Graham A, Heptinstall LE, Powell SJ, Summers C 1989. Analysis of any point mutation in DNA. The amplification refractory mutation system (ARMS). Nucleic Acids Res., 17: 2503-2516.

Nichols WC, Liepnieks JJ, McKusick VA, Benson MD 1989. Direct sequencing of the gene for Maryland/German familial amyloidotic polyneuropathy type II and genotyping by allele-specific enzymatic amplification. Genomics, 5: 535-540.

O'Brien SJ 1991. Molecular genome mapping: lessons and prospects. Curr Opin Genet Dev 1:105–111.

Orita M, Iwahana H, Kanazawa H, Hayashi K, Sekiya T 1989. Detection of polymorphisms of human DNA by gel electrophoresis as single-strand conformation polymorphisms. Proceedings of the National Academy of Sciences USA 86: 2766-2770.

Peura T, Hyttien JM, Turunen M, Janne J 1991. Theriogenology, 35: 547–555.

Piper LR, Bindon BM 1982. The Booroola Merino and the Performance of Medium Non-peppin Crosses at Armidale (eds Piper, L. R. et al.), CSIRO, Melbourne, pp. 9–20.

Plotsky Y, Kaiser MG, Lamont SJ 1995. Genetic characterization of highly inbred chicken lines by two DNA methods: DNA fingerprinting and polymerase chain reaction using arbitrary primers. Anim. Genet. 26:163–170.

Pollak EJ 2005. Application and impact of new genetic technologies on beef cattle breeding: a 'real world' perspective. Australian Journal of Experimental Agriculture, 45 (7-8): 739-748.

Saiki R, Gelfand DH, Stoffel S, Scharf SJ, Higuchi R 1988. Primer-directed enzymatic amplification of DNA with a thermostable DNA polymerase. Science, 239: 487-491.

Sarkar G, Cassady J, Bottema CDK, Sommer SS 1990. Characterization of polymerase chain reaction amplification of specific alleles. Anal Biochem, 186: 64-68.

Sax K 1923. The association of sizes differences with seed coat pattern and pigmentation in *Phaseolus vulgarus.* Genetics, 8: 552-560.

Smith C, Simpson SP 1986. The use of genetic polymorphisms in livestock improvement. J Anim Breed Genet 103: 205-217.

Smith C, Smith DB 1993. Anim. Breed. Abstr., 61:197–204.

Soller M, Beckmann JS 1987. Consultation Report submitted to FAO, Rome, March 1987.

Teneva A, Petrovic MP 2010. Application of molecular markers in livestock improvement. Biotechnology in Animal Husbandry 26(3-4): 135-285.

Thibier M, Nibart M 1995. Theriogenology, 43:71–80.

Vaiman, D, Scibler L, Ourgeosis F, Oustry A, Amigues Y, Cribiu, E.P. 1996. Genetics, 144:279–305.

Vaiman M, Cotinot C, Kirszenbaum M, Leonard M, Chense P, Heyman Y, Stinnakre MG, Bishop C, Fellous M 1988. Proceedings of the 3rd World Congress on Sheep and Beef Cattle Breeding, 19–23 June, Paris, 1:93–105.

Vignal A 2002. Denis Milan, Magali Sancristobal and Andre Eggen. A review on SNP and other types of molecular markers and their use in animal genetics. Genet. Sel. Evol. 34: 275-305

Vos P, Hogers R, Bleeker M, Reijans M, Van De Lee T, Hornes M, Frijters A, Pot J, Peleman J, Kuiper M, Zabeau M 1995. AFLP: a new technique for DNA fingerprinting. Nucleic Acids Res., 23: 4407-4414.

Weller JI, Yanai A, Blank Y 1995. Proceedings of the Third International Mastitis Seminar, Tel Aviv, Israel, 1: 3–13.

Welsh J, McClelland M 1990. Fingerprinting genomes using PCR with arbitrary primers. Nucleic Acids Res., 18:7213-7218.

Williams JGK, Kublelik AR, Livak KJ, Rafalski JA, Tingey SV 1990. DNA polymorphism's amplified by arbitrary primers are useful as genetic markers. Nucleic Acids Res., 18: 6531-6535.

Yadav BR, Mitra A 1999. One day symposium on emerging DNA Technologies for the Next Millennium, CCMB/CDFD, Hyderabad, 23 February, 1999 Abstr., p. 9.

Zabeau M 1993. Selective restriction fragment amplification: A general method for DNA fingerprinting. European Patent Application, No. 0-534-858-A1.

Zhu KY, Clarke JM 1996. Addition of a competitive primer can dramatically improve the specificity of PCR amplification of specific alleles. Biotechniques 21:586-587.

3

Bioinformatics in Livestock Production and Management

Dhirendra Kumar, Pranav Kumar and Vibha Raj Shanti

Bioinformatics is the study of analysis of biological data using computer programming, mathematics and statistics. Bioinformatics is the application of computer science and information technology to the field of biological sciences. Bioinformatics involves databases and information systems, web technologies, artificial intelligence and soft computing, information and computation theory, software engineering and statistics. The one of the most important achievement of bioinformatics is the human genome project which was mapped in 2001. Most of the work in this field is based on the research. Scientists make databases and develop different tools to analyze different gene expressions. Within last few decades, a number of research in the field of molecular biology, along with advances in genomic technologies, have generated a huge number of biological information Similar is the case in the field of animal production where many advances have occurred. The day to day advances in bioinformatics and computational science have led development in the field of animal as well as veterinary science. Bioinformatics generates new knowledge in the field of Animal Sciences, and improving & discovering new models of computation for the advanced livestock production.

Bioinformatics deals with the creation and maintenance of a database to store biological information for further utilization.

Current research in bioinformatics can broadly be classified into.

(i) *genomics* – sequencing and comparative study of genomes to identify gene and genome functionality

(ii) *proteomics* – identification and characterization of protein related properties and reconstruction of metabolic and regulatory pathways,

(iii) cell visualization and simulation to study and model cell behavior, and

(iv) application to the development of drugs and anti-microbial agents.

Livestock bioinformatics research can be classified under three major approaches:

1. analysis based upon the available experimental wet-lab data,
2. the use of mathematical modeling to derive new information, and
3. an integrated approach that integrates search techniques with mathematical modeling.

The major impact of bioinformatics research has been to automate the genome sequencing, automated development of integrated genomics and proteomics databases, automated genome comparisons to identify the genome function, automated derivation of metabolic pathways, gene expression analysis to derive regulatory pathways, the development of statistical techniques, clustering techniques and data mining techniques to derive protein-protein and protein-DNA interactions, and modeling of 3D structure of proteins and 3D docking between proteins and biochemical's for rational drug design, difference analysis between pathogenic and non-pathogenic strains to identify candidate genes for vaccines and anti-microbial agents, and the whole genome comparison to understand the microbial evolution. The development of bioinformatics techniques has enhanced the pace of biological discovery by automated analysis of large number of microbial genomes. We are on the verge of using all this knowledge to understand cellular mechanisms at the systemic level. The developed bioinformatics techniques have potential to facilitate

(i) the discovery of causes of diseases,

(ii) vaccine and rational drug design, and

(iii) to improved cost effective agents for bioremediation by pruning out the dead ends.

Three approaches have been used in bioinformatics:

(i) use of computational search and alignment techniques to compare new genome against the set of known genes to annotate the structure and function of genes in a newly sequence genome,

(ii) the use of mathematical modeling techniques such as data mining, statistical analysis, neural networks, genetic algorithm, and graph matching techniques to identify common patterns, features and high level functions, and

(iii) an integrated approach that integrates search techniques with mathematical modeling.

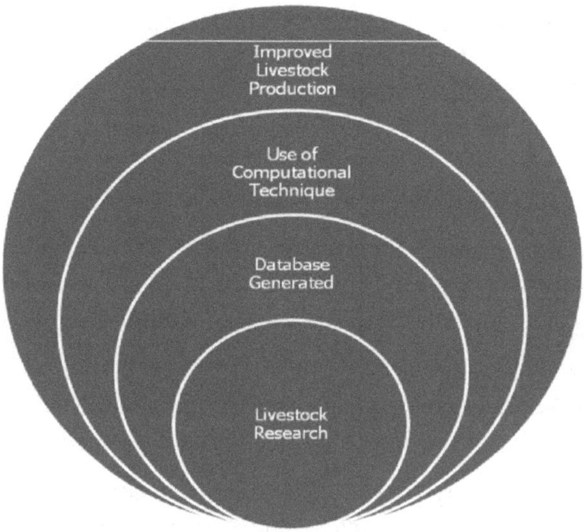

Fig : Bioinformatics in Livestock Improvement

Different kind of bioinformatics

1. Computational biology
2. Protein bioinformatics
3. Cheminformatic
4. Molecular modeling and systems biology
5. Systems bioinformatics
6. Pharmaceutical bioinformatics
7. Research informatics
8. Laboratory informatics
9. Toxic informatics
10. Health information data

Databases in livestock production

With the increasing amount of data being generated in different field of animal sciences, it has become important for animal scientist to store data in proper way that makes it easily accessible and also allow integration of data. Databases on livestock and poultry breeds characterization along with information on the environment and management practices are required for decision making on conservation of indigenous animal genetic resources and for further study in improved livestock production. Computer databases are created and maintained with the software called database management system (DBMS). DBMS allows creation of database, adding, changing, and deleting data in the database, sorting and retrieving data, and creating forms and reports for database users.

Databases are of two types

Primary database: Primary database include all of the repositories of the primary output from the experimental work such as GenBank. These repositories generally house information that is submitted by the scientist who generated it.

Secondary database: Secondary database are less all –inclusive than primary databases but concentrate on data quality and include additional information.

A database contains many files, a file contains records, a record contains fields, and a field contains characters. Field is the smallest unit that a data user accesses. Data type specifies kind of data contained in the field.

Data on breeds is sourced from surveys on breeds in the breeding tracts. Data is collected from the breeding tracts and analyzed to obtained breed descriptors.

Various databases on animal genetic resources are available. Such as - **Animal Genetic Resources - Information System (AGRI-IS)** developed at NBAGR, Karnal.

Which contains following information as-

1. District-wise livestock & poultry census data from 1961 with age, sex and utility details.
2. Information on various breeds of livestock and poultry species (General information, utility, Geographical distribution, Population, Morphology, Performance parameters, Management practices) along with two photographs of animals of each breed.
3. District-wise infrastructure facilities from 1990 onwards (No of Veterinary hospitals, extension centres for poultry, sheep & wool, pig; livestock

farms of all species, gaushalas, villages & area per veterinary institute, and A.I. done in cattle & buffalo).

4. District-wise production statistics from 1990 onwards (Milk production, wool, eggs, no of slaughter houses, species-wise no of animals slaughtered and meat production).
5. State-wise per capita availability and utilization of animal products from 1990 onwards (milk, egg and dung).
6. Livestock farm details (location, breed, herd strength).
7. Information about semen banks (location, breed, number of bulls, semen doses, etc.).
8. Vaccine production (centre, vaccine and quantity).
9. Import & export of animals and animal products for the country.

Domestic Animal Diversity Information System (DAD-IS) developed at the Food and Agriculture Organization (FAO).

1. General information: Species, geographical location of a breed, local name, synonyms, international name of the breed, source of information.
2. Origin and development of breed: Origin of breed, wild or primitive, population size, risk status, use of A.I. and storage of semen.
3. Breed description: Coat color, horn description, wool type, adult weight and size, genetic characters (marker genes, chromosomal aberrations).
4. Use and qualities of breeds: Main usc, special quality like disease resistance, adaptability to harsh climate.

Information System for Genebank Management (ISGerm) developed for management of genebank.

Database at NCBI

The **National Center for Biotechnology Information** (**NCBI**) is part of the United States National Library of Medicine (NLM), a branch of the National Institutes of Health. The NCBI houses genome sequencing data in Gen Bank and an index of biomedical research articles in PubMed Central and PubMed, as well as other information relevant to biotechnology. All these databases are available online through the Entrez search engine.

Major research areas through bioinformatics for the advancement in livestock production

1. Sequence analysis

2. Genome annotation
3. Computational evolutionary biology
4. Analysis of gene expression
5. Analysis of regulation
6. Comparative genomics
7. Analysis of protein expression
8. Modeling biological systems
9. Parentage
10. Breed signature
11. Breed diversity estimation
12. Breed clustering- OTU- target of conservation
13. Rationalization of conservation priority indices
14. Degree of admixture

Current progress in genomics, transcriptomic and proteomics technologies generate a large amount of livestock molecular data. Databases, tools and computational programs are required for proper storage, utilization and accessing the wealth of livestock molecular data. Numerous databases and computational tools are available on web to provide agricultural community access to a range of molecular data.

Databases and tools for DNA and protein sequence analysis

S.No	Databases Sequence Databases	Description	Web Address
1.	Ncbi	For nucleotide and protein sequence information	www.ncbi.nlm.nih.gov/
2.	Embl	For nucleotide and protein sequence information	http://www.ebi.ac.uk/embl/
3.	Web Cutter	Generate a map of enzyme sites	http://users.unimi.it/~camelot/tools/cut2.html
4.	Neb Cutter	This tool will take a DNA sequence and find the large, non-overlapping open reading frames	http://tools.neb.com/NEBcutter2/

Metabolic Pathway

A metabolic pathway is a graphical representation of metabolism in a cell. Metabolic pathways analysis in livestock is a necessary step in understanding the evolutionary origins

S.No	Name	Description	Web address
1.	Kegg	Kyoto Encyclopedia of Genes and Genomes, provides a collection of manually created maps of biochemical pathways, and precomputed results of comparative sequence analysis, pattern/motif searching, and orthologous gene clusters, among others.	http://www.genome.jp/kegg
2.	Cattlecyc	Cattle-specific pathway genome database is used for the reconstruction of metabolic pathways in cattles.	http://biocyc.org/CATTLE/organism- Summary

Genome mapping tools and databases

Genomic research in agriculturally important species is largely concerned with mapping genes that influence economically important traits.

S.No	Name	Description	Web address
1.	Cattle EST Gene Family Database	Est family database through Gbrowse	http://genomes.arc.georgetown.edu/drupal/bovine/?q=cattle_est_database
2.	GENETPIG	For functional and comparative mapping of the pig genome	http://www.infobiogen.fr/services/Genetpig/.

Phylogenetic analysis/representation tools

Phylogenetic analysis has many applications in animal science, this study is required to find out the divergence among different subgroups of animal, for genetic diversity studies, gene frequencies can be determined for several loci in several breeds or populations

S.No	Name	Description	Web address
1.	Phylip	Phylogeny Inference Package	http://bioweb2.pasteur.fr/phylogeny/intro- en.html

Microsatellite markers in animal sciences

Microsatellites have been recommended by FAO as first priority molecular tools for the Measurement of Domestic Animal Diversity (MoDAD). The term microsatellite was introduced by Litt and Luty (1989) to characterize the tandemly repeated simple sequence motifs, one to six nucleotides (mono, di, tri, tetra, penta and, hexanucleotides tandem repeats) in length

1. Biodiversity analysis
2. Parentage testing

3. Population bottleneck
4. Mapping of qtl

SNP (single nucleotide polymorphism) markers

SNPs are stable genetic markers and have low mutation rate in comparison with other genetic markers. The frequency of occurrence of single base difference in genomic DNA is very high when two equivalent chromosomes are compared. SNPs are evolutionary stable from generation to generation and thus, are useful in population studies and in tracking down the inheritance of any gene that contributes to a disease. SNPs are also useful in epidemiological studies, gene discovery, gene mapping, candidate gene polymorphism testing, diagnostics and risk profiling, and homogeneity testing.

SNP commonly known as *Snip* is just a single base change in a DNA sequence, with a usual alternative of two possible nucleotides at a given position. For such a base position with sequence alternative within genomic DNA to be considered as an SNP, the least frequent allele should have a frequency of 1% or greater.

Types of SNP

a. Direct sequencing
b. Restriction enzyme cutting (PCR-RFLP)
c. Single strand conformation analysis (SSCP)
d. Primer extension
e. Oligonucleotide ligation assay (OLA)
f. Exonuclease detection (TaqMan)

Database	Emphasis	Details
dbSNP http://www.ncbi.nlm.nih.gov/projects/SNP	SNPs from the complete genome	More than 6 million validated SNPs
HapMap http://hapmap.org/cgi-perl/gbrowse	Whole genome SNP in four populations	More than 1 million SNPs
HGVbase http://hgvbase.cgb.ki.se/	SNPs from the complete genome	More than 2.8 million SNPs

DNA microarray

A **DNA microarray** (also commonly known as gene chip, DNA chip, or biochip) is a collection of microscopic DNA spots attached to a solid surface. Scientists use DNA microarrays to measure the expression levels of large numbers of genes simultaneously or to genotype multiple regions of a genome. Each DNA spot contains picomoles (10"^{12} moles) of a specific DNA sequence, known as *probes* (or *reporters*). These can be a short section of a gene or other DNA element that are used to hybridize a cDNA or cRNA sample (called *target*) under high-stringency conditions. Probe-target hybridization is usually detected and quantified by detection of fluorophore-, silver-, or chemiluminescence-labeled targets to determine relative abundance of nucleic acid sequences in the target.

Microarrays have now been evolved as a very powerful technology that enables large numbers of genes, up to the order of tens of thousands, to be evaluated simultaneously. The objective of a microarray experiment might be to investigate genes which are differentially up or down regulated in cells.

Types of microarrays

1. Microarray expression analysis.
2. Microarray for mutation analysis.
3. Comparative genomic Hybridization

Applications of Microarrays in Livestock Improvement

Gene discovery: DNA Microarray technology helps in the identification of new genes, know about their functioning and expression levels under different conditions.

Disease diagnosis: DNA Microarray technology helps researchers learn more about different diseases such as heart diseases, mental illness, infectious disease and especially the study of cancer

Drug discovery: Microarray technology has extensive application in *Pharmacogenomics*. Pharmacogenomics is the study of correlations between therapeutic responses to drugs and the genetic profiles of the patients

Toxicological research: Microarray technology provides a robust platform for the research of the impact of toxins on the cells and their passing on to the progeny.

Why are microarrays so important?

Microarrays are a significant advance both because they may contain a very large number of genes and because of their small size. Microarrays are therefore useful when one wants to survey a large number of genes quickly or when the sample to be studied is small. Microarrays may be used to assay gene expression within a single sample or to compare gene expression in two different cell types or tissue samples, such as in healthy and diseased tissue. Because a microarray can be used to examine the expression of hundreds or thousands of genes at once, it promises to revolutionize the way scientists examine gene expression.

A DNA microarray experiment

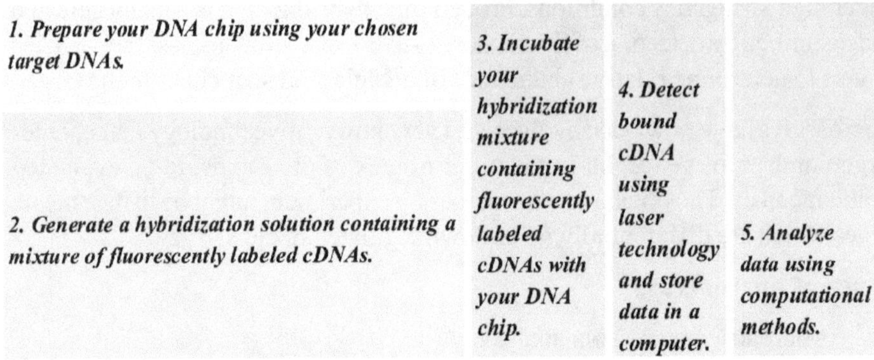

Next generation sequencing

The transforming aspect of the Human Genome Project was not the completion of the genome sequence itself, but rather the technologies that enabled, and were enabled by, the sequencing of that first reference genome. The evolution of 'omic science through microarray transcriptomics, metabolomics, proteomics, and wholegenome SNP-omics has in many ways come full circle with a new focus on genomics and genome sequencing. Next-generation sequencing technologies have begun to revolutionise genomics and their effects are becoming increasingly widespread.The 1000 genomes project (http://www.1000genomes The automated Sanger method is considered as a 'first-generation' technology, and newer methods are referred to as next-generation sequencing (NGS). These newer technologies constitute various strategies that rely on a combination of template preparation, sequencing and imaging, and genome alignment and assembly methods.

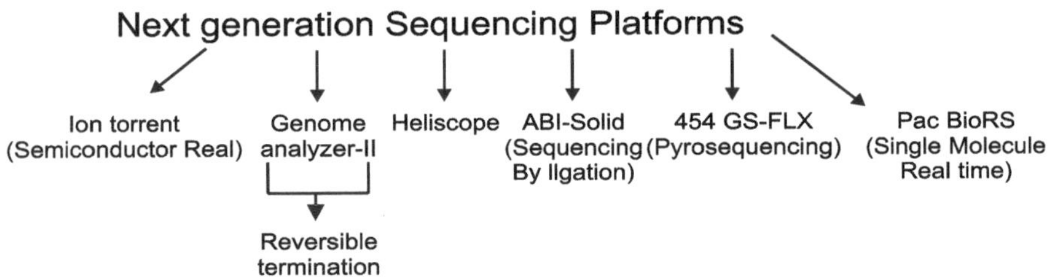

Bioinformatics in microbial genomics

In the molecular analyses, the researchers target both functional genes and the 16 or 18S rRNA gene. Both functional and phylogenetic analyses are performed. In the last decade, bioinformatics has been used for the microbial biotechnology in many ways: computationally analyzing the wet-lab data, genome sequencing, identification of protein coding segments and genome comparison to identify the gene function], the development of genomic and proteomics databases and inference of phenotypes (higher level functions) from genotypes (gene level functions).

1. Genome sequencing
2. Automated identification of genes
3. Identifying gene function: searching and alignment
4. 3D structure modeling and docking
5. Pair-wise genome comparison
6. Reconstructing metabolic pathways

Role of bioinformatics in drug discovery

Bioinformatics has made it possible to sequence the genome of various organisms and there are almost hundred organisms whose genome has been mapped so far. The databases of these organisms are increasing in size day by day because every day they get new information about any organism. As other fields of science have made use of bioinformatics, similarly, pharmaceutical industry has also adopted bioinformatics and genomics for the drug targets and drug discovery. The understanding of molecular biology has made it possible to design and develop the drugs. In the beginning, whole animals were used to test the synthetic organic molecules in the organ preparation. In the recent years, bioinformatics has made it easy for the researchers that they can now easily target the molecules in the in vitro environment. Now the screening of newly developed compounds can be done against the molecules of the proteins or genetically modified cells and it gives very efficient results. This way of drug development has eased the identification of the disease in an organism.

Bioinformatics in disease diagnosis

It is widely known that genetic mutations cause disease. What are largely unknown are the mechanisms by which these mutations wreak havoc at the molecular level, giving rise to clinically observable symptoms in patients. Now a new study using bioinformatics, reports the ability to predict the molecular cause of many inherited genetic diseases. These predictions involve tens of thousands of genetic disease-causing mutations and have led to the creation of a web-based tool available to academic researchers who study disease. We now have a quantitative model of function using bioinformatic methods that can predict things like the stability of the protein and how its stability is disrupted when a mutation occurs. Traditionally people have used a very time consuming process based on evolutionary information about protein structure to predict molecular activity.

The research was done in the contexts of inherited single gene diseases, complex diseases such as cardiovascular and developmental disorders and mutations in cancerous tumors. The study focused on amino acid substitutions (AAS), which are genetically driven changes in proteins that can give rise to disease, and utilized a series of complex mathematical algorithms to predict activity stemming from the mutations.

Bioinformatics provide platform to combines the genetic and geographic data of a pathogen to reconstruct its history and to identify the migration routes through which the strains spread regionally and internationally. Bioinformatics assist biosurveillance and early warning to address and solve the inefficiencies in traditional surveillance as well as the need for more timely and comprehensive infection monitoring .Through bioinformatics tool the microbial profiling and biosurveillance can be improved which enable integrated infectious disease outbreak detection and response.

References

Ahmed F, Ifeanyi, Kenneth DE 2005. Farm animal genomics and informatics: an update. Nucleic Acids Res 2005, 33(19):6308-6318.

Arora R 2012 Genetic Diversity Based on Microsatellite Data Training programme on Advances in Animal Bioinformatics', Feb 21 to March 3 2012.

Bhatia AK 2012 Databases for Management and Conservation of Livestock Biodiversity Training programme on 'Advances in Animal Bioinformatics', Feb 21 to March 3 2012.

Chitty M 2012. Bioinformatics in drug discovery & Development www.healthtech.com. Visited march 14.

Clark MS, Schlitt T 2007. Navigating sequenced genomes. Scion ,U.K.15-38p

Hall N, Schriml 2007. Database recourses for wet – bench scientists. Scion ,U.K.1-13p infectious disease control BMC Bioinformatics. 2009; 10(Suppl 2): S10. Published online 2009 February 5. doi: 10.1186/1471-2105-10-S2-S10.

Pandey A, Sharma A, Bhatia AK 2012. Web Resources for Livestock Information Training programme on 'Advances in Animal Bioinformatics', Feb 21 to March 3 2012.

Sharma U, Vijh RK 2012. Next Generation Sequencing -Concept and Strategies Training programme on 'Advances in Animal Bioinformatics', Feb 21 to March 3, 2012.

4

Laboratory Animals in Animal Breeding: A Review

Bharat Bhushan, Manjit Panigrahi and G K Gaur

Introduction

Animals used for experimental research purpose, which show results resembling to human beings are known as lab animals. They are mostly used because of their small size, lesser feed and space requirement. Furthermore, they are not natural host for most of the animal / human pathogens. The laboratory animal research is conducted at universities, medical colleges, pharmaceutical companies, farms, defense establishments and commercial facilities that provide animal-testing services to industry. It includes basic research in the area of genetics, developmental biology, behavioural science etc., as well as applied research in the area of biomedical science, xenotransplantation and drugs including cosmetics testing. Laboratory animals are also used for education, breeding and defense research. The common laboratory animals are mice, rats, hamsters, guinea pigs and rabbits. However, cats, dogs, monkeys, goats, sheep, cattle, pig, horses, fishes etc. can also be utilized as laboratory animals for experimental purpose in biomedical sciences.

Mouse

A mouse (plural mice) is a rodent that belongs to one of numerous species of small mammals. The best known mouse species is the common house mouse (*Mus musculus*). It is found nearly in all countries, serves as an important model organism in biology and is also a popular pet. The American white-footed mouse (*Peromyscus leucopus*) and the deer mouse (*Peromyscus maniculatus*) also sometimes live in houses. These species of mice live commensally with humans. Although they may live up to two years in the lab, the average mouse in the wild lives only about 5 months, primarily due to heavy predation. Cats, wild dogs, foxes, birds of prey, snakes and even certain kinds of insects have been known to prey heavily upon mice. Nevertheless, due to its remarkable adaptability to almost any environment, and its ability to live commensally with humans, the mouse is regarded to be the third most successful mammalian species living on Earth today, after humans and the rat.

Kingdom	Animalia
Phylum	Chordata
Class	Mammalia
Order	Rodentia
Superfamily	Muroidea
Family	Muridae
Subfamily	Murinae
Genus	Mus

Mice can be harmful pests, damaging and eating crops and spreading diseases through their parasites and feces. In the Western United States, breathing dust that has come in contact with mouse feces has been linked to the deadly Hantavirus. The original motivation for the domestication of cats is thought to have been for their predation of mice and their relatives, the rats.

Behaviour

The laboratory mouse is a docile animal and can be easily handled. Animals can be grouped soon after weaning usually coexists peacefully. However, some strains of mice (i.e. BALB/cJ, SJL/J and HRS/J) will begin to fight even if grouped at weaning. Breeding males that have been removed from breeding cages and caged together usually fight. Wounds on the tail are a good sign of aggression between cage-mates.

Chromosome number	40
Birth weight	1.5 gm
Life span	1.5-3 years
Adult body weight: male	20-40 gm
Adult body weight : female	25-90 gm
Breeding onset : male	50 days
Female	35-60 days
Gestation period	19-21 days
Body temperature : male	36-37^{0}C

Contd.

Female	36.5-38.0^{0}C
Heart rate: male	500-600 beats per minute
Female	325-780 beats per minute
Respiratory rate: male	84-230 per minute
Female	60-220 per minute
Estrous cycle	4-5 days
Estrous period	10 hours
Litter size	11

Breeding system in mice

Breeding onset is at about 50 days of age in both females and males, although females may have their first estrus at 25-40 days. Mice are polyestrous and breed year round; ovulation is spontaneous. The duration of the estrous cycle is 4-5 days and estrus itself lasts about 12 hours, occurring in the evening. Vaginal smears are useful in timed mating to determine the stage of the estrous cycle. Mating is usually nocturnal and may be confirmed by the presence of a copulatory plug in the vagina up to 24 hours post-copulation. The presence of sperm on a vaginal smear is also a reliable indicator of mating.

Female mice housed together tend to go into anestrus and do not cycle. If exposed to a male mouse or the pheromones of a male mouse, most of the females go into estrus in about 72 hours. This synchronization of the estrous cycle is known as the Whitten effect. The exposure of a recently bred mouse to the pheromones of a strange male mouse may prevent implantation (or pseudo pregnancy), a phenomenon known as the Bruce effect.

The average gestation period is 20 days. A fertile postpartum estrus occurs 14-24 hours following parturition, and simultaneous lactation and gestation prolongs 3-10 days due to delayed implantation. The average litter size is 10-12 during optimum production, but is highly strain dependent. As a general rule, inbred mice tend to have longer gestation periods and smaller litters than out bred and hybrid mice. The young are called pups and weigh 0.5-1.5 grams at birth, are hairless, and have closed eyelids and ears. Cannibalism is uncommon, but females should not be disturbed during parturition and for at least 2 days postpartum. Pups are weaned at 3 weeks of age; weaning weight is 10-12 grams. If the postpartum estrus is not utilized, the female resumes cycling 2-5 days post weaning.

Newborn male mice are distinguished from newborn females by noting the greater anogenital distance and larger genial papilla in the male. This is best accomplished by lifting the tails of litter mates and comparing perineums.

Rat

Rats are various medium sized rodents. “True rats” are members of the genus

Rattus, the most important of which to humans are the black rat, *Rattus rattus*, and the brown rat, *R. norvegicus*. Many members of other rodent genera and families are also called rats and share many characteristics with true rats. A rat has an average life span of 2-3 years. Rats are distinguished from mice by their size; rats generally have bodies longer than 12 cm (5 inches).

Kingdom	Animalia
Phylum	Chordata
Class	Mammalia

Behaviour

The rat can become accustomed to handling providing they are not upset by the experience. Rats will bite without warning, but not repeatedly. Unlike mice, groups of the same sex can be housed together without fighting. Rats are active primarily during the night at which time they feed; the light hours are used primarily for rest, sleep and digestion. Handling animals during the night phase can be more difficult due to this increase in activity. The diurnal rhythm can be changed by a 12 hour shift in the light cycle. It takes approximately two weeks for rats to adjust to this shift.

Chromosome No.	42
Life span	2-3.5 years
Birth weight	5-6 gm
Adult body weight: male	450-520 gm
Adult body weight: female	250-320 gm
Breeding onset: male	65-110 days
Female	65-110 days
Gestation Period	21-23 days
Estrous cycle	5 days
Estrous period	13-15 hours
Body Temperature: male	38-39^0 C
female	35.9-37.5^0 C
Heart rate: male	320-480 beats per minute
female	250-450 beats per minute
Respiratory rate: male	85-110 per minute
female	70-115 per minute
Litter size	8-12

Uses

The albino rats (*Mus norvegicus albinus*) is the most commonly used lab animal. They live for around three years and begin breeding in about three months of age. The rat is resistant to usual infections and is particularly satisfactory for surgical experiments.

Breeding system in rats

(1) Moderate Inbreeding (the "harem" system)

Polygamous or Harem breeding utilizes on male to breed several females. It is the most efficient for space and utilization of males, but it makes record keeping more difficult. Pregnant females must be separated from the group prior to delivering their litters. Under the system up to four females are regularly mated to one male but as soon as pregnancy is definitely established the female is removed to a separate cage and allowed to deliver and nurse the young on her own. No advantage is taken of the postpartum estrus and litters result only every 7-8 weeks.

(2) Monogamous pair mating

A monogamous system consists of one male and one female permanently housed together. It will produce the maximum number of litters per female in the shortest time frame, by utilizing the female's post-partum estrus (~12h post partum). Record keeping is easy with this method. However, it requires a larger population of males to be maintained. Mice mated under this system are left together for the rest of their useful breeding life.

(3) Trio mating

One male is mated with two females under this system. Again mice can be permanently allocated to the breeding cage.

(4) Other mating systems

The production of large numbers of pregnant rats is obtained by caging one male with up to four females. Mating is confirmed by vaginal lavage. Females are removed and replaced at the detection of mating. Males may be changed at weekly or less frequent intervals.

Breeding systems can be classified as follows

1. Random breeding

True random breeding cannot be achieved by indiscriminate mating of a finite number of animals but must be planned by use of the tables of random sampling numbers or other comparable randomizing devices.

2. Inbreeding

The word 'inbreeding' has unpleasant connotations to many breeders. It is often thought of as being scary and something to condemn; a practice that only produces unfit individuals. Such prejudice often originates in the fallacious belief that inbreeding is simply the random mating of close relatives. This, however, is simply not the case.

In brother- sister mating type of inbreeding litter mates who possess a useful character is mated. Under pen mating type of inbreeding one or more males is mated with two or more females. All the progeny are reared together and no distinction is made between litters when mates are chosen to produce next generation.

3. Crossbreeding

Controlled crossbreeding of inbred strains is a means to restore hybrid vigor.

Hamster

Hamsters are rodents belonging to the subfamily Cricetinae. The subfamily contains about 24 species, classified in six or seven genera. Because they are easy to breed in captivity, hamsters are often used as lab animals and kept as pets. The word is rather frequently misspelled as "hampster". This may have been popularized by The Hampster Dance meme. The name *hamster* is derived from the German verb *hamstern,* which means "to hoard" because hamsters have expandable cheek pouches lined with fur to store their food.

Kingdom	Animalia
Phylum	Chordata
Class	Mammalia
Order	Rodentia
Suborder	Myomorpha
Superfamily	Muroidea
Family	Cricetidae
Subfamily	Cricetinae

Behavior

The Syrian Hamster or Golden Hamster, *Mesocricetus auratus*, is a very well-known member of the rodent subfamily Cricetinae, the hamsters. Syrian hamsters are generally solitary and may fight to the death if put together, whereas some of dwarf hamster species may get along with others of the same species. Hamsters are primarily considered crepuscular because they live underground during most of the day, only leaving their burrows about an hour before sundown and then returning when it gets dark.

Chromosome No.	44
Life span	2-3 years
Birth weight	2 gm
Adult body weight: male	90-120 gm
Adult body weight: female	95-140 gm
Breeding onset: male	60 days
Female	60 days
Gestation Period	16-18 days
Estrous cycle	4-5 days
Estrous period	20 hours
Body Temperature	36.6^{0}C
Litter size	6-10

Uses

The first use of Syrian hamsters was in 1930 when they were used as experimental animal for kala-azar. It was used as experimental animal for viruses for polio, ND, influenza, rabies and others. Because of its ability to develop dental caries the hamsters are used in dental research. Since hamsters hibernate in response to cold weather they are well suited to studies of natural and induced hypothermia and its effects upon mammalian physiology. The presence in these animals of a naked cheek pouch has given scientists a living lab for study of microcirculation.

Breeding system in hamsters

Breeding hamsters should be housed either in separate quarters or with some degree of isolation from adult males and non pregnant females. Pseudo pregnancy and anestrus are induced by crowding females in small or large groups (Lee Boot effect). Pregnancy may be terminated if recently bred females are housed with strange males (Bruce effect) or with too many females. These are due to action of pheromones. Group breeding is practiced in some colonies in that several females and an equal no of males may be housed in a single cage for nine days i.e. for two estrus cycles and then the females are placed in individual cages for gestation and lactation.

Trial breeding is used when timed mating is needed for special purposes like collection of ova, embryos and teratological studies. A female hamster is test bred for at least nine days before she is discarded as a non breeder. If females to be bred are examined each morning for vaginal discharges it is possible to schedule breeding several days in advance. When hamsters are maintained in inbred lines the easiest method of breeding is to leave litters together and observe females daily until copulation plugs are observed. At that time the animals are placed in separate maternity cages. In group breeding four females and one male are housed in five cages with connecting passages. There are two major methods used i.e. test breeding and pen breeding.

Guinea pig

The domestic Guniea pig is originated in South America. It is related to the wild, semi-domesticated Guniea pig (*Cavia aperea*) found in the mountains and grasslands of Peru, Argentina, Brazil, and Uruguay. It has been domesticated for over 400 years and is well known as a pet and laboratory animal. The Guinea Pig, *Cavia porcellus*, is a mammal of the order Rodentia, sub-order Hystricomorpha and family Caviidae. Three basic breeds of guinea pigs exist, the English (short-haired), Peruvian (long-haired) and Absynninian, which has a rosette hair pattern. Guinea pigs have several unique biological characteristics. Guinea pigs are herbivores and unlike most laboratory animals, except nonhuman primates, guinea pigs require a nutritional source of Vitamin C.

Kingdom	Animalia
Phylum	Chordata
Class	Mammalia
Order	Rodentia
Superfamily	Muroidea
Family	Cavidae
Genus	*Cavia*

The English short hair is the preferred animal for pet and laboratory usage. The Dunkin-Hartley strain developed in 1926 is an outbred. Two inbred strains, numbers 2 and 13, were developed also and are lesser used due to small numbers and higher costs. The Abyssinian is easily identified by its characteristic rosettes or whorls in the hair coat. The Peruvian is distinguishable by its long silky hair.

Behaviour

Guinea pigs are very docile and rapidly become accustomed to gentle handling, in fact guinea pigs rarely bite. Aggression between females is uncommon and is more likely to occur between males in competition for a female in estrous. Guinea pigs are easily alarmed and will often "freeze" for extended periods (30 minutes) when startled. Group housed guinea pigs may stampede when startled which may result in injury to young guinea pigs, orthopedic injuries, and abortion in pregnant dams. Guinea pigs are considered crepuscular animals.

Chromosome No.	64
Birth weight	90 gm
Life Span	4-5 years
Adult body weight: male	800-1200 gm
Adult body weight: female	250-320 gm
Puberty: male	56-70 days
Puberty: female	67.8 + (21.5 SD) days
Gestation Period	65-72 days

Contd.

Estrous cycle	16.5 days
Estrous period	6-11 hours
Body Temperature	37.2-39.5 C
Heart rate	230-380 beats per minute
Respiratory Rate	42-104 per minute
Litter size	1-6

Uses

Historically the guinea pig was sought after as a culinary delicacy in its native habitat and invading Spaniards who discovered those in South America took them to Europe for fanciers. Today, however, the guinea pig is well known for its use as a pet, show, and research animal. In the research arena guinea pigs have found usage in studies of human infections, immunology, gnotobiology, genetics, otology, and nutrition. Although the term laboratory animal is often associated with guinea pigs it is interesting to note that guinea pigs comprise only about 2% of the total usage of research animals.

Breeding system in guinea pigs

1. Group breeding

A ratio of eight females to each male is near optimum for a cage of app. 0.75 m^2. 18 days after the introduction of the male into a breeding cage all the females should be pregnant. In practice this seldom happens. However, before discarding the female it is reasonable to wait until she has had no fewer than two estrus cycles in the breeding cage. Any non pregnant female should be discarded from the breeding programme at this time and replaced with another female. Once the breeding cage is established the primary task is to monitor them to assure a continued high level of reproductive performance in each group. Many commercial breeders replace a colony breeder after a specified length of time usually 12-18 months. Some females prove to be more productive at two years than others.

2. Timed mating

In many research programs the need will arise for timed pregnancies. When such need arise it is to the advantage of the colony to have a system for timed pregnancies. The following two methods are used:-

A. open cages method

The detection of females in estrus is facilitated by the "open cages method" of housing. Adult females are kept in open cages without a male present. Several hours after onset of the dark period females begin coming into estrus. We have to find the estrus females by stroking each female in turn. In open cage system the task is extremely simple. After the female is found in heat she is taken to a 61 by 61 by 30 cm cage and placed with a male. Most males will ejaculate within a 10 minute period with a fully receptive female if they have not ejaculated within the last 24 hours. After the ejaculation is over the female is taken from the cage and placed in a special cage reserved for females mated in this manner. It is best to use different males for any additional females found in estrus. When males are housed individually in cages 61 by 61 30 cm, the mating process becomes highly efficient. Keeping sexually aggressive males under these conditions reduces ejaculation latency to 1-2 minutes. Eight males caged individually can breed approximately 128 females. By eliminating females in which estrus cannot be detected from the breeding program a highly efficient system for obtaining timed pregnancies can be developed.

B. Vaginal membrane method

An alternate method of determining the date of conception involves a daily check of the vaginal membrane of each female housed with a male. A vaginal smear is performed on any female with a ruptured vaginal membrane. The presence of sperm indicates the date of the conception. She may be moved to a separate cage with other pregnant females.

Rabbit

Rabbits are small mammals in the family Leporidae of the order Lagomorpha, found in several parts of the world. There are seven different genera in the family classified as rabbits, including the European rabbit (*Oryctolagus cuniculus*), cottontail rabbit (genus *Sylvilagus*; 13 species), and the Amami rabbit (*Pentalagus furnessi*, endangered species on Amami Ôshima, Japan). There are many other species of rabbit, and these, along with cottontails, pikas, and hares, make up the order Lagomorpha. Rabbits generally live between four and 20 years.

Kingdom	Animalia
Phylum	Chordata
Class	Mammalia
Order	Lagomorpha
Family	Leporidae
Genus	*Oryctolagus*

Behaviour

The rabbits are gentle animals and if care is taken become accustomed to handling if they are not upset by the experience. Inappropriate handling can result in toenail scratches to the handler from the toe nails on the powerful rear legs. Rabbits are herbivores and have the unique characteristic of reingesting soft fecal pellets directly from the anus. This process of "pseudorumination" allows the acquisition of B vitamins that have been produced by microorganisms in the cecum.

Chromosome number	44
Birth weight	60 gm
Adult body weight: male	2-5 kg
Adult body weight: female	2-6 kg
Breeding onset: male	6-7 months
Female	5-6 months
Life span	5 to 15 years
Gestation Period	31 days
Estrous cycle	-
Estrous period	10-11 hours
Rectal Temperature	38.5-39.5C
Heart rate	205-235/minute
Respiratory rate	30-60 per minute
Litter size	8

Uses

Of the many breeds of the domesticated European rabbit (*Oryctolagus cuniculus*), the albino New Zealand White is the most common breed utilized in biomedical research. The visibility of the peripheral vasculature in albino rabbits is advantageous to the biomedical use of this rabbit. Non-albino breeds such as the black and white Dutch belted rabbit tend to be preferred where pigmentation is required (i.e. ophthalmological research).

Breeding system in rabbits

1. Natural breeding in captivity

Provide one buck for 8 to 10 does. The doe should be taken to the buck's hutch. Copulation occurs within a few minutes and is considered successful if the buck falls on his side. If they fail to mate within a few minutes, take her to a different buck. If this fails, try again the next day, but don't leave the doe with the buck all day or even an hour in an attempt to solve a mating problem. A mature vigorous buck can be used for 2-3 successive days.

2. Post partum breeding

Most breeders mate does 6-8 weeks after kindling when the young from the previous litter are weaned, but a doe can be bred immediately after parturition. A doe should be bred for the first time while she is still young. With the gestation period of 31 days and a suckling period of eight weeks she can produce four litters a year. If the litter is in good condition she may be rebred 3-4 days after kindling. In order to obtain as many offspring as possible a doe may be bred repeatedly and her young are given to foster mothers. In this way as many as 11 litters can be obtained per year. Inbreeding, line breeding and crossbreeding are used in commercial colonies. Inbreeding and line breeding are used to reduce the number of ancestors and to retain the characteristics of an outstanding buck or doe. In crossbreeding purebred animals from two stocks are mated. The following criteria should be considered in selection: general health, vigor, high productivity, good mothering ability, tameness and ease of handling.

The type and quality of rabbits can be improved through careful selection of breeding rabbits and their use in a definite breeding plan. The important factors to be taken into account for selection are as under:-

1. Body weight and size
2. Wool quality
3. Reproductive efficiency
4. Feed conversion and weight gains
5. Mothering ability
6. Disease resistance

Reproductive efficiency, viz. conception rate, litter size at birth, litter size at weaning and mothering ability, are also important factors considered for selection but with low heritability. Disease resistance, better feed conversion efficiency, lesser tendency to moult and matting of wool are other important factors to be considered for selection.

Optimizing Breeding Performance

Factors to consider

To optimize the breeding performance of laboratory mice, observe the following practices:

1. Replace breeders before their reproductive performance declines: Maintain breeder pairs of various ages by replacing a percentage of them monthly or weekly. A colony of mixed-aged breeders produces a more consistent number of pups than does a colony of even-aged breeders.
2. Replace non-productive breeders: The following signs indicate that breeders are non-productive:
   - If they produce no litter within 60 days of mating (longer may be acceptable if delayed breeding is a strain characteristic).
   - If they produce no litter within 60 days of their last litter.
   - If they produce litters but do not wean pups for two to three litters.
3. Mate mice early: Mate mice when they are six to eight weeks old. Younger mice generally breed better than do older ones.

4. Use experienced males: Pairing young females with older males often improves breeding performance.
5. Keep meticulous and accurate breeding records: To evaluate the breeding performance of a mouse colony, meticulously maintain accurate records and examine them regularly. The sooner a problem is detected, the sooner it can be corrected. Adopt a detective-like demeanor and cultivate habits such as the following:
   - Investigate deviations in breeding performance and phenotype immediately.
   - Compare your colony's breeding performance to that characterized by your supplier. If mutant strain breeding data are not available, use data for the inbred strain background. All mouse facilities are different: strains that breed well in one facility may not breed well in another.
   - Keep a colony's environmental conditions suitable and stable.
   - Periodically verify the genotypes of pedigreed breeders for colonies of induced mutants (including those with no visible phenotype) with allele-specific molecular probes or other diagnostic assays.

Miscellaneous precautions

- A mutation's genetic background may affect phenotype, including breeding performance. If you transfer a mutation to a different background, maintain several generations of each background until you are sure that the second background does not affect phenotype, especially reproductive performance and survival.
- Try to avoid inadvertently placing selective pressure on your mouse colony. For example, if you pick only offspring from the best breeding females to perpetuate your colony, you may select for genes that change your strain's characteristics and inadvertently develop a sub-strain.

Development of genetically controlled lab animals

Experimental animals are classified into germ-free animals, specific pathogen free animals, Gnotobiotic animal and common animals, each according to the presence of microbes in a certain environment. Germfree is the highest quality level of laboratory animals in which there are not any detectable microorganisms in contrast to those commonly known as specific pathogen free (SPF) which is merely free of specific pathogens. Germfree animals are especially useful in the researches of genetic engineering, cancer, normal intestine flora, immunology and nutrition. Specific Pathogen Free animals are bred uniquely for the production of antibodies. These animals are housed in a pathogen free environment – all entering air, food, water is guaranteed to not contain foreign organisms that could provoke an immune response. The study of organisms or environmental conditions that have been rendered free of bacteria or contaminants or into which a known microorganism or contaminant has been introduced for research purposes is known as gnotobiotic animal. This can be bred in a controlled environment like Isolator. These gnotobiotics are used as a source of foundation stock for the production of highly defined barrier bred specific pathogen free animals. Gnotobiotic animals may be derived from germ free animals.

The genetic manipulation of animals has revolutionized our understanding of biology, by making it possible to test gene expression and function at the whole-animal level. The ability to introduce DNA into the germ line of mice for the production of transgenic mice is one of the greatest achievements of the twentieth century and has paved the way for the transformation of other mammals. Genetically modified mammals have been used not only to study gene function and regulation, but also as bioreactors producing valuable recombinant proteins, e.g. in their milk. Transgenic animals are animals in which every cell carries new genetic information.

- Abnormally large mice, with an inserted rat growth hormone gene.
- Onco mice, with an activated oncogene, so as to significantly increase the incidence of cancer.
- Doogie mice, with enhanced NMDA receptor function, resulting in improved memory and learning.

Gene targeting (also, replacement strategy based on homologous recombination) is a genetic technique that uses homologous recombination to change an endogenous gene. The method can be used to delete a gene, remove exons, add a gene, and introduce point mutations. Gene targeting requires the creation of a specific vector for each gene of interest. However, it can be used for any gene, regardless of transcriptional activity or gene size. Knockout mice are produced by the technique of gene targeting. The technique is essentially the opposite of Gene Knock-in that involves the insertion of a protein coding cDNA sequence at a particular locus in an organism's chromosome. Typically, this is done using a murine (mouse) model.

What is a knockout mouse?

A knockout mouse is a laboratory mouse in which researchers have inactivated, or "knocked out," an existing gene by replacing it or disrupting it with an artificial piece of DNA. The loss of gene activity often causes changes in a mouse's phenotype, which includes appearance, behavior and other observable physical and biochemical characteristics. For example, Fat mice are prone to obesity due to a carboxypeptidase E-deficiency and strong mice are produced with a disabled myostatin gene.

Why mice are used for knockout?

Mice are able to adopt new genes. They are cheaper, smaller and have shorter life span. Its genome is completed and 99% of its genes are homologous to humans. A large proportion of the mouse and human genomes are "syntenic" *i.e.*, they possess chromosomal regions with the same order of genes.

The first knock out mouse was created by Mario Renato Capecchi, Oliver Smithies and Martin Evans in mid 1980's. They were awarded Nobel Prize in Medicine and Physiology in 2007. Martin Evans identified and isolated the embryonic stem cell of the early embryo, the cell from which all cells of the adult organism are derived. He established it in cell culture, modified it genetically, and reintroduced it into foster mothers in order to generate a genetically modified offspring. Mario Capecchi and Oliver Smithies, independently of each other, discovered how homologous recombination between segments of DNA molecules can be used to target genes in the mammalian genome and developed methods to generate genetically modified mice.

Method of producing knockout mouse

A gene to be knocked out is isolated from gene library. A new DNA is engineered very similar to original gene and its immediate neighbour sequence. Gene is disabled by inserting a new gene into one of coding exon. Provision for positive-negative selection is added (insertion of marker genes). Embryonic stem cells (ESC) from a mouse blastocyst are used as target cells for insertion of vectors. ESC containing new DNA is then injected into blastocyst of another mouse and the blastocyst is implanted into a foster mother. The chimaeric progenies are tested for the presence of artificial gene. Heterozygous animals having the gene are crossed together to get the null mouse.

Why do we study knockout?

Knocking out the activity of a gene provides valuable clues about what that gene normally does. Knockout models can discover the function of a protein, gene for which is sequenced but its function is not known. Humans share number of genes with mice so these knockout models can help understand human diseases better.

Diseases	Mouse models
Cystic Fibrosis	*Cftr* ko mouse
Glaucoma	DBA/2J mouse
Type 1 Diabetes	NOD mice
Type 2 Diabetes	*Cpefat, Lepob*
Epilepsy	*swe* mouse
Atherosclerosis	*Apoe* ko mouse
Muscular dystrophy	*Dmd mdx* mouse
Ovarian Tumors	SWR and SWXJ mouse

Genetic control and monitoring

Genetic monitoring of the animals is necessary particularly among the inbred and special strains to ensure genetic homogenicity of desired characters in a particular strain. The inbred strains particularly need such monitoring at regular time interval to ensure freedom from genetic contamination. The different methods, used in detecting the status of inbred strains for their homogenicity are:-

1. **Histocompatibility or Skin grafting:** It is a very convenient and reliable method for routine testing of all inbred strain against genetic contamination. Skin grafts exchanged among members of the same inbred strain or F_1 hybrid or from either parent to an F_1 hybrid should be accepted.

2. **Electrophoresis or Biochemical Markers:** Inbred strains differ at many genetic loci. Some of these loci code for proteins and enzymes that may be identified by a number of biochemical techniques, the most important of which is electrophoresis. Typically, the samples of body fluids or organ homogenates are electrophorosed on starch cellulose acetate or polyacrilamide gels with specified buffer systems and under controlled conditions. After an appropriate time the gels are stained.

3. **Immunological Markers:** Inbred strains differ in the alloantigens that they carry and these immunological markers can be used effectively in routine genetic quality control. For immunological markers, there are two basic methods for demonstrating allo-antibodies. The first is the haemagglutination technique in which the erythrocytes are the test cells. The second is the cytotoxic test in which the lymphocytes are typically the test cells.

4. **Coat Colour Studies:** The coat colour genes carried by most inbred strains of mice and many inbred strains of rats are known. Some pigmented strains have an unusual or unique coat colour, which may be sufficient both to type the strain and to serve as an indicator in the event of genetic contamination. A cross with virtually any other strain would result in

offspring that would not have the parental coat colour, and the occurrence of coat colour variants would immediately suggest genetic contamination.

5. **Mandibular biometry:** Strain differences in the morphology of the skeleton have been known for many years. One method of genetic quality control using skeleton morphology is based on mandible shape. The technique is highly sensitive. All strains seem to differ in mandible shape and even closely related sublines can be distinguished, by measuring different parameters on the mandibles of 10-15 random animal samples taken from the group.

Record keeping

Good quality animals are those which are free from disease. The results of regular monitoring of parameters of genetic purity must be scrupulously recorded. Proper record-keeping is extremely important and vital for an animal facility. The forms should be simple but complete and preferably computer compatible. Too exhaustive and unnecessary recording should be avoided as these are not useful. Records of breeding and experimentation and deaths of all experimental animals at various stages are essential. Receipt and issue of food and other stores should be recorded. Log books of various machines such as incinerator, boilers, air-conditioning plant should be maintained. Monthly and annual reports of the activities should be prepared and reviewed for evaluation of work and future planning. The animal house should maintain following records:

(i) Animal House plans, which includes typical floor plan, all fixtures *etc.*

(ii) Animal House staff record - both technical and non - technical

(iii) Health record of animals

(iv) Breeding, stock, purchase and sales records

(v) Records of institute Animals Ethics Committee Meetings

(vi) Records of experiments conducted with the number of animals used

(vii) Death Record

(viii) Clinical record of sick animals

(ix) Training record of staff involved in animal activities

(x) Water analysis report *etc.*

Laboratory animal ethics

All scientists working with laboratory animals must have a deep ethical consideration for the animals they are dealing with. The three R's—replacement, refinement, and reduction—continue to be the core foundation of the Guide for scientific laboratory animal use. From the ethical point of view it is important that such considerations are taken care at the individual level, at institutional level and finally at the national level. A number of animals are used in this country for conducting a variety of experiments. There has been a long felt need to prescribe guidelines and procedures for animal experimentation covering all aspects. Good Laboratory Practices (GLP) for animal facilities is intended to assure quality maintenance and safety of animals used in laboratory studies while conducting biomedical and behavioral research and testing of products. Committee for the Purpose of Control and Supervision on Experiments on Animals (CPCSEA) laid guidelines for laboratory animal facility.

Ethical guidelines for use of animals in scientific research

1. Animal experiments should be undertaken only after due consideration of their relevance for human or animal health and the advancement of knowledge.
2. The animals selected for an experiment should be of an appropriate species and quality, and minimum number should be used to obtain scientifically and statistically valid results.
3. Investigators and other personnel should treat animals with kindness and should take proper care by avoiding or minimizing discomfort, distress or pain.
4. Investigators should assume that all procedures which would cause pain in human beings may cause pain in other vertebrate species also (although more needs to be known about the perception of pain in animals).
5. Procedures that may cause more than momentary pain or distress should be performed with appropriate sedation, analgesia or anaesthesia in accordance with accepted veterinary practice. Surgical or other painful procedures should not be performed on unanaesthetized animals.
6. At the end of, or when appropriate during an experiment, the animal that would otherwise suffer severe or chronic pain, distress, discomfort, or disablement that cannot be relieved or repaired should be painlessly killed under anaesthesia.
7. The best possible living condition should be provided to animals used for research purpose. Normally the care of animals should be under the supervision of a veterinarian or a person having adequate experience in laboratory animal care.

8. It is the responsibility of the investigator to ensure that personnel conducting experiment on animals possess appropriate qualifications or experience for conducting the required procedures. Adequate opportunities have to be provided by the institution for in-service training for scientific and technical staff in this respect.

9. In-vitro systems to replace or reduce the number of animals should be used wherever possible.

References

Baker HJ, Lindsey JR, Weisbroth SH 1979. The Laboratory Rat, vols 1 and 2. Academic Press, New York.

Capecchi MR 1989. The new mouse genetics: Altering the genome by gene targeting. Trends Genet 5: 70-76.

Chitty D, Southern HN 1954. Control of rats and mice, vol. 1-3. Clarendon Press, Oxford.

Davies B, Morris T 1993. Physiological Parameters in Laboratory Animals and Humans. Pharmaceutical Research 10 (7): 1093–1095.

Encyclopedia Britannica (Standard ed). 2007. Chicago: Encyclopedia Britannica, Inc.

Fritzsche P 2008. Hamsters: A Complete Pet Owner's Manual. Barron's Educational Series Inc., NY.

Gill TJ, Smith GJ, Wissler RW, Kunz HW 1989. The Rat as an Experimental Animal. Science 245: 269–276.

Guidelines to Promote the Well-being of Animals Used for Scientific Purposes: The Assessment and Alleviation of Pain and Distress in Research Animals. 2008. National Health and Medical Research Council, Australian Government. Available at www.nhmrc.gov.au/_files_nhmrc/file/publications/synopses/ea18.pdf, accessed January 24, 2010.

Hafez ESE 1970. Reproduction and Breeding Techniques for Laboratory Animals. Lea and Febiger, Philadelphia.

Hawk CT, Leary SL 1995. Formulary for Laboratory Animals. Iowa State University Press. Institute for laboratory animal research http://dels.nas.edu/ilar/

International Council for Laboratory Animal Science 1987. ICLAS Guidelines on the Selection and Formulation of Diets for Animals in Biomedical Research. London, International Council for Laboratory Animal Science.

Laboratory Animal Breeders Association of Great Britain Limited/Laboratory Animal Science Association 1993. Guidelines for the Care of Laboratory Animals in Transit. Laboratory Animals 27: 93–107.

Loeb WF, Quimby FW 1989. The Clinical Chemistry of Laboratory Animals. Pergamon Press, New York.

Nakajima T, Wang R, Ito Y, Aoyama T, Kamijima MS 2005. A Review of Hazardous Chemical Toxicity Studies Utilizing Genetically-Modified Animals —Their Applications for Risk Assessment, Industrial Health 43: 615–622

National Academy of Sciences 2011. Guide for the Care and Use of Laboratory Animals. Natl Acad Sci, Washington, DC.

National Research Council 1996. Rodents . National Academy Press, Washington, D.C.

Nebert DW, Duffy JJ 1997. How knockout mouse lines will be used to study the role of drug-metabolizing enzymes and their receptors during reproduction and development, and in environmental toxicity, cancer, and oxidative stress. Biochem Pharmacol 53: 249-254.

Nowak, Ronald M 1999. Walker's Mammals of the World, 6th edition. Johns Hopkins University Press.

Painful dilemmas: The ethics of animal-based pain research. 2009. Magalhaes-Sant'Ana M, Sandoe P, Olsson IAS. Anim Welf 18:49-63

Spotorno AE, Marín JC, Manríquez G, Valladares JP, Rico E, Rivas C 2006. "Ancient and modern steps during domestication of guinea pigs (Cavia porcellus L.)" Journal of Zoology.

Svendsen P, Hau J 1994. Handbook of Laboratory Animal Science, Vol. 1. CRC Press, Inc., Boca Raton.

The Jackson Laboratory http://www.jax.org/The Three Rs: A journey or a destination? 2000. Richmond J. ATLA 28:761-773. The use of non-human animals in research: a guide for scientists. 2004. The Royal Society, p1.

Use of Laboratory Animals in Biomedical and Behavioral Research 1988. Institute for Laboratory Animal Research. The National Academies Press.

Williams, Christine SF 1976. Practical Guide to Laboratory Animals. The CV Mosby Co, St Louis.

Zambrowicz BP, Sands AT 2003. "Knockouts model the 100 best-selling drugs-will they model the next 100?" Nature Review Drug Discovery 2 (1): 38–51.

Zutphen LFM, Baumans V, Beynen AC 1993. Principles of Laboratory Animal Science: A Contribution to the Humane Use and Care of Animals and to the Quality of Experimental Results. Elsevier Science Publishers, Amsterdam.

5

Impact of Nutrition on Reproductive Efficiency of Small Ruminants

V P Maurya, V Sejian, Gyanendra Singh and Mihir Sarkar

Introduction

Sheep and goats have a vital economic importance in India, inhabitant in semi-arid region and mostly rose under harsh environmental conditions (high ambient temperature, scarcity of feed and water). The arid and semi-arid regions of India are drought prone due to low and uneven rainfall. The vegetation of arid and semi-arid tract is more acute during summer when the quality of the pastures becomes vulnerable which in turn decreases the body condition score of animals and reproductive efficiency is being compromised (Maurya *et al.* 2009a & 2010). The small ruminants are widely distributed in comparison of other domesticated species. These species of livestock is also a source of employment for marginal and land-less farmers particularly in the arid and semi-arid region of India. Most of the Indian farmers have limited resources and small ruminants production offers an opportunity to alleviate financial conditions and improve living standards.

The reproductive efficiency and performance adaptability of sheep inhabiting the semi-arid regions of India is relatively low (Maurya *et al.* 2004).The nature of nutritional resources and management problems differs with climate, soil and vegetation. The physical process which governs the animals' reproductive performance is the same; the female must reach puberty, show behavioral estrus and shed one or more ova which have then to be fertilized. There after, the resultant embryos have to be maintained throughout pregnancy. Combined stress affects growth, blood metabolites and hormonal changes in sheep (Sejian *et al.* 2010a & b). Dietary nutrition promotes the programming and expression of the metabolic pathways that enable animals to achieve their genetic potential for reproduction.

Nutrition has an important impact on the reproductive performance in sheep and goats but the magnitude of the effect on reproduction may vary with the season (White *et al.* 1983). Sheep and goats are prone to neglect compared to cattle and often suffer from lack of feed. This leads to decrease in body condition score, body weight and loss of reproductive functions. The nutritional stress appears to be a prime cause of acyclicity in sheep and goats. The low birth rates may be caused by acyclicity, early embryonic mortality or abortions not noted by the farmer. It has been reported that the depleted body condition during period of energy deficiency also reduces heat tolerance (Hooda and Naqvi 1990), which in turn affects the reproductive potential of sheep and goats. It is, therefore, important for farmers to know when these critical periods of nutritional stress occur which could have an impact on the reproductive activities of their animals. In this chapter we are going to discuss impact of nutrition on reproductive efficiency of Indian sheep and goats.

Nutrition, sexual behaviour, incidence of estrus and ovulation rate

Puberty, the age at first ovulation, occurs approximately at 7 to 9 month of age. The onset of puberty in sheep is influenced by genetic and environmental factors such as breed and strain differences, the nutrition plane and time of birth. Prerequisites of the onset of puberty in the female sheep include the achievement of a critical body size. In sheep (Foster *et al.* 1988; Dyrmundsson 1987) restricted nutrition during the early life and failure to achieve the necessary live weight before the end of the first breeding season, after birth, results in delay in onset of puberty at least at the following breeding season when the animal has usually reached the critical live body weight. Any alteration of nutrition before puberty alters the rate of growth of young animals and which affects the attainment of puberty (Riera 1982, Mc Call *et al.* 1989). Adequate nutrition is necessary for the proper pubertal development. Under-nutrition at the time of prepubertal development retards the growth of animals. High plane of nutrition advances the puberty considerably (Moustgaard 1969). The dietary energy level affects the blood profile of ewes (Maurya *et al.* 1998a) which indirectly affects the reproductive potential of ewes. At the ovarian level, low feed intake that delay puberty is accompanied by reduced follicular development which in case of heifer is associated with smaller dominant follicles (Bergfeld *et al.* 1994). Blood borne metabolites that reflect nutritional status and are involved in the control of GnRH release are subjects of speculation. The observation of Hall *et al.* (1992) demonstrating a stimulatory effect of the abomasal infusion of tyrosine on LH pulse frequency in growth restricted ewe lambs imply that this amino acid may function as a nutritional signal influencing the central nervous system control on GnRH release.

Under-nutrition and body condition score affects the incidences of estrus very badly (Maurya *et al.* 2004 & 2011) and sometimes estrus is being completely inhibited (Hafez 1952) or breeding season can be prematurely terminated (Knight *et al.* 1983). Sheep are frequently raised in areas where feed resources are limited, suppression of estrus as a result of low level of nutrition, may be an important determinant of reproductive performance in ewes. Energy restriction suppresses the increase in LH that is necessary for growth of ovarian follicles in the pre-ovulatory stage (Schillo 1992). Flushing is generally recognized as a significant regulator of reproduction and it can be accomplished either by allowing animals to graze lush nutritious pasture or by feeding energy-rich supplements (Webb and Mamabolo 2004). Furthermore, flushing has been reported to be positively affecting the body condition, fertility and ovulation rate of dams with poor nutritional status just before or during mating (Chowdhury *et al.* 2002, Karikari and Blasu 2009).

It is well established that the follicular development in the many mammals is affected by nutrition. This effect can either be indirect, acting by way of hypothalamic pituitary axis to alter the secretion of gonadotrophin, or can be direct acting on the ovary to mediate the action of gonadotrophin on the follicles. Allison (1977) reported that greater number of follicle >2mm diameter and larger mean follicle diameters in ewes of high live weight compared with ewes of low live weight. It has also been shown that a higher population of large follicle in the high condition animal compared with the low condition animal was estrogenic and potentially ovulatory (Mc Neilly *et al.* 1987).

Ovulation rate is one of the most important determinants of reproductive performance in sheep, which depends on the age (Knight *et al.* 1975), genotype (Wheeler and Land 1977) and stage of breeding season (Davis *et al.* 1976, Maurya *et al.* 1998[b]) as well as nutrition (Gunn 1983). The ewes raised under tropical conditions are generally maintained under condition of minimal management in a climatically harsh environment and have lower ovulation and lambing rate, which can be increased by providing them good nutrition (Gunn 1983, Naqvi *et al.* 2002). In Australia, Downing *et al.* (1990 & 1995) showed that the infusion of branched chain amino acids over a five-day period in the late stage of estrous cycle would significantly increase ovulation rate in sheep. The body weight of ewes, at mating, representing the static effect, has been shown to influence subsequent litter size (Coop 1966). This effect is mainly the result of different ovulation rate. Cumming (1977) reported that heavier ewes within the flock were found to have more ovulations than the lighter ones, showing about 2.5-3.0% increase for each 1.0 Kg increase in live weight. The net nutritional status of ewes also affects the ovulation rate. Lindsay (1976) suggested that ovulation rate in ewes depends on the net nutritional status of the ewes.

According to him, heavy ewes given poor feeding may still show a good ovulation rate because they have a reasonable endogenous source of energy and protein. On other hand, poor ewes temporarily well-fed will also ovulate well because the contribution of exogenous nutrients. Effect of nutrition on follicle development and ovulation rate could potentially be mediated through changes in circulating blood metabolites (glucose, non-etherified fatty acids, amino acids etc), metabolic hormones (insulin , growth hormone etc.), gonadotrophin (LH and FSH) or a combination of several of these factors (Robinson 1996).

Nutrition, embryo mortality, fetal and neonatal metabolism

After the ovum is shed, one of the first determinants of the rate of reproductive success is the fertilization rate. Extremes in the level of feeding are detrimental to embryo survival. The retinoids are the main metabolites of vitamin A and are involved in cell proliferation and differentiation, expression of growth factor, steroidogenesis, which play important role in embryo survival. Folic acid, for which there is evidence of improved prenatal survival, is essential for nucleic acid synthesis, and vitamin C appears to enhance luteal function.

Recently, attempts have been made to identify the mechanism involved in mediating the adverse impact of high protein diet on the fertility rate of dairy cow. Elord and Butler (1993) found that high intake of rumen degradable protein (RDP), leading to excess rumen ammonia production, were associated with reduction of pH of the uterine environment. This disturbance in the uterus environment leads to the embryonic death. It is well known that progesterone plays a crucial role in maintaining pregnancy in ewe (Denamur and Martinet 1955) and some workers have used exogenous progesterone to improve embryo survival in sheep. Nutrition in early pregnancy and peripheral progesterone concentration may be inversely related (Williams and Cumming 1982). Further Parr *et al.* (1987) demonstrated that sheep fed high energy ration after mating, had reduced progesterone levels and showed an increase in embryo mortality. High plane of feeding increases the metabolic clearance rate of progesterone (Parr *et al.* 1993) and the concomitant reduction of progesterone concentration at Day 11 and Day 12 of post mating in ewes (Parr 1992). Dietary induced suppression of circulating progesterone during oocyte maturation in superovulated ewes primed with a single CIDR device (0.3 g progesterone) can impart a legacy of developmental retardation which leads to decreased embryo survival (McEvoy *et al.* 1995). Severe under-nutrition significantly decreases the pregnancy rate of sheep (Abecia *et al.* 1994, Maurya *et al.* 2004). It is also evident that under-nutrition of female sheep during prenatal life and early in the postnatal period may reduce life time reproductive performance of ewes. Studies reported by Borwick *et al.* (1994) led to the conclusion that under-nutrition of ewe lambs at mating significantly retarded ovarian development in terms of oogonia differentiation.

Advance pregnancy is the most important period of pregnancy with fetal mass increasing in a curvilinear manner such that in the last 8,4,2 weeks of gestation the fetus is gaining equivalent to 85, 50 and 25% of the final birth weight (Robinson 1983) respectively. This is the period when pregnancy nutrition can have the greatest effect on total potential lamb production. The degree of under-nutrition can also be observed with the help of plasma glucose and plasma free fatty acids. Under-nutrition decreases uterine blood flow and accompanying decrease in fetal insulin and IGF-I coupled with increase in growth hormone, adrenocorticotrophin corticosterone impair fetal growth and development. Fetal urea concentration increases during the under-nutrition, reflecting increase in gluconeogenesis by fetus from amino acids (Bell 1993). In new born lambs, low insulin, high corticosterone and a low selenium and /or iodine status inhibit thermo genesis from brown adipose tissue (Robinson 1990[b], Robinson and Symonds 1995) which leads to a reduced viability at birth. Under nutrition during pregnancy significantly reduces the live birth weight of lambs (Maurya *et al.* 2004).

Nutrition, early stage of pregnancy and in later stage of development

The animals which are exposed to under-nutrition during fetal life have reduced subsequent litter size (Robinson 1990[a]). Recent studies have demonstrated an effect of under-nutrition on the concentration of oogonia in the ovine ovary as early as day 47 of fetal life (Borwick *et al.* 1994) and an associated postponement in the arrest of ovarian metabolic activity on day 62 (Borwick *et al.* 1995). Thompson *et al.* (1994) found that birth weights of lamb from embryos cultured in synthetic oviduct fluid medium supplemented with human serum were greater than those from spontaneous ovulating ewes or those derived from embryos cultured in synthetic oviduct fluid supplemented with amino acids and bovine serum albumin. Lamming (1969) points out that inadequate nutrition can have adverse effect on the pituitary function i.e. on the pattern of synthesis and/or release of pituitary gonadotrophin. He also points out that level of nutrition may influence the response of target organs to either gonadotrophin or gonadal hormone. Rearing of ewe lamb on high plane of nutrition will normally promotes both faster growth and development of reproductive organs (Kuliev 1965). The plane of nutrition and body condition score has positive correlation and affects fetal growth of animals (Sejian *et al.* 2008, Maurya *et al.* 2010a). Thus owing to the close association between general body growth and sexual development, raising the level of nutrition will not only result in more ewes lamb attain puberty during the first breeding season, but also in ewe lambs growing at faster rate exhibiting their first estrus and more likely to conceive at lower age and heavier body weight than ewe lambs growing at slower rate. McKelvey *et al.* (1988) reported that high plane of nutrition during the post mating period can adversely affect the embryo survival and that this was mediated through the decline in plasma progesterone concentration; low plane of nutrition at this time may affect the embryo survival. However, not all studies have shown that a high plane of nutrition after ovulation influence the embryo survival, although there may be significant decrease in the peripheral progesterone concentrations (Wallace *et al.* 1994).

Nutrition and male reproductive efficiency

In male sheep, reproductive activity is affected by a range of external factors, including socio–sexual cues, temperature, photoperiod and nutrition. Among these factors affecting rams, nutrition is arguably the most powerful. The responses of ram to nutritional manipulations can be divided into short-term effects that act mainly on the neuroendocrine system controlling testicular activity (Martin *et al.* 1994, Blache *et al.* 2000) and long-term effects that act on testicular growth and sperm production (Oldham *et al.* 1978). Under extensive management system, level of nutrition may have a greater influence on reproductive activity of rams than photoperiod (Perez *et al.* 1997). Energy balance, appetite and reproductive functions are tightly linked and alterations in energy and appetite-endocrine axes will lead to altered feedback and feed-forward in the hypothalamo-pituitary-gonadal axis. Sperm production and testicular size in rams has been found to be influenced by the plane of nutrition (Cameron *et al.* 1988, Murray *et al.* 1990, Maurya *et al.* 2010). The male breeding animal makes an important contribution to the reproductive performance and genetic improvement in a herd/flock. It is therefore important to pay attention to the factors which may influence male fertility such as libido, mating ability and semen quality.

An adequate plane of nutrition is clearly important for the early sexual development of the male lambs. Nutritive deficiency specially low energy intake, will retard sexual development and delay the attainment of puberty (Foote 1969).There is a lack of gonadotrophin from the hypophysis though the testis will usually continue to produce testosterone and androgenic function of the testis is retarded more markedly than is spermatogenesis (Mann 1964). Conversely, in well fed and rapidly growing lambs it would appear that androgenic function is perfectly advanced (Dyrmundsson 1987). In rams there is 6 to 7 week delay in the response of spermatozoa's number to diet, reflecting the time it takes for the development of spherical spermatids in the germinal epithelium to fully mature spermatozoa in the distal cauda epididymus. Evidence for the direct effect of the nutrition on testis growth and production of spermatozoa in the absence of alteration in the GnRH gonadotrophin system (Martin and Wlkden Brown, 1995) to develop the concept of both GnRH independent and dependent pathway that are involved in nutritional effect on spermatogenesis. Salaman (1964) reported that ram fed on a high protein diet displayed a more intense sex drive than ram fed on a low protein supplement.

It is evident from the results that animals, which have undergone a period of under-nutrition, are capable to regaining body weight loss and reproductive function. It appears that animals fed *ad libitum* could overcome the harsh climate effect on their body weight and reproduction, while animals exposed to thermal and nutritional stress simultaneously have reduced body reserves and reproductive efficiency. The losses incurred on body weight and reproduction during coupled stresses (thermal and nutritional) could not be compensated during a post stress period.

References

Abecia JA, Rhind SM, McMillen SR 1994. Effect of under nutrition on luetial function and the distribution of progesterone in endometrial tissue in ewes. ITEA, Production Animal 90 A: 63-71.

Allision AJ 1977. Effect of nutritionally induced live weight differences on the ovulation rates and the production of ovarian follicles in ewes. Theriogenology 8:19-24.

Bell AW 1993. pregnancy and foetal metabolism. In: Forbes, JM and France, J (edts.)Aspects of ruminant digestion and metabolism. CAB, International ,Wallingford. pp 405-431.

Bergfeld EGM, Kojima FN, Cupp AS, Kinder JE 1994. Ovarian follicular development in prepubertal heifer is influenced by level of dietary energy intake. Biology of Reproduction.51:1051-1057.

Blache D, Chagas LM, Blackberry MA, Vercoe Pe, Martin GB 2000. Metabolic factors affecting the reproductive axis in male sheep. Journal of Reproduction and Fertility 120: 1–11.

Borwick SC, Rhind SM, McMlillen SR1995. Effect of under nurtition from the time of mating on the ovarian; development in foetal sheep at 62 d of gestation. Journal of Reproduction and Fertility. Abstract Series No. 15, p.52.

Borwick SC, Rhind SM, McMlillen SR1994. Ovarian steroidogenesis and development in fetal Scottish Blackface ewes, undernourished in utero from conception. Journal of Reproduction and Fertility Abstract Series No. 14, p.14.

Cameron AWN, Murphy PM, Oldham CM 1988. Nutrition of rams and output of spermatozoa. Proc Aust Soc Anim Prod 17: 162–165.

Chowdhury SA, Bhuiyan MSA, Faruk S 2002. Rearing Black Bengal goat under semi-intensive management. 1 Physiological and reproductive performance. Asian-Australasian J Anim Sci, 15(4): 477-484.

Coop IE 1966. Effect of flushing on reproductive performance of ewes. Journal of Agricultural Science, Cambridge 67: 305-321.

Cumming IA 1977. Relationship in the sheep of ovulation rate with live weight, breed , season and plane of nutrition. Australian Journal of Experimental Agriculture and Animal Husbandry, 17:234-241.

Davis IP, Kenny PA, Cummins IA 1976. Effect of time of joining and rate of stocking on the production of Corredale ewes in southern Victoriya II Ovulation rate and embryonic survival. Australian Journal of Experimental Agriculture and Animal Husbandry,16:11-18.

Denamur R, Martinet J 1955. Efects de, I'ovariectomise chez in brebis pendant in gestation. Cr. Senic Social Biolology140:2105-2107.

Downing JA, Scaramuzzi RJ, Joss J 1990. Infusion of branched chain amino acids will increase ovulation rate in the ewe. Proceedings Australian Society Animal Production 18:472.

Downing JA, Joss J, Scaramuzzi RJ 1995. A mixture of branched chain amino acids leucine, isoleucine and valine increases ovulation rate in ewes when infused during the late luteal phase of the estrous cycle: an effect that may be mediated by insulin. Journal of Endocrinology 145:315-323.

Dyrmundsson OR 1987. Advancement of puberty in male and female. In: Marai IFM, Woven JB (edt.) New Techniques in sheep production, Butterworths, London. pp 65-76.

Elrod CC, Butler WR 1993. Reduction infertility and alteration of uterine pH in heifers fed access ruminally degradable protein. Journal of Animal Science.71:694-701.

Foote RH1969. In: Reproduction in Domestic Animals. Cole, H.H. and Cupps, P.T (edts.) Academic Press,London. pp. 313-333.

Foster DL, Ebling EBJ, Vannerson LA, Veenvilet BA 1988. Modulation of gonadotrophins secretion during development by nutrition and growth. Proceedings of 11th International Congress on Animal Reproduction and Artificial Insemination. 5:101-108.

Gunn RG 1983. In:Haresign, W. (edt.) Sheep Production. Proceedings of the 35th Easter School in Agricultural Sciences University of Nottinghom, London. Butterworths. pp 99-110.

Hafez ESE 1952. Studies on the breeding season and reproduction of the ewe. Journal of Agricultural Science Cambridge. 42: 189-265.

Hall JB, Schillo KK, Hileman SM, Boling JA 1992. Does tyrosine acts as a nutritional signal mediating the effect of increased feed intake on leutenizing hormone patterns in growth restricted lambs. Biology of Reproduction. 46:573-579.

Hooda OK, Naqvi SMK 1990. Effect of thermal load and feed restriction on relative adaptability of Malpura and Avikalin sheep in semi-arid region. Indian Journal Animal Sciences, 60: 608-611.

Karikari PK, Blasu EY 2009. Influence of Nutritional Flushing Prior to Mating on the Performance of West African Dwarf Goats Mated in the Rainy Season. Pakistan Journal of Nutrition. 8(7):1068-1073.

Knight TW, Hall DRH, Wilson ID 1983. Effect of teasing and nutrition on the duration of the breeding season in Romney ewes. Proceedings of the New Zealand Society of Animal Production, 43:17-19.

Knight TW, Oldham CM, Smith JF, Lindsay DR 1975. Studies in ovine infertility in Agricultural region in western Australia: analysis of reproductive wastage. Australian Journal of Experimental Agriculture and Animal Husbandry, 15:183-188.

Kuliev GK 1965. Morpho-Biological principles in growth and developments of ovaries of ewes at deferent levels of nutrition. Izvstiya Akedemii Nauk. SSSR. Biologicheskaya 3:51-58.

Lamming GE 1969. Nutrition and reproduction. In: Cuthbertson (edt.) The Science of Nutrition of Farm Livestock. Pt I Pergamon, London pp 411-453.

Lindsay DR 1976. The usefulness to the animal producer of research finding in nutrition on reproduction. Proceedings Australian Society Animal Production 18: 217-224.

Mann T 1964. The biochemistry of semen and of the male reproductive track. Methan & Co. Ltd. London. pp. 493.

Martin GB, Tjondronegoro S, Blackberry MA 1994. Effects of nutrition on testicular size and the concentrations of gonadotrophins, testosterone and inhibin in plasma of mature male sheep. J Reprod Fertil 101:121-128.

Martin GB, Walkden-Brown SW 1995. Nutritional influence on reproduction in mature male sheep and goat. Journal Reproduction and Fertility. Suppl. 49:437-494.

Maurya VP, Naqvi SMK, Mittal JP 1998a. Reproductive behaviour of Malpura and Bharat Merino ewes lamb under semi-arid condition. Indian Journal of Animal Sciences. 68:533-535.

Maurya VP, Naqvi SMK, Mittal JP 1998b. Influence of dietary energy level on circulating blood constituents of sheep in hot semi-arid region of India. 8th World Conference on Animal

Production Seoul.Korea.ON4-10.

Maurya VP, Naqvi SMK, Mittal JP 2004. Effect of dietary energy level on physiological responses and reproductive performance of Malpura sheep in the hot semi-arid regions of India. Small Ruminant Research 55:117-122.

Maurya VP, Kumar S, Kumar D, Gulyani R, Joshi A, Naqvi SMK, Arora AL, Singh VK 2009. Effect of Body Condition Score on Reproductive Performance of Chokla Ewes. Indian Journal of Animal Sciences 79:1136-1138.

Maurya VP, Prince LLL, Naqvi SMK, Sejian V, Kumar D, Joshi A 2010a. Effect of body condition score on productive and reproductive performance of Avikalin ewes. Indian Vet Journal 87:189-190.

Maurya VP, Sejian V, Naqvi SMK 2010b. Transabdominal ultrasonography for early pregnancy diagnosis in Malpura sheep. Indian Veterinary Journal, 87:1097-1099.

Maurya VP, Sejian V, Naqvi SMK 2011. Body condition score and sexual behavior of Malpura ewes. Indian Veterinary Journal, 88(12):18-20.

Maurya VP, Sejian V, Kumar D, Naqvi SMK 2010c. Effect of induced body condition score differences on sexual behavior, scrotal measurements, semen attributes, and endocrine responses in Malpura rams under hot semi-arid environment. Journal of Animal physiology and Animal Nutrition, 94: e308-e317.

McCall DG, Clayton JB, Dow BW 1989. Nutrition effect on live weight and reproduction on Cashmere Doe hogget's. Proceedings of the New Zealand Society of Animal Production 49:157-161.

McEvoy TJ, Robinson JJ, AitkenFidley PA, Plmer RM, Robertson IS 1995. Dietary induced suppression of preovulatory progesterone concentrations in superovulated ewes impairs the subsequent in vivo and in vitro development of their ova. Animal reproduction Science. 39:89-107.

Mc Kelvey WAC, Robinson JJ, Iitken RP 1988. The use of reciprocal embryo transfer to separate the effect of pre and post mating nutrition of embryo survival in growth of the ovine concetus. Proceedings of the 11th International Congress Animal Reproduction and AI (Dulbin) 2. Paper 176.

McNeilly AS, Jonassen JA, Rhind SM 1987. Reduced ovarian follicular development as a consequence of low body condition in ewes. Acta Endocrinologica.115:75-83.

Moustgaard J 1969. In: Reproduction in domestic animals. Cole, H.H. and Cupps, P.T (edts.) Academic Press, London. pp. 489-516.

Murray PJ, Rowe JB, Pethick DW, Adams NR 1990. The Effect of Nutrition on Testicular Growth in the Merino Ram. Aust J Agric Res 41:185-195.

Naqvi SMK, Gulyani R, Joshi Anil, Das GK, Mittal JP 2002. Effect of dietary regimens on ovarian response and embryo production of sheep in tropics. Small Ruminant Research. 46:167-171.

Oldham CM, Adams NR, Gherardi PB, Lindsay DR, Mackintosh JB 1978. The Influence of Level of Feed Intake on Sperm-producing Capacity of Testicular Tissue in the Ram. Aust J Agric Res 29:173-179.

Parr RA 1992. Nutrition progesterone interaction during early pregnancy in sheep. Reproduction Fertility Development.4:297-300.

Parr RA, Davis IF, Fairclough RJ, Miles MA1987. Overfeeding during early pregnancy reduces peripheral progesterone concentration and pregnancy rate in sheep. Journal of Reproduction and fertility.80:317-320.

Parr RA, Davis IF, Miles MA, Squires TJ 1993. Feed intake affects metabolic clearence rate of progesterone in sheep. Research in Veterinary Science.55:306-310.

Perez CR, Lopez A, Castrillejo A, Bielli A, Laborde D, Gastel T, Tagle R, Queirolo D, Franco J, Forsberg M, Rodriguez-Martinez H 1997. Reproductive seasonality of corriedale rams under extensive rearing conditions. Acta Vet Scand 38:109-117.

Riera S 1982. Reproductive efficiency and management in goat. Proceedings of the 3rd International Conference in Goat Production and Diseases Tucson, Arizona, USA. pp.162-174.

Robinson JJ 1983. Nutrition of the pregnant ewes. In: Haresign, W.(edt.) Sheep Production. Butterworths . London.pp.111-131.

Robinson JJ 1990. Nutrition in the reproduction of farm animals. Nutritional Research Reviews.3:253-276.

Robinson JJ 1990. The postural animal industries in the 21st century. Proceedings of the New Zealand Society Animal Production 50:345-359.

Robinson JJ 1996. Nutrition and reproduction. Animal Reproduction Science.42:25-34.

Robinson JJ, Symonds ME 1995.Whole body fuel selection: "Reproduction" Proceeding of Nutritional Society. 54:283-299.

Salamon S 1964. The effect of nutritional regimen on the potential semen production of rams. Australian Journal of Agricultural Research. 15:645-646.

Schillo KK 1992. Effect of dietary energy on control of luteineizing hormone secretion in cattle and sheep. J Anim Sci, 70: 1271 -1 282.

Sejian V, Maurya VP, Naqvi SMK 2010a. Adaptability and growth of Malpura ewes subjected to thermal and nutritional stress. Tropical Animal Health and Production, 42:1763-1770.

Sejian V, Maurya VP, Naqvi SMK 2010b. Adaptive capability as indicated by endocrine and biochemical responses of Malpura ewes subjected to combined stresses (thermal and nutritional) under semi-arid tropical environment. International Journal of Biometeorology, 54:653-661.

Thompson JG, Gardner DK, Pugh P, AmcMillan WH, Tervit HR 1994. Lamb birth weight following transfer is affected by the culture system used for pre elongation development of embryos. Journal of Reproduction and Fertility. Abstract Series.13:25.

Wallace JM, Airken RP, Cheyne MA 1994. Effect of post ovulation nutritional status in ewes on early conceptus survival and growth in vivo and luterotophic protein secretion in vitro. Reproduction Fertility and Development 6:253-259.

Webb EC, Mamabolo MJ 2004. Production and reproduction characteristics of South African indigenous goats in communal farming systems. South African J. Anim. Sci., 34(5):236-239.

Wheeler AG, Land RB 1977. Oestrus and ovarian activity of Finnish Landrace, Tashmanian Merino and Scottish Blackface ewes. Animal Production 24:363-376.

White DH, Bowman PJ, Morley FHW, Mc Manus WR, Filan SJ 1983. A simulating model of the breeding ewe flock. Agricultural System 10:149-189.

Williams AH, Cumming AI 1982. Inverse relationship between concentration of progesterone and nutrition in ewes. Journal of Agriculture Science, Cambridge 98:517-522.

6

Cactus (*Opuntia*) as Strategic Source of Feed for Farm Animals in Arid and Semi-arid Regions

Abed M Al-Bial, Jai Singh and Ram Niwas

Introduction

Agricultural activity in the arid and semi-arid divided into two main components are plant production and animal production. Animal production is associated with a component of a large proportion of workers in the agricultural sector, either individually or as part of agricultural activity Integrative. The livestock are plays major role in the lives of the rural population to the availability of several important products such as meat and dairy products.

Farm animals in the arid and semi-arid depend on their feeding mainly on crop residues and natural pastures that do not meet their needs due to scarcity of resources in these regions. Such scarcity is caused by a number of factors; the most important factor is limited agricultural land devoted to cultivation of fodder crops, low rainfall and the deterioration of natural pastures. However, there are many studies in different countries suggesting the possibility of widespread exploitation of cactus as a strategic resource, especially in the areas of low rainfall and scarce feed resources (FAO 2007). Those studies recommend the use Cactus to feed cows, goats, sheep and rabbits without causing any problems to the animals' health. In addition, it can be used as medicine and cosmetics for humans and for many other uses.

The plant has become important fodder in many parts of the world. It is used all year round, or as emergency feedstock during drought. In many arid areas (Mexico, South Texas, South Africa, Tunisia, etc.) farmers use cactus extensively as emergency forage.

Cactus occupy the great economic importance in a number of countries, which established an International Network of Technical Cooperation in 1993, under the auspices of the Food and Agriculture Organization (FAO) to enhance cooperation between the countries producing the crop which include 14-country.

Therefore, this topic is highlighting the importance of the development and exploitation of resources in rural areas through the use of Cactus (*Opuntia*) as fodder to feed livestock. This can be done through supplying livestock with food, creating sustainable economic opportunities and activities, increasing the income of male and female workers in rural sectors and reducing the environmental degradation of pastures and farmland.

Cactus as a fodder bank

Opuntias used for animal feeding are abundant, easy and cheap to grow, palatable and drought tolerant (Shoop et al. 1977). Such characteristics make them a potentially important feed supplement for livestock, particularly during periods of drought and low feed availability. A large portion of opuntia plant biomass is vegetative material rather than fruits, and it can be fed to livestock as fresh forage or stored as silage for later feeding (Castro et al. 1977). The idea of using cactus to feed livestock is not recent. Griffiths and Hare (1996) was certain that feeding cactus to livestock started in the USA before the Civil War, and before and after the war; there was extensive freight transportation of cactus pads between Brownsville, Indianola, San Antonio and Eagle Pass in Texas. The plant has become important fodder in many parts of the world, based on natural and cultivated populations. It is cultivated in Africa, Argentina, Bolivia, Brazil, Chile, Colombia, Israel, Italy, Mexico, Spain, USA and Peru (Barbera et al. 1992, Hawkins 1970, Brutsch 1984, Pimienta 1990, Saenz 1985, Curek M and Ozen 2001, Cushnahan 2006).

Large areas are found in Algeria, northeast Brazil, Mexico and South Africa. The opuntia is used all year round, or as emergency feedstock during drought. In many arid areas (Mexico, South Texas, South Africa, Tunisia, etc.), farmers use cactus extensively as emergency forage that is harvested from both wild and cultivated populations to prevent the disastrous consequences of frequent and severe droughts (Le Houérou 1992).

From the early 1900s in North Africa, several strategies were introduced to reduce water and wind erosion and rangeland degradation, using shrubs (*Acacia cyanophylla*, *Atriplex nummularia* and *A. halimus*) and cacti (*Opuntia ficus-indica f. inermis*). Large areas have been planted in Algeria, Morocco and

Tunisia since the 1950s. It is estimated that in low rainfall areas, some 0.7 to 1 million ha plantings are helping to combat erosion and desertification and provide feed for livestock during drought.

Chemical composition

Hoffman and Walker (Table 1) analysed nutrient content in *Opuntia* in 1912. These early investigations indicate that spiny and spineless cactus has practically the same chemical composition and are of equal value for feeding purposes (Nefzaoui *et al.,* 2001). Cladodes have high water content (90%), ash value for feeding purposes (Nefzaoui *et al.,* 2001). Cladodes have high water content (90%), ash (20% DM), Ca (1.4% DM), soluble carbohydrate and vitamin A. They have low contents of crude protein (CP) (ca. 4% DM), crude fibre (CF ca. 10% DM), and phosphorus (P ca. 0.2% DM) (Nefzaoui *et al.,* 2001).

Table 1: Average chemical composition (as % of DM) of *Opuntia engelmanni* and *O lindheimeri*

Fraction	Content	Fraction	Content
Water	85	Phosphoric acid	0.33
Crude protein	1.4-4.4	Potassium	3.04
Nitrogen-free extract	7.85	Magnesium	1.6
Fat	1.55	Calcium	2.84-13.85
Crude fibre	8.65		

Source: Nefzaoui *et al.* 2001

Table 2: Average chemical composition of *Opuntia ficus-indica* cladodes produced in Tunisia

	DM(%)	Chemical content (as % of DM)							
		Ash	CP	CF	NFE	P	Ca	K	Na
Average	13.48	27.41	3.84	8.55	58.16	0.04	8.66	1.09	0.05
Minimum	8.95	23.11	1.90	7.39	52.60	0.02	7.56	0.43	0.001
Maximum	21.48	33.70	7.51	10.60	63.79	0.07	10.62	1.92	0.17
Standard deviation	4.50	3.77	1.48	1.03	4.02	0.02	1.09	0.45	0.05

Key: DM = dry matter. CP = crude protein. CF = crude fibre. NFE = nitrogen-free extract. P = phosphorus. Ca = calcium. K

Use of cacti as fodder

Opuntia is not a balanced feed and should rather be considered as a cheap source of energy. Cladodes have low crude protein content and consequently should be supplemented with protein sources. They are also low in phosphorus and sodium.

Intake

Generally, cacti are highly palatable. Jersey cows fed on opuntia and supplemented with 1 kg of concentrate feed/day, ate 50.6 kg/day of fresh cactus. Reynoldset *et al.* (2001) obtained similar results, with cows consuming a voluntary intake of 60 kg when cactus was fed alone. Viana (1965) obtained higher values, with an average voluntary intake of 77.3 kg and 117 kg/day maximum.

Zeeman (2005) observed higher intakes with sheep fed with *Opuntia ficus-indica* (11 kg/day) than with *Opuntia robusta* (6.5 kg/day). Nefzaoui *et al.* (2001) reported intake values ranging from 2.5 to 9 kg/day. It is also reported that higher intakes are observed when water content of pads is higher. Similar results were observed in the authors' work (Barbera *et al.* 1992; Hawkins 1970, Brutsch 1984). It is well established that besides its low feed value, straw intake is low. Combining straw with cactus increases straw intake and consequently animal performance.

Sheep fed with straw were able to consume up to 560 g DM of spineless cactus. This level represents nearly half of the total diet. This beneficial effect of spineless cactus could be explained by the improvement in rumen fermentation conditions. Animals receiving diets containing up to 500 g of spineless cactus did not show any digestive disturbance, supporting earlier findings (Hawkins 1970). The response pattern to spineless cactus supply with respect to straw intake is in agreement with those generally observed with soluble-carbohydrate-rich diets. Earlier reports (Reynoldset *et al.* (2001) indicated that supplementation of poor quality roughages with molasses increased their palatability. Thus, spineless cactus may have a similar effect. Absence of a negative effect of spineless cactus supply on straw intake presumably lies in the high digestibility of spineless cactus in the rumen and in the rapid outflow rate of this feed from the rumen, as it is rich in water.

Digestibility

Opuntia cladodes are highly digestible. *In vivo* average values obtained with sheep varied from 60 to 65%, 60 to 70%, 35 to 70% and 40 to 50%, for DM, OM, CP and CF, respectively. These coefficients are similar to those observed with common forage crops. Since *opuntia* should not be used alone to feed animals, digestibility is calculated by difference, assuming no interaction between diet components.

The main difference between cactus and other forage crops is nutrient degradability in the rumen. While forage crops potential degradability in the rumen is often reached after 48 hours, cactus nutrients are rapidly degraded (in between 6 and 12 hours), so it can be assumed that no significant nutrient extraction is operating after 24 hours (Reynoldset *et al.* 2001).

According to Shoop *et al.* (1977), 80% of the total digestion of Great Plains opuntia (*O. polyacantha*) occurred during the first 16 hours of a 48-hour incubation period, whereas only 73% and 71% of total digestion for hay pellets and alfalfa hay, respectively, occurred during the initial 16 hours.

Effect of feeding cactus on rumen fermentation pattern

The effect of spineless cactus supply on digestion of wheat straw was studied in rumen-cannulated sheep. Animals received wheat straw *ad libitum,* with graded levels of cactus (0, 150, 300, 450 or 600 g DM/day). When the level of cactus in sheep diet increased, fibrous feed intake, rumen volatile fatty acids concentration, rumen protozoa number and rumen ammonia concentration increased, while water intake, rumen cellulolytic activity and rumen acetic acid/propionic acid ratio decreased (Reynoldset *et al.* (2001).

Rumen pH

Rumen pH remained in the range of 6.80 to 7.13, even when animals received the highest level of spineless cactus, so pH of rumen fluid was not affected by the presence of spineless cactus in the diet. Even though spineless cactus is rich in highly fermentable carbohydrates, no differences were observed. Consumption of large volumes of spineless cactus probably enhanced salivation as a result of the high level of mineral salts and the abundance of mucilage in cactus, which might explain the discrepancy. Efforts to quantify effects of mucilage on saliva production and rumen buffer would

Cactus feeding helps to solve the problem of watering Animals in arid environments

Water is scarce in arid zones and smi-arid regions, and animal watering is a real problem during summer and drought periods. Animals expend a lot of energy to reach watering points, and rangeland degradation is a serious risk in the area surrounding watering points. Therefore, the high water content of cactus pads could help mitigate the problem of watering animals in dry areas. Research clearly shows that water intake is nil when cactus intake by sheep is about 300 g of dry matter. The volume of water consumed by animals decreased from 2.4 litres for the control diet to 0.1 litre when the level of spineless cactus consumption exceeded 300 g DM. reported similar findings.

Feeding cactus helps to solve the problem of animal watering. Sheep fed for a long period (400 to 500 successive days) with large amounts of cactus stopped drinking (Hawkins 1970, Cushnahan 2006) with Jersey cows reported similar results. However, Cushnahan (2006) suggested that it is not possible to suppress water in cattle fed on cactus.

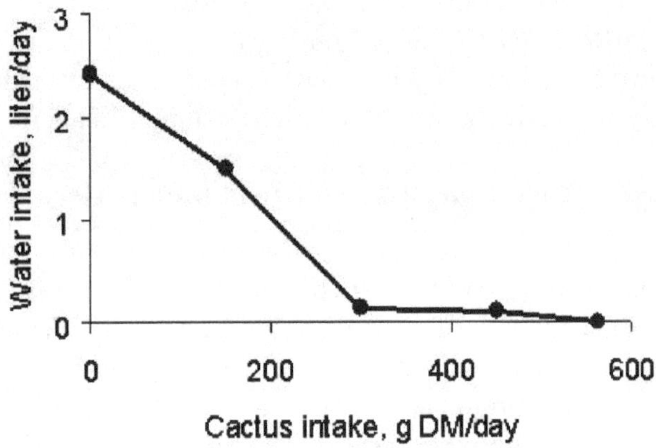

Fig. 1: Feeding cactus helps solving livestock watering problem

Some practical considerations

The method of utilization of spineless cactus will differ from farm to farm according to circumstances, such as available labour, facilities, volume of spineless cactus, etc. It is often recommended to use opuntia for feeding livestock by.

- Grazing of cladodes *in situ*. Although this is the simplest method, it is not the most efficient, and care should be taken so that the animals do not overgraze and destroy the plants.
- Cutting harvested cladodes into small pieces or strips and feeding them in a confined area to limit unnecessary wastage.
- Making silage. The cladodes are cut into small pieces and mixed with hay or low quality alfalfa. If no fruits are included it is necessary to add molasses. The silage container must be airtight.
- Supplementation in case of emergency. Cactus, fed in any form, will keep the animals alive for long periods. Tegegne (2005) emphasized the desirability of supplementing opuntia with a protein-rich supplement of alfalfa or hay (200 g in winter and 100 g in summer) with cactus fed *ad libitum*. A lick of equal parts by mass of bone meal, salt and fodder lime is recommended by De Kock (1980) to supplement the phosphate and sodium.

Spines

Cactus pads are valuable feed provided the spines are singed off first, usually through the use of a propane weed burner (Shoop *et al.* 1977). Other practices were discussed by Tegegne (2005). Steaming to moisten the spines and chopping of the large pads were, and are, very efficient practices to facilitate the use and maximize the amount of cactus eaten by livestock. According to the same author, tools and machines have been built for these purposes.

While in some countries (Mexico and USA) the whole standing plant is burned before grazing, in North Africa individual pads are burned and chopped into small pieces with hand tools or appropriate cutting machines.

References

Barbera G, Carimi F, Inglese P 1992. Past and Present role of the Indian-fig prickly pear (Opuntia ficus-indica (L) Miller, Cactaceae) in the agriculture of Sicily. Economic Botany 46:10-22.

Brutsch MO 1984. Oruckly pear (Opuntia ficus-indica) cultivation in Southern Africa. In : Symposium of agricultural use of the Cactaceae. Prospects and problems. International Organization for Succulent Plant Study, Frankfurt.

Castro J, Pérez S, Riquelme E 1977. Evaluation of thornless prickly pear silages as a feedstuff for ruminants. Proc. Western Section, American Soc Animal Sci, 28.

Curek M, Ozen N 2001. Feed value of Cactus and cactus silage. Turk J Vet Anim Sci 28: 633-639.

Cushnahan A 2006. Silage analysis. College of Agric, Food and Rural enterprise (CAFRE). Greenmount.

Griffiths D, Hare RF 1906. Pricky pear and other cacti as food for stock. II. New Mexico College of Agric. Mechanic Arts Agric Exp Station Bulletin 60:125.

Hawkins D R, Henderson H E & Purser, DB, 1970. Effect of dry matter levels of alfalfa silage on intake and metabolism in the ruminants. J Anim Sci 31 : 617.

López-García JJ, Fuentez-Rodríguez JM, Rodríguez RA 2001. Production and use of Opuntia as forage in Northern Mexico. In: Cactus (Opuntia spp.) as forage Eds.

Mondragón-Jacobo C, Perez-Gonzalez S. FAO Plant Production and Protection Paper 169. Food and Agriculture Organization of the United Nations, Rome pp 29-36.

Mondragón-Jacobo C, Perez-Gonzalez S. FAO Plant Production and Protection Paper 169. Food and Agriculture Organization of the United Nations, Rome pp 73-89.

Nefzaoui A, Ben Salem H 2001. Opuntia – A strategic fodder and efficient tool to combat desertification in the WANA region. In: Cactus (Opuntia spp.) as forage. Eds.

Paper 169. Food and Agriculture Organization of the United Nations, Rome. pp. 1-4.

Reynolds SG, Arias E 2001. Introduction In: Cactus (*Opuntia* spp.) as forage Eds. Mondragón-Jacobo C, Perez-Gonzalez S. FAO Plant Production and Protection

Shoop MC, Alford EJ, Mayland HF 1997. Plain prickly pear is good for cattle. J Range Management, 30: 12-17.

Zeeman DZ 2005. Evaluation of sun-dried Opuntia ficus-indica var. Algerian cladodes in sheep diets. M.Sc. Agric. thesis. University of the Free State, Bloemfontein.

7

Computation of Ration for Cattle and Buffaloes

Shalik Gram Shukla

As an engine requires fuel to run, similarly the animal body requires food to run its vital activities. Food, after its digestion, absorption and oxidation, provides energy to the body. The energy thus obtained is utilized by the animal for maintenance, growth, reproduction and production. Hence, a good quality of food should be provided to the animal so that it may fulfil all the requirements of the animal. Food contains water, carbohydrates fats, proteins, vitamins and minerals which are essential constituents of the body and indispensable for life. Water being an essential constituent of protoplasm carries out many other important and inevitable functions in the body. Water helps the body absorb nutrients from the food and throw out wastes from body in the form of sweat and urine. Carbohydrates, fats, and proteins provide energy to the body. Fat provides 2.25 times more energy than carbohydrate and proteins. Besides providing energy, proteins are also essential for body building, growth and repair of wear and tear of the body tissues. Vitamins protect the body against diseases and help in keeping eyes, gums, skin, teeth and bones healthy. Vitamins are required by the body in small quantities. Vitamins are essential for normal growth, health and reproduction. Prolonged deficiency of vitamins causes characteristic diseases. Minerals required in small quantities are essential for normal growth of body and maintaining good health. Skeleton system of the body contains principally Calcium and Phosphorus with small amount of Magnesium. Calcium takes part in formation of blood clot and plays an important role in maintaining the acid-base equilibrium in the body. Sodium, Potassium and Chlorine maintain the osmotic pressure and acid-base equilibrium in the body. Iron and copper are essential for formation of hemoglobin of blood. Iodine is essential for synthesis of thyroxine hormone which plays an important role in controlling metabolism. Cobalt is a constituent of vitamin B_{12}. Animal body contains about 0.2% Sulphur. Fibers of food are also known as roughage. Roughage is an essential component of foods and adds to the bulkness of the food and helps the body get rid of undigested food.

Animals need food for maintenance, growth, reproduction, work and production. While feeding cows and buffaloes for production of milk, percentage of fat in milk also should be considered. Milch animal producing milk with higher percentage of fat requires more nutrients than that producing milk with lower percentage of fat. Hence, the food to be given to the animal should contain all the nutrients in the amounts and proportions as required by the body and fulfils all the requirements of the animal.

Ration

The amount of feed given to an animal during a day of 24 hours is called ration. The ration provided to an animal should be balanced and ideal.

Balanced Ration

A ration which provides all the essential nutrients for the animals is called as balance ration.

Ideal Ration

A ration possessing the following qualities is said to be an ideal ration:

1. Should be balanced

An ideal ration should be balanced. It should have adequate quantity of the nutrients required by the body.

2. Easily and Locally Available

The ration should be easily available in the locality where it is to be provided to the animals.

3. Cheap and Good

The ration should be economical in cost and good in the quality.

4. Non-Toxic

The ration to be given to the animal should be non-toxic.

5. Fresh, Palatable and Easily Digestible

The ration to be provided to the animal should be fresh, tasty and easily digestible. The ration should be as per liking of the animal.

6. Succulent, Green and Bulky

The ration should contain adequate quantity of succulent and green fodders in order to provide bulkiness to the ration. Bulkiness of the ration provides satisfaction to the animal and facilitates digestion of the food.

7. Variety

The ration should have a variety of feeds which makes the ration balanced, tasty, appetizing and nutritious and health giving. It provides adequate quantity of proteins, vitamins and minerals.

8. Quality

The ration should be of good quality. Poor quality reduces the feeding value of the ration. The ration should be clean, wholesome and free from moulds and foreign matters.

9. Laxative

The feeds given to the animal should be laxative in nature in order to facilitate the digestion. If the food is not laxative, it may not get completely digested.

10. Properly prepared

The food before it is given to an animal should be properly prepared. The green and succulent fodders should be chaffed into small pieces. Dry fodders should be moistened. Hard cereal grains should be crushed. Seeds, cakes, crushed hard grains, Chunis, cotton seeds and husks should be soaked in water preferably overnight to make them soft, palatable and easily digestible.

Not too bulky

The ration to be given to an animal should not be too bulky as it may cause the animal fail to receive all its nutrients from the food.

Types of ration

1. Maintenance ration

The amount of balanced ration which is given to an animal to run the normal body processes without any gain or loss in the body weight so as to keep the animal alive and healthy.

2. Production ration

The amount of balanced ration which is given to an animal for growth, production or work over and above the maintenance ration.

3. Ration for milk production

The amount of ration to be given for milk production depends upon the quantity of milk produced and the percentage of fat in milk. The animal producing milk with higher percentage of fat requires more nutrients to be supplied than that producing milk with lower percentage of fat. A cow needs 0.045 kg. of DCP and 0.316 kg. of TDN for every kg. of milk produced with 4% fat.

4. Gestation ration

A lactating and pregnant cow or buffalo requires pregnancy allowance from fifth month of pregnancy onward to be given over and above the ration given for maintenance and milk production for proper growth and development of the foetus and keep herself healthy after calving. 0.140 kg. of DCP, 0.70 kg. of TDN, 12 gms. of Calcium, 7 gms. of Phosphorus and 30 mgs. of Vitamin A, should be given to a cow as pregnancy allowance.

In order to meet the aforesaid requirements 1.50 kg. of concentrate mixture to zebu and 1.75 kg. of concentrate mixture to crossbred cows/buffaloes should be given.

5. Ration for bulls

A bull requires an extra allowance over and above his maintenance requirement for maintaining good health and extra energy required for reproduction. The requirement of a bull depends upon the age and weight of the bull. A bull weighing 600 kg. requires 0.53 kg. of DCP, 4.2 kg. of TDN, 13 gms. each of Calcium and Phosphorus.

6. Ration for working animals

A bullock requires an extra amount of nutrients over and above his maintenance requirements. The requirement of a bullock also depends upon the nature of the work, whether the work done is normal or heavy. A bullock of the same age doing heavy work requires more nutrients or energy than that doing the normal work. A bullock weighing 400 kg. and doing normal work requires 0.45 kg. of DCP and 4.0 kg. of TDN against 0.57 kg. of DCP and 4.8 kg. of TDN required for the bullock weighing 400 kg. and doing heavy work.

How to feed the animal

Feeding of animals is an art which depends upon the feeding experience of the animal farmers. The art of feeding developed gradually with the practical experience of the animal farmers. The scientific feeding started only a century ago. Since then, the animals are being fed for their various requirements on the basis of their body weights. While computing ration for the different groups of animals (for dry animals, lactating animals, lactating and pregnant animals, growing animals, working animals, breeding bulls etc.), their requirements or the purpose for which they are to be fed should be properly considered. The ration of the animal should contain all the nutrients in the amounts and proportions as required by the animal body. It is not possible for a single feed to supply all the nutrients. Hence, the animals have to depend on a varied of feeds.

Principles of rationing for cattle and buffaloes

1. Ration should be balanced, economical and easily available in the locality.
2. Feeding should be done twice a day at fixed times at an interval of 8-10 hours.
3. Liberal feeding should be allowed to the animals and an extra allowance should be given for the wastage of feeds.
4. Individual feeding should be done to high yielding animals according to their production.
5. If the ration includes adequate quantity of high quality leguminous fodders like Berseem, Lucern or Cow-pea, there is no need to give concentrate mixture upto 10 litres of milk produced.
6. There is no need to supply carotene to pregnant animals or lactating animals if the ration includes sufficient quantity of green fodders.
7. Sudden change in the rationing pattern should not be done. If a change is essential it should be done gradually.
8. The ration should contain green roughages, dry roughages and concentrate mixture in the quantities as per the rationing rules. The ration containing excess quantity of green and succulent roughages may become unable to meet the total requirement of dry matter.
9. Feeding at the time of milking is claimed to stimulate the let down of milk.
10. Feeding green and succulent roughages alone and in excess may lead to bloat formation. Hence, green and succulent roughages should always be given along with dry roughages and concentrates.
11. Ration of lactating animals should contain 17% DCP and 70% TDN.
12. Dry matter requirement of the cattle and buffaloes should be met on

the basis of their body weights. Indian cattle should be given 2.0 to 2.5 kg. of dry matter (DM) for every 100 kg. of their body weights. Buffaloes and crossbred cattle eat slightly more. Hence, they should be provided 2.5 to 3.0 kg. of dry matter for every 100 kg. of body weight.

13. 1/3rd of the total dry matter requirement should be met through concentrate mixture and 2/3rd through roughages. Of the total quantity of roughages 1/3rd should be given through green roughages it may be reduced to 1/4th if the sufficient quantity of high quality leguminous green fodders like Berseem, Lucern or Cow-pea are available) and 2/3rd through dry roughages (it may be increased to 3/4th if the sufficient quantity of high quality leguminous green fodders are available).

14. Lactating cows should be given 1.0 kg. of concentrate mixture for every 3.0 kg. of milk produced whereas, buffaloes should be given 1.0 kg. of concentrate mixture for every 2.5 kg. of milk produced, because of higher percentage of fat in buffalo milk.

15. The quantity of concentrate mixture should be given to the different groups of animals as follows:-

For maintenance	-	1.0 to 2.0 kg.
For growth	-	1.5 to 2.0 kg.
For breeding	-	1.0 to 2.0 kg.
For pregnancy	-	1.0 to 2.0 kg.
For work	-	1 .5 to 2.0 kg.

Computation of ration

The following two methods are used for computing ration for cattle and buffaloes:

1. Thumb Rule Method
2. DCP and TDN Method or Scientific Method or Conventional Method

1. Thumb rule method

This method is simple and does not require any calculation. Hence, most of the Indian farmers use this method for feeding their cattle and buffaloes. This method is used particularly where individual care of the animals and formulation of ration on the basis of body weight is impractical. While formulating the ration for an animal, the various requirements for which the animal is to be fed are properly considered. Say for example the animal in question is to be fed whether for maintenance, production or pregnancy. For feeding by this method, practical experience of the farmer is more important than that done by the scientific method.

Example 1. Prepare a balanced ration for an Indian cow weighing 400 kg. and producing

6 litres of milk. The locally available feed stuffs are as follows:

1. Bhusa
2. Berseem
3. Groundnut cake
4. Muize (Crushed)
5. Rice bran

Solution

1. Dry matter requirement @ Rs. 2.5 kg/100 kg. body weight = 2.5 × 400 / 100 = 10 kg.

2. Amount of concentrate mixture :

(i) For maintenance = 1.0 kg.

(ii) For milk production = 2.0 kg.

@ 1.0 kg. for every 3.0 kg. of milk = 3.0 kg.

Dry matter obtained from concentrate $\frac{90}{100} \times 1.83 = 2.0$ kg

Hence, the quantity of dry matter to be given through roughages = 10.0 - 2.7 = 7.3 kg.

The quantity of dry method to be given through dry roughage (Bhusa) = 7.3 x ¼ = 1.83 kg.

The quantity of DM to be given through green roughages (high quality roughages) = 7.3 x 3/4= 5.47 kg.

3. Quantity of roughages :

(i) Bhusa = $\frac{100}{90} \times 1.83 = 2.0$ kg.

(ii) Berseem = $\frac{100}{30} \times 5.47 = 18.23$

4. Quantity of each ingredirent in 3.0 kg. of the concentrate mixture :

(i) Quantity of Groundnut cake in 3.0 kg. of the concentrate mixture @ 40 parts

$$= \times 3 = 1.20 \text{ kg. } \frac{40}{100}$$

(ii) Quality of crushed Maize

@ 30 parts of the concentrate mixture $= \frac{30}{100} \times 3 = 0.090$ kg

(iii) Quantity of Rice bran $= \frac{30}{100} \times 3 = 0.090$ kg

@ 30 parts of the concentrate mixture

Total = 3.0 kg.

Mineral Matter	= Chalk = 60 gms.
Salt	= 50 gms.
Water	= as per requirement

Example 2. Prepare a balanced ration for a Murrah buffalo weighing 450 kg. and producing 8.0 kg. of milk daily. The buffalo is in advanced stage of pregnancy. The locally available feed-stuffs are as follows:

1. Wheat Bhusa
2. Berseem
3. Mustard cake
4. Crushed Maize and
5. Wheat bran

Solution

1. Total Dry matter (DM) requirement of the buffalo @ of 3.0 kg. per

3.0

100 kg. of body weight

100

$$= \times 450 = 13.50 \text{ kg. } \frac{3.0}{100}$$

2. Requirement of concentrate:

(i) For maintenance = 1.5 kg.

(ii) For milk production = 3.2 kg.

@ 1.0 kg. for every 2.5 kg of milk

3.For Pregnancy = 2.0 kg.

Total = 6.7 kg.

DM supplied through concentrate @ 90% DM $\frac{100}{90}$

DM to be supplied through roughages = 13.5 – 6.0 = 7.5 kg.

Here, we have two roughages viz., Bhusa and Berseem. Supply half of 7.5 kg. of DM through Bhusa and the rest half through Berseem.

Quantity of Bhusa $= \frac{100}{90} \times 3.75$

(@ 90% DM in Bhusa) = 4.17 kg.

Quantity of Berseem $\frac{100}{30} = \times 3.75$

(@ 30% DM in Berseem) = 12.5 kg.

Quantity of mustard cake, crushed maize and wheat bran separately in the concentrate mixture @ 35%

40% and 25%, respectively.

Mustard Cake $= \frac{35}{100} \times 6.7$

= 2.3 kg.

Crushed Maize $= \frac{40}{100} \times 6.7$

= 2.7 kg.

Wheat Bran $= \frac{25}{100} \text{ x } 6.7$

= 1.7 kg.

Total = 2.3 + 2.7 + 1.7 = 6.7 kg.

Common Salt = 50 gm.

Chalk = 60 gm.

Water = as per requirement

2. DCP and TDN method

This method of computing ration is also known as scientific method or conventional method or orthodox method. Being modern and easier, this is at present most commonly used by the Govt. and organized dairy farms.

At present, Prof. F.B. Morrison's Feeding Standard after making slight modification in it is used all over the world for computing ration for different groups of animals. In India, Morrison's Feeding Standard as modified by Dr. K.C. Sen and the recommendations of the National Research Council made in its report are used for computing the ration for cattle and buffaloes which are based on Dry Matter (DM), Digestible Crude Protein (DCP) and Energy or Total Digestible Nutrients (TDN), Calcium, Phosphorus and Carotein requirements of the different groups of the animals. Carotein is not required to be supplied along with the ration if the ration includes adequate quantity of green roughages.

The various requirements of buffaloes are similar to the requirements of cattle.

Example 1. Find out the total requirements of DM, DCP and TDN for an Indian cow weighing 350 kg. and producing 8 litres of milk daily with 4.50% fat.. The cow is in advanced stage of pregnancy.

Sl.No.	Requirments	DM (Kg.)	DCP (Kg.)	TDN (Kg.)
1.	For maintenance	@ 2.5 kg. per 100 kg. of body	0.227	2.70
2.	For milk production	Weight	0.048x8=0.384	0.34x8=2.72
3.	For Pregnancy	= 2.5/100x350 = 8.75	0.140	0.70
Total requirements		8.75	0.751	6.12

Example 2. Find out the total requirements of DM, DCP and TDN for a Murrah buffalo weighing 450 kg. and producing 10 litres of milk daily with 7% fat.

Solution

S.No.	Requirements	Nutrients to be given in Kg.		
		Dry Matter	DCP	TDN
1.	For maintenance	3.0/100x450	0.282	3.37
2.	For milk production	=13.50	0.063x10=0.630	0.458x10=4.58
	Total requirements	13.5	0.912	7.95

Example 3. Compute a balanced ration for a crossbred cow weighing 450 kg and producing 12 kg of milk with 4.0% fat. The cow is pregnant for about 6 months. The following feed-stuffs are locally available.

Sl. No.	Feed-stuffs	% Dry Matter	% DCP	% TDN
1.	Wheat bhusa	90.00	0.0	48.00
2.	Berseem	30.00	1.5	12.00
3.	Maize (crushed)	90.00	7.00	87.00
4.	Groundnut cake	90.00	38.00	77.00
5.	Wheat bran	90.00	8.00	77.00

Solution

Requirements of the Cow

Sl. No.	Requirements	Nutrients in Kg.		
		Dry Matter	D.C.P.	T.D.N.
1.	For maintenance	3.00/100x450	0.282	3.37
2.	For milk production	=13.50	0.045x12=0.540	0.34x12=4.08
3.	For pregnancy		0.140	0.70
	Total requirements	13.5	0.962	8.15

Total Dry matter requirement = 13.50 kg.

Dry matter to be given through concentrate = 13.50x1/3=4.5 kg.

Amount of concentrate = 100/90x4.5=5.00 kg.
(90% DM in concentrate)

Dry matter to be given through roughages = 13.5x2/3=9.00 kg.

Dry matter to be given through dry roughage = 9x3/4 = 6.75 kg.
(Wheat bhusa)

Dry matter to be given through green roughage = 9 x ¼ = 2.25 kg.
(Berseem)

Quantity of wheat bhusa and berseem

S. No.	Roughages	Amount to be given	Obtained (kg.)		
		(kg.)	DM	DCP	TDN
1.	Wheat bhusa	100/90 × 6.75 = 7.50	6.75	0.0	48/100 × 7.50 = 3.60
2.	Berseem	100/30 × 2.25 = 7.50	2.25	1.5/100 × 7.50= 0.112	12/100 × 7.50 = 0.90
	Total	15.00	9.00	0.112	4.50

Quantities of Dry matter, DCP and TDN to be given through concentrate mixture

Nutrients	Total requirements	Supplied through roughages (kg.)	To be given through concentrate (kg.)
Dry matter	13.50	9.00	4.50
DCP	0.962	0.112	0.85
TDN	8.15	4.50	3.65

4.50 kg. Dry Matter, 0.85 kg. DCP and 3.65 Kg. TDN are to be given through concentrate mixture

Computation of ingredients of concentrate mixture

$$\% \text{ DCP in the concentrate mixture} = \frac{\text{Quantity of DCP to be given through Concentrate mixture}}{\text{Total quantity of concentrate}} \times 100$$

$$= \frac{0.85}{5.00} \times 100$$

$$= 17.0\%$$

Now, we have three ingredients of the concentrate mixture viz., groundnut cake, crushed maize and wheat bran with 38%, 7% and 8% DCP, respectively. Groundnut cake is containing the highest percentage of DCP. Here, we have to find out the average DCP content in the rest two ingredients as follows.

$$\% \text{ DCP in crushed maize and wheat bran} = \frac{7+8}{2} = 7.2\%$$

Now, with the help of Pearson Square the quantity of each ingredient is computed as follows:

Place the value of DCP content (17.0) in the concentrate mixture in the middle, percent DCP in groundnut cake on the top left and average percent DCP in

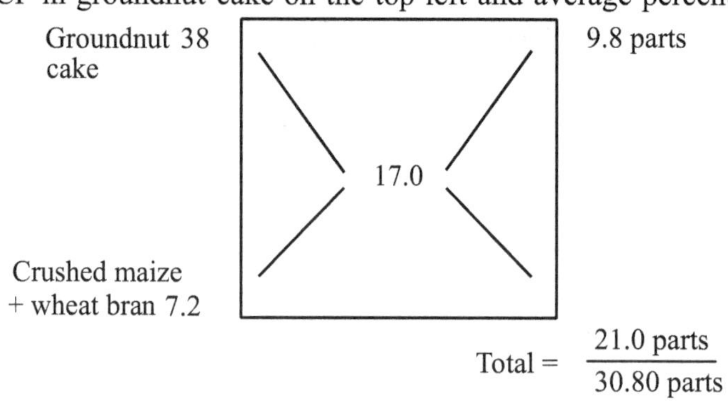

crushed maize and wheat bran (7.2) on the bottom left corners of the square. Now, subtract the lower values from the higher values diagonally and write them on the top right (9.8) and bottom right (21.0) corner, respectively.

Amount of groundnut cake in the concentrate mixture $= \frac{9.8}{30.80} \times 5 = 1.59$ kg.

Amount of crushed maize and wheat bran $= \frac{21}{30.80} \times 5 = 3.41$ kg.

Amount of crushed maize $= 3.41 \div 2 = 1.70$ kg.

Amount of wheat bran $= 3.41 - 1.70 = 1.71$ kg

$= 7.2\%$

Nutrients obtained from the ingredients of the concentrate mixture

Sl.No.	Concentrate ingredients	Quantity given (kg.)	Obtained (kg.)		
			DM	DCP	TDN
1.	Groundnut cake	1.59	90/100x1.59 =1.43	38/100 x1.59 =0.604	77/100x1.59 = 1.22
2.	Crushed maize	1.70	1.53	7/100x1.70 =0.119	87/100x1.70 =1.48
3.	Wheat bran	1.71	1.54	8/100x1.71 =0.137	77/100x1.71 =1.32
	Total	5.0	4.50	0.860	4.02

Sl.No.	Feed-stuffs	Quantity given (kg.)	Obtained (kg.)		
			DM	DCP	TDN
1.	Wheat Bhusa	7.50	6.75	0.0	3.60
2.	Berseem	7.50	2.25	0.112	0.90
3.	Ground nut cake	1.59	1.43	0.604	1.22
4.	Maize (crushed)	1.70	1.53	0.119	1.48
5.	Wheat bran	1.71	1.54	0.137	1.32
	Obtained	20.00	13.50	0.972	8.52
	Recommended		13.50	0.962	8.15
	Difference		Nil	+0.010	+0.37

Salt and mineral mixtures should be added to the above concentrate mixture @ 1.0 % each.

Related terms

1. Concentrates

A feed-stuff which contains less than 18 percent crude fiber and 60 percent or more TDN is called concentrates e.g. cereal grains, oil seed cakes, brans etc.

2. Digestible crude protein (DCP)

Crude Protein (CP) includes both true protein and non-protein nitrogenous (NPN) compounds. It is calculated by using the formula.

$CP = \% N \times 6.25$

Where, N = Nitrogen

The amount of crude protein (CP) actually absorbed by the animal is known as Digestible Crude Protein.

3. Dry matter

Total weight of feed minus the weight of water in the feed expressed as a percentage. It may also be referred to 'on dry basis'.

4. Gestation period

The time in which a foetus develops, beginning with fertilization and ending at birth is called gestation period.

5. Net energy

Net energy is metabolizable energy minus heat increment of feeding. Heat increment of feeding is the heat produced when feed is ingested and utilized.

6. Roughages

The feed-stuff which contains more than 18 percent crude fiber and less than 60 percent TDN is known as roughages e.g. Wheat Bhusa, Berseem etc.

7. Total digestible nutrients (TDN)

% TDN = % Dig. Total Carbohydrate + % DCP + (% Dig. EE. x 2.25)
= % Dig. NFE + % Dig. CF. + % DCP + (% Dig. EE. x 2.25)

Where, Dig = Digestible
NFE = Nitrogen Free Extract
CF = Crude Fiber
EE = Ether Extract

Table 1: Nutrients required for maintenance of adult cattle per head per day (Sen and Ray 1971)

Body weight (kg.)	DCP (kg.)	TDN (kg.)	Calcium (g)	Phosphorus (g)
150	0.102	1.27	4	4
200	0.148	1.66	5	5
250	0.168	2.02	6	6
300	0.197	2.36	7	7
350	0.227	2.70	8	8
400	0.254	3.03	9	9
450	0.282	3.37	10	10
500	0.296	3.69	11	11
550	0.336	3.71	12	12

Table 2. Nutrients required for growing cattle per head per day (Sen and Ray 1971)

Body weight (kg.)	DCP (kg.)	TDN (kg.)	Calcium (g)	Phosphorus (g)
45	0.15	0.80	7	6
70	0.22	1.30	12	10
100	0.28	1.90	13	10
150	0.35	2.6	14	12
200	0.40	3.0	13	12
300	0.47	4.0	13	12
450	0.48	5.0	12	12

Table 3: Nutrients required for production per kg. of milk (to be added to the maintenance allowance) (Sen and Ray 1971)

% Fatin milk	DCP (kg.)	TDN (kg.)	Calcium (g)	Phosphorus (g)
3	0.040	0.269	2	1.4
4	0.045	0.316	2	1.4
5	0.051	0.363	2	1.4
6	0.057	0.411	2	1.4
7	0.063	0.458	2	1.4

The actual milk yield may be converted into the milk with 4% fat by using the Gainer formula as follows.

FCM = 0.4 M + 15 F

where, FCM = Fat Corrected Milk

M = Weight of milk

F = Weight of fat in weight of milk

Table 4: Nutrients required for breeding bull per head per day

Body weight (kg.)	DCP (kg.)	TDN (kg.)	Calcium (g)	Phosphorus (g)
400	0.38	2.7	9	9
500	0.45	3.4	11	11
600	0.53	4.2	13	13

Table 5: Nutrients required for working animals per head per day (Sen and Ray, 1971)

Body weight (kg.)	Normal Work		Heavy Work	
	DCP (kg.)	TDN (kg.)	DCP (kg.)	TDN (kg.)
200	0.24	2.0	0.25	2.7
300	0.33	3.1	0.42	4.0
400	0.45	4.0	0.57	4.8
500	0.56	4.9	0.71	6.4

References

Banarjee GC, A Textbook of Animal Husbandry (Eighth Edition), pp. 380-386, 661-675
ICAR publication 2008. Handbook of Animal Husbandry, pp. 424-432.
Pandey Devnarayan 2011. Animal Husbandry and Veterinary Science (Hindi Version), pp. 33-45.
Prasad, Jagdish. Pashu Palan Avam Dugdh Vigyan (Hindi Version), pp. 77-99.
Singh KS 1997. Animal Nutrition. Pp. 189-196.
Singh TB, Shukla SG 2006. Principles of Animal Nutrition (Hindi Version), pp. 117-122.

8

Minerals: An Important Micronutrient in Livestock Enterprises

Ravindra Kumar Tiwari and Sunil Kumar

There are six major classes of nutrients: carbohydrates, fats, fibre, proteins, minerals and vitamins. Water is not a nutrient but play important role in livestock industry. These nutrient classes can be categorized as either macronutrients (needed in relatively large amounts) or micronutrients (needed in smaller quantities). The macronutrients are carbohydrates, fats, fiber, and proteins. The micronutrients are minerals and vitamins.

The term minerals is used by nutritionists to refer to inorganic chemical elements, but is used by many other scientists to mean a homogeneous crystalline compounds. Mineral elements are not inert materials. Mineral elements may change valence states and are transferred from one chemical compound to another. Minerals are active participants in enzymatic reactions, have specificity of function, and are critical to life.

All animal and plant tissues contain widely varying amounts and proportions of mineral elements, which largely remain as oxides, carbonates, phosphates and sulfates in the ash after ignition of organic matter.

Large number of livestock in many parts of the world consumes diets that do not meet the exact mineral requirements causing acute to severe mineral deficiency or toxicity and expressed merely as unthrift ness, or unsatisfactory growth, production and fertility.

In 1981, 22 mineral elements were believed to be essential for the higher forms of animal life (Underwood 1981). These comprised the seven major or macro minerals (Calcium, Phosphorus, Magnesium, Sodium, Potassium, Chlorine and Sulphur) and 15 trace or micronutrient mineral elements (Iron, Copper, Zink, Cobalt, Molybdenum, Manganese, Iodine, Selenium, Chromium, Tin, Vanadium, Flourine, Silicon, Nickel and Arsenic). Since then, Aluminium, Lead and Rubidium have been shown to be beneficial in some circumstances. Minerals are essential for maintenance and normal functioning of the animal body and are present in all organs and tissue of the body. The functions of the inorganic minerals are extremely diverse. They range from structural function in some tissues to a vide variety of regulatory function of other tissues (NRC 1980, Mc Dowell 1992, Underwood and Suttle 1999). To better understand the role of minerals in animal production it is important to recognize that mineral elements are functional components of numerous metabolic events. Mineral functions can be described by four broad categories (Underwood and Suttle 1999) structural, physiological, catalytic and regulatory. Structural function refers to minerals forming structural components of body organs and tissue. An example is the contribution of calcium and phosphorus to the body skeleton. Physiological function occurs when minerals in body fluids and tissues act as electrolytes to maintain osmotic pressure, acid base balance, and membrane permeability. Catalytic function is the chief function of trace minerals as it refers to catalytic role of metallo enzymes in enzyme and hormone systems. Upon removal of the trace element or lack of adequate trace mineral levels the enzyme activity is lost. There are numerous metalloenzymes that are required for a wide range of metabolic activities such as energy production, protein digestion, cell replication, antioxidant activity and wound healing. Regulatory function is exemplified by the role of zinc to influence transcription and iodine serving as constituent of thyroxine, a hormone associated with thyroid function and energy metabolism.

The net mineral requirements may under estimate the dietary needs of livestock for the minerals, because ingested minerals are incompletely utilized, due to limits upon their absorption from the gut. Therefore, apart from the requirement for various minerals their availability should also be known to prevent deficiency or oversupply. Much attention has to be paid to make a proper comparison between different mineral sources based on the bioavailability. Absorption and utilization of major minerals may be, apart from passive transport through the gut wall, mediated by hormonal control that is primarily based on their concentration in the extracellular fluid. Before absorption by te enterocytes from the gastrointestinal tract can take place, the minerals must become available in ionic form (as cations and anions), which is suitable for uptake and transport. In principal, the transepithelial transport consists of both, an active transcellular component, which can be regulated, and / or a passive paracellular component which depends on chemical and electrical gradients existing across the intestinal wall (Nys and Mongin 1980, Schroder *et al.* 1996, Jongbloed and Mroz 1997). The highly-soluble monovalent minerals, such as sodium, potassium and chlorine can be transported easily. However, the solubility of various other minerals is often low at neutral pH. Their solubility is dependent on the presence of other compounds as they can relatively easily precipitate or form non-absorbable complexes. Well-known complexing food components are phytic acid and oxalic acid (Harland 1989). In addition, there are several interactions among various minerals (e.g. calcium and phosphorus; calcium and zinc; copper and zinc; copper, molybdenum and sulphur), which complicate an easy understanding of absorption (Pointillart *et al.* 1987, Ashmead and Zunino 1994, Van der Klis 1994, Jongbloed *et al.* 1995). Also, viscosity of the chime in the intestinal tract may negatively affect absorption of minerals (Van der Klis 1993).

Mineral requirements

The functions performed by minerals can only be fulfilled if sufficient amounts of the ingested mineral are absorbed and retained to keep pace with maintenance, growth, production, reproduction and to replace minerals that are 'lost' either as products, such as milk or eggs, or insidiously during the process of living.

Mineral supplementation

Inorganic mineral supplements are added routinely to home-mixed and purchased rations as an insurance against the inclusion of natural feeds that deviate from the norm of mineral or antagonist composition. In some localities, supplements of specific minerals are always necessary because the pastures or feeds are abnormal in their mineral composition as a consequence of local soil and climatic effects. The use of industrial nitrogen sources to provide protein for ruminants increases the need for minerals from other dietary components. Urea supplementation is increasingly used to improve the nutritive value of low-quality roughages and crop by-products in developing countries, but may only do so if accompanied by additional phosphorus and sulfur (Suttle 1991). Successful procedures have been developed for the prevention and control of all disorders caused by mineral deprivation and for many of those caused by mineral toxicities. Methods can be divided into three categories: indirect methods that increase the mineral composition of pastures and feeds while they are growing; and continuous or discontinuous direct methods, which involve the administration of minerals to animals. The procedure of choice varies greatly with different elements, climatic environments, conditions of husbandry and economic circumstances.

Macro minerals

Calcium and phosphorus

Calcium is the most abundant mineral found in the animal body (1-2%), with 99% of it occurring in bone and teeth. Calcium is generally deficit in grains but abundant in most forages. Legume forages are good source of calcium. Phosphorus is second most abundant mineral found in te animal body and about 80-85% is in bones and teeth. Calcium and phosphorus are considered together because they constitute the major part of the mineral content of bone and are closely related. They are found in insufficient amounts in common feedstuffs to mete requirements of livestock. Calcium deficiency occurs in animals fed on ruminants. Phosphorus plays a major role in the structure and function of living cells. It is an integral component of nucleic acids, nucleotides, phospholipids, and proteins, and a key component of many coenzymes. These comounds function in cell division and growth in the transport and metabolism of fats, and in the absorption and utilization of carbohydrates, fatty acids, and proteins. A deficiency or excess of one will interfere with the proper utilization of the other.

Physiological functions

Ca and P make about 70% of the total mineral elements in the body. The ratio in bone is nearly constant and somewhat greater than 2:1 (Hegeted 1973). Most of the Ca and P in bones are in the form of calcium phosphate and hydroxyapatite. The calcium phosphate of bone is deposited with in a sort fibro-organic matrix composed of collagen fibers and to a much less extent of mucopolysaccharide gel. The protein matrix in bone is calcified when the proper levels of Ca, P, Mg and other minerals are present. Bone undergoes a continuous progress of resorption and formation with mobilization and restorage of Ca and P occurring throughout life. There is a continuous interchange of Ca and P between bone, the blood and other parts of the body. The animal continuously needs Ca, P and other nutrients to maintain bone in strong condition. During heavy lactation, bone is the source of minerals secreted in the milk. The body's regulatory mechanism, which allows minerals deposited in the bone to be drawn during emergency needs, is an efficient one (Cunha 1990)

Deficiency of calcium and phosphorus

Ca deficiency mostly occurs in areas of high rainfall and humidity (Sharma *et al.* 2003). Ca deficiency frequently occurred in high producing grazing dairy cows and has led to osteodystrophic diseases and low productivity in other livestock. In high yielding dairy cow an acute calcium deficiency still strikes many animals at calving with no signs of bone disorders. Ca deficiency may be directly or indirectly responsible for infertility, but the disturbed Ca: P ratio has a blocking action on the pituitary gland and consequently on ovarian action (Herrick 1977), prolonging the interval of first ovulation. According to Mosdol and Waage (1981), hypocalcaemia rarely affects ewes in their first pregnancy, and it is physiologically similar to milk fever in cattle in that it is caused by disequilibrium in the supply and demand for calcium (Underwood and Suttle 1999). At the time of parturition all older cows show a sharp but limited decrease in plasma calcium, which may recur at roughly 9 days as they readjust to the demands of lactation (Hove 1986). Parturient paresis is a metabolic disease of bovines associated with parturition and the initiation of lactation. The changes those appear in association with parturition and with milk fever in the dairy cow involve an acute hypocalcaemia. Prolonged calcium deprivation in domestic animals eventually causes lameness, enlarged and painful joints, deformation and fracture of the pelvis and long bones arching of the back with posterior paralysis causing low productivity (Miller and Miller 1962). The key to control calcium deficiency is to restrict the post parturient fall in plasma calcium to level of 1.75 mmol. That is done by single intravenous injection of 8-10 g Ca as calcium borogluconate. High dose with vitamin D3 (2-30 million units a day for 5-7 days before calving) has been successful (Hibbs and Pounden 1955).

Phosphorus deficiency is the most common in several developing countries due to the low phosphorus content in soils and plants (McDowell *et al.* 1986). Out of all mineral deficiencies phosphorus deficiency is most widespread and economically important deficiency as far as reproduction is concerned (Underwood 1966).Phosphorus is critically deficient for most classes of productive ewes throughout the year and being most deficient during the period of early lactation. A craving for and a consumption of abnormal materials often parallel the loss of appetite caused by phosphorus deprivation. Deficiency of phosphorus results in the delayed outset of puberty and sexual maturity. Call *et al.* (1986) reported that in cows pregnant with their second calf, given diets low in phosphorus from the seventh month of gestation and for 7.5- 10 months of lactation, showed maximum depression of milk yield (32-35%) between 18-34 weeks of lactation and proportionate reductions in food intake. Dietary imbalance of phosphorus results in depressed growth, bone development and reproduction. Dutta *et al.* (2001) observed significantly higher overall mean values for serum inorganic phosphorus in normal cyclic heifers and lower in anoestrus heifers. Phosphorus supplementation during late pregnancy and early lactation coupled with the early weaning of lambs, would appear to be one of the management strategies which should be evaluated to improve the condition and performance of breeding ewes (Cook and Fadlalla 1987).Phosphorus can be given directly to grazing livestock in phosphatic saltlicks and blocks and also in water supply which may be given to the growing cattle 5 g/head/day and109 for breeding stock (Miller *et al.* 1990). Underwood (1981) estimated that 1.5-1.7 g P/kg is required by young growing sheep. In grazing livestock, deficiencies can be prevented by direct treatment of the animal through supplementing the diet or the water supply or indirectly by appropriate fertilizer treatment of the soils.

Requirements

The amount of Ca and P that must enter the extra cellular compartment for maintenance, growth, pregnancy and lactation is fairly well known. Ca and P requirements are highest during lactation, early growth, and for reproduction. High-producing cows need much more Ca and P than cows producing at average or low levels. Milk contains about 0.12% calcium and 0.095% phosphorus. Calcium and phosphorus requirements of different species are given in Table 1.

Table 1. Calcium and phosphorus requirements (%)

Species	Calcium (%)	Phosphorus (%)
Dairy cattle		
Young stock	0.50	0.31
Lactating	0.70	0.48
Sheep (all classes)	0.20-0.82	0.16-0.38

Magnesium

Magnesium is an extremely important element for all stock. It is defined as a macro element, which means it is required in large quantities. Magnesium is vital for a number of different functions in the body; including relaxation of muscles and nerves, utilization of calcium and converting sugars to energy. Magnesium is not stored well as a reserve in the body, therefore stock need their daily requirement from feed intake.

Function

- Bone formation
- Cell replication
- Hormone regulation
- Nerve impulse transmission (stress) and muscle contraction
- Fertility
- Resistance to infection (immune system, vitamin E)

Deficiency

Calves maintained on a simulated milk diet for a longer period induces hypo magnaesaemic tetany (Blaxter *et al.* 1954). In Australia a high incidence of hypomagnaesaemic tetany in breeding ewes was reported during the periods with rapid winter growth of pastures (Blumer *et al.* 1939). Thus Mg deficiency led to reduced reproduction efficiency in animals.

Requirement

The dietary requirements of Mg in livestock vary with the species and breed of the animal and its age and rate of growth or production but mostly with the absorbability of the mineral in the diet of ruminants (Sharma *et al.* 2002a). In cattle requirement for Mg is 10-15 mg/kg body weight (O'Dell 1960). Braak *et al.* (1987) suggested that prepartum serum Mg should be more than 0.85 mmol.

Potassium

Potassium has been recognized as an essential nutrient in animal nutrition since its importance was pointed out by Sidney Ringer in 1883. Potassium is essential for life. Young animals will fail to grow and will die within a few days when the diet is extremely deficient in K. Potassium is the third most abundant mineral element in the animal body, surpassed only by calcium (Ca) and phosphorus (P). Potassium concentrations in cells exceed the concentration of sodium (Na) by 20 to 30 times. Outside the cell the reverse is true. Potassium comprises about 5 percent of the total mineral content of the body. Muscle contains most of the total K in the bodies of animals. Potassium is contained almost entirely within the cells and is the most plentiful ion of the intracellular fluids. Potassium is found in every cell. It is present in tissues and cells only in ionic form (K+).

Functions of potassium

The major functions of K animal body are to:

- maintain water balance
- maintain osmotic pressure
- maintain acid-base balance
- activate enzymes
- help metabolize carbohydrates and proteins
- regulate neuromuscular activity (along with Ca)
- help regulate heartbeat.

Potassium deficiency

There are several causes of K deficiency: inadequate amounts of K in diet, K losses in digestive secretions caused by vomiting and diarrhea, high intake of Na, increased urination, and stress conditions. Potassium deficiency may commonly be manifested by depressed growth, muscular weakness, stiffness, decreased feed intake, intracellular acidosis, nervous disorders, reduced heart rate, and abnormal electrocardiograms. The first sign of K deficiency is reduced feed intake. Many of the other signs stem from reduced feed intake. Potassium must be supplied in the daily ration because it is a mobile nutrient and there are not any appreciable reserves.

Importance of potassium

Potassium is especially important in diets of chickens and turkeys during the first 8 weeks. During heat stress, or if there is any diarrhea, the needed levels may be higher. Adequate K in the ration of laying hens assures good egg production, egg weight, and shell thickness. In starter chicks and turkey poults, adequate K increases weight gain, improves feed efficiency, and reduces mortality. Swine K requirement is higher for young pigs than for older ones. It ranges from about 0.33 percent (dry matter basis) in rations of small pigs weighing up to 8 lb, to 0.19 percent in rations of pigs weighing more than 180 lb. The K requirement for gestating and lactating sows is 0.20 percent. Potassium requirement increases in diets with higher Na and chloride (Cl) levels. Ruminants have a higher K requirement than non ruminants. Potassium is essential for rumen microorganisms. The single most consistent effect of suboptimal K in the ration of ruminants is decreased feed intake. Lactating dairy cattle, particularly high producing cows, require the highest levels of dietary K. Under heat stress, their optimal level of dietary K can be as high as 1.9 percent, but the normal National Research Council (NRC) recommendation is 1.0 percent of dietary dry matter. Less K (0.65 percent) is recommended for dry cows, calves and heifers. During the last three to four weeks before calving, excessive K in the dry cow diet can increase the incidence of milk fever and retained placentas. This can lead to reduced milk production during the subsequent lactation. The maximum amount of K desirable in the dry cow diet depends on the use of anionic salts and other factors, but generally forage K should be less than 2.5 percent.

Sodium and chlorine

Common salt contains both sodium and chloride and is also called sodium chloride. Salt is unique in that animals have a much greater appetite for the sodium and chloride in salt than for other minerals. Because most plants provide insufficient sodium for animal feeding and may lack adequate chloride content, salt supplementation is a critical part of a nutritionally balanced diet for animals. Even though the body only contains about 0.2% sodium, it is essential for life and is highly regulated. About half of the sodium in the body is in the soft tissues of the body; the other half in bones. Sodium makes up about 93% of the basic mineral elements in the blood serum and is the chief cation regulating blood pH. Sodium plays major roles in nerve impulse transmission and the rhythmic maintenance of heart action.

Dietary sources

Sodium is the minerals most likely to be deficient when ruminants receive no mineral supplementation. Forages and grains generally contain level of sodium below dietary requirements of ruminants. The content of sodium in forages can be increased by fertilization, but direct supplementation of the diet with sodium is generally preferable. All dietary sodium sources are thought to have good bioavailability.

Deficiency

Sodium deficiencies in ruminants are most likely to occur during lactation or rapid growth, in tropical or semi-arid environments and in ruminants grazing pasture which have been heavily fertilized with potassium. Growth, milk production, feed intake and body condition of the animals are reduced in sodium deficiency. As the deficiency progress, there is loss of body weight, shivering, in coordination, general weakness, cardiac arrhythmia and death.

Requirements

Generally 0.5-1% NaCl or a trace mineralized salt mix is added to ruminants rations. When fed free-choice in loose salt form, the consumption of salt varies from 20-40 g/head/day for adults animals.

Sulpur

Recognition that sulfur is an important dietary component stems from the discovery in the 1930s that the essential amino acid, methionine, contains one atom of sulfur per molecule (McCollum 1957). Subsequent investigations showed that sulfur is an integral part of the majority of tissue proteins, comprising 0.5–2.0% by weight. Other sulfur amino acids such as cystine, cysteine, cystathionine and taurine contribute to protein sulfur, but each can be derived from methionine and they are not essential constituents of the diet.

Functions

The functions of sulfur are as diverse as the proteins of which it is a part. Sulfur is frequently present as highly reactive sulfhydryl (SH) groups or disulfide bonds, maintaining the spatial configuration of elaborate polypeptide chains and providing the site of attachment for prosthetic groups and the binding to substrates that are essential to the activity of many enzymes. Hormones such as insulin and oxytocin contain sulfur, as do the vitamins thiamine and biotin. The conversion between homocysteine and methionine facilitates many reactions involving methylation. Cysteine-rich molecules such as metallothionein play a vital role in protecting animals from excesses of copper, cadmium and zinc, while others influence selenium transport and protect tissues from selenium toxicity. Sulfur is present as sulfate (SO_4) in the chondroit in SO_4 of connective tissue, in the natural anticoagulant heparin and is particularly abundant in the keratin-rich appendages (hoof, horn, hair, feathers, wool fibre and mohair).

Sulfur sources

Sulfur in forages

The sulfur concentrations found in pasture and conserved forages range widely from 0.5 -5.0 g kg dry matter, depending mainly on the plant species, availability of soil sulfur, nitrogen and phosphorus, and on maturity of the sward. Any area away from the coast and industrial activity, where the annual rainfall is moderate (>500 mm) and the soils derive from weathered volcanic material, is likely to have leached soils, low in sulfur and swards that respond to supplementation. In many countries, sulfur deficiency limits forage production. Areas of deficiency have been delineated in Australia and in the south-east, north-east, mid-west, west and north of the USA (Beaton *et al.* 1971, Underwood and Suttle 1999). There is little published information on seasonal variations in sulfur concentrations, but sulfur and protein are highly correlated and protein concentrations decline at a rate of 1–3 g/ kg DM/ day, depending on species and growth conditions (Minson 1990) with the approach of maturity; this is equivalent to a decline in sulfur of 0.1–0.3 g kg"1 DM week"1.

Sulfur in other foodstuffs

Cereal grains tend to be low in sulfur (approximately 1 g /kg) (Todd 1972), cereal straws slightly richer (1.4–4.0 g/ kg) (Suttle 1991), protein supplements rich (2.2–4.9 g kg"1 DM) (Qi *et al.* 1994) and brassica crops too rich (4.8–9.0 g kg"1 DM) (Cornforth *et al.* 1978) in sulfur. Leafy brassicas such as kale and the leaves of root crops such as swedes and turnips contain the most sulfur. Among by-products, molasses contains 3–7 g S /kg, while animal by-products are moderate sources.

Sulfur requirements of ruminants

Dietary sulfur requirements (NRC) as a percent of ration dry matter are as follow:

Growing heifers and bulls	0.16%
Dry pregnant cows	0.17%
Lactating cow	0.20%
Mature bull	0.11%
Young lambs	0.18-0.26%
Mature ewes	0.14-0.18%

Needs for hair and wool production

Sheep and goats differ from cattle in that hair and wool growth adds significantly to their need for sulfur. Since hair and wool growth is inevitable and is related to body size and growth rate, net sulfur and sulfur amino acid needs increase with body size and growth rate and are considered part of the maintenance need (ARC 1980, AFRC 1993). In double-coated, cashmere producing goats, there is an additional need during the cashmere-growing season in autumn (Souri *et al.* 1998).

Trace minerals

Trace minerals – zinc, copper, manganese, selenium, cobalt, iron, iodine etc. are essential nutrients for all animals. They have a wide range of activities and functions within the body, being involved individually or collectively in general metabolism, reproduction, the immune system, growth, development and repair of various tissues and so on. But, as the name suggests, trace minerals are required in very small amounts in the diet and their uptake from the digestive tract can be impaired by other dietary components or the presence of 'antagonists'. Natural feedstuffs such as corn, wheat, soybean meal, etc. contain essential trace elements, which are required by animals. However, these trace elements are often in a form which renders them unavailable to the animal. Also, even if the elements were totally available, in many cases, they would not be in adequate concentrations to meet the animal's requirement. Therefore, when deficiencies of one or more of the trace mineral elements exist in a diet, they are usually provided to the animal in an inorganic or organic supplemental form. The requirement for trace minerals by an animal is affected by many factors including genetics, age, maintenance, growth, reproduction, and lactation.

Adequate mineral status is also needed during gestation to provide the fetus with sufficient trace minerals for development and/or tissue accretion. Supplementation of the pregnant cow with adequate trace minerals is essential. As Hidiroglou and Knipfel (1981) stated, the fetus relies entirely on the dam for a sufficient supply of trace minerals and other nutrients needed for growth and development. As the trace mineral status in the calf declines, immunity and enzyme functions are compromised first; this is followed by a reduction in maximum growth; and finally normal growth decreases prior to clinical deficiency (Wikse 1992).

For newborn calves colostrums is the main source of minerals for the calf and mineral content of colostrums is largely affected by mineral supplementation to the cow during the transition period. Therefore, supplemental trace minerals should be fed to the dam during the prepartum transition period to increase the trace mineral content of colostrums as well as in fetal tissues to enhance immunity in the calf.

Cobalt

The essentiality of cobalt in the forages of sheep and cattle emerged from investigations of two naturally occurring disorders: 'coast disease' of sheep in South Australia (Lines 1935, Marston 1935) and 'wasting disease' of cattle in Western Australia (Underwood and Filmer 1935). The mode of action of cobalt was not recognized until 1948, when the anti-pernicious anaemia factor, subsequently designated vitamin B_{12}, was found to contain cobalt (Smith 1948).

Functions of cobalt

The essentiality of cobalt for mammals is linked to two distinct forms of vitamin B_{12} with contrasting coenzyme functions. As methyl cobalamin (MeCbl), cobalt assist a number of methyl transferase enzymes by acting as a donar of methyl groups and is thus involved in one carbon metabolism. MeCbl is important for microbes as well as mammals and is needed for methane, acetate and methionine synthesis by rumen bacteria (Postan and Stadman 1975)

Sources of cobalt

Cobalt concentrations in pastures and forages vary widely between plant species and with soil conditions (Minson 1990). In early Scottish studies, cobalt concentrations in mixed pasture ranged from 0.02 to 0.22 mg kg"1 DM and legumes were richer in cobalt than grasses grown in the same conditions, with red clover and ryegrass containing 0.35 and 0.18 mg kg"1 DM, respectively (Underwood and Suttle 1999). Cereal grains are poor cobalt sources with concentrations usually within the range 0.01–0.06 mg kg"1 DM, and are used to produce cobalt-deficient diets (Kennedy *et al.* 1992). Like most trace elements, cobalt is highly concentrated in the outer fibrous layers, so that bran and pollard are much richer in cobalt than the whole grain. Feeds of animal origin, other than liver meal, are mostly poor sources of cobalt. Cow's milk normally contains 0.5–0.9 mg Co/litre, but it is possible to increase this values several folds by heavy supplementatation with cobalt salts (Grace *et al.* 2006).

Deficiency

Symptoms of cobalt deficiency include anemia, rough hair coat, depressed appetite, increased susceptibility to heat stress and reduced conception rates

(Puls 1994). Quirk and Norton (1987) reported that cobalt deficiency during early pregnancy in ewe lead to parturition of fewer lambs. In their study 5 out of 9 supplemented cows had failed to conceive, whereas in each of 3 cobalt supplemented groups of 9 to 11 cows, only 2 or 3 had not conceived. The poor reproductive performance was not attributable to anorexia because ewe live weight and body condition were not affected while infertility was always arisen as a secondary consequence of debilitating conditions such as severe cobalt deprivation. According to Judson *et al.* (1997) infertility may occur in the absence of ill-thrift in beef cows. Ruminants appear to be more sensitive to vitamin B_{12} deficiency than non-ruminants largely because they are dependent on gluconeogenesis for meeting tissue glucose needs. Clinical signs of Co deficiency are delay in onset of puberty, anemia, general unthrift ness, delayed uterine involutions and decreased conception rate (Hidiroglou 1979). Young animals are more sensitive to dietary cobalt insufficiency because they have lower liver vitamin B_{12} reserves (Underwood and Suttle 1999). Vitamin B_{12} level in the blood is the most common diagnosis of cobalt deficiency and the level of above 0.5 mg per milli litres to be considered as adequate. Ewes should receive cobalt supplements throughout late pregnancy to avoid loss of appetite and risk of pregnancy toxemia. Quirk and Norton (1988) stated that in pregnant heifers Co supplementation could produce massive increase in the milk concentration of vitamin B_{12} Intramuscular injections of vitamin B_{12} (100 Jlg500 Jlg) were reported to be highly effective (Judson *et al.* 1988). Oral administration of vitamin B_{12} is not effective because of poor absorption in rumen of ruminants.

Cobalt requirements

The foregoing differences in vitamin B_{12}, energy and sulfur amino acid metabolism probably cause cobalt requirements to vary between species, with the type and amount of cereal in the ration, stage of the reproductive cycle and the criterion of adequacy used. Dietary requirements for Co by ruminants have been established at 0.1 ppm for dairy cattle and sheep.

Copper

Copper was first shown to be essential for growth and hemoglobin formation in laboratory rats by Hart and Elvejhem in 1928 and subsequently for the prevention of a wide range of clinical and pathological disorders in several monogastric species. A number of naturally occurring wasting diseases in grazing animals were soon found to respond to copper therapy, such as 'salt-sick' of cattle in Florida (Neal *et al.* 1931), 'lecksucht' of sheep and cattle in the Netherlands (Sjollema 1933) and 'copper pine' in UK calves (Jamieson and Allcroft 1947). An ataxic disease of newborn lambs ('swayback') in Western Australia was attributed to copper deficiency in the ewe during pregnancy (Bennetts and

Chapman 1937). Subsequently, extensive copper-deficient areas, affecting both crops and livestock, were discovered throughout the world.

Functions of copper

Copper is presenting in an essential for the activity of numerous enzymes, cofactors and reactive proteins. The essentiality of copper for processes such as reproduction and bone development is presumed to depend on such functions, but only two processes have unequivocal dependency on a specific enzyme: pigmentation (tyrosinase) and connective tissue development (lysyl oxidases) (Prohaska 2006). Copper plays various important biological roles in animals through several copper dependent enzymes (Sharma *et al.* 2005a). It is required in the development of nervous system of growing and unborn calves (Marston 1999). There is existence of an interaction between estrogen and Cu (Hidiroglou and Knipfel 1981). Suttle (1987) stated that Cu is required for reproduction, bone development, growth, and connective tissue development and pigmentation of the skin appendages. Copper concentration is significantly higher on the day of estrus than during the other stages of reproduction (Bhattacharya *et al.* 1995).

Deficiency

Production losses associated with Cu deficiency mostly occur when Mo is responsible for the reduction in the Cu status (Humphries *et al.* 1983,). Functional activity of the ovary is reduced due to copper deficiency and the plasma level of this mineral in cow might be of importance in relation to the time of occurrence of postpartum estrus. Impaired reproductive performances due to both primary and induced Cu deficiencies were reported by Van Niekerk (1989). Prenatal mortality and early embryonic losses are associated with hypocuprosis (Mc Chowell 1968). Puls (1994) reported that marginal copper deficiency causes reduced growth rates and reduced feed efficiencies, fertility and increased incidence of retained placenta. Phillipo *et al.* (1987) showed reduced conception rates and disrupted estrus activity in cattle suffering from a primary copper deficiency. Sanders and Sanders (1983) reported that young calves are unable'to suckle, suffer from in coordination, stiff gait, and opisthotonos whose dams were Cu deficient. Initially calves are dependent on the Cu in the colostrums as there is little placental transfer of Cu (Ullrey *et al.* 1977). Delayed or depressed estrus and infertility was found to be associated with cow grazing on Cu deficient pastures (Annekov 1981). Symptoms of Cu deficiency related to reproduction include embryonic deaths; resorption of the embryo, increased retained placenta and necrosis of the placenta (Puls 1994). Adverse changes in reproductive organs can arise due to inadequate level of copper in the diet (Tohami *et al. 1997).* Due to Cu deficiency progesterone secretion is impaired during the estrus cycle and especially late pregnancy in ewes (Van Niekerk 1989b). NRC

(2001) recommends that cows in gestation of more than 225 days, should have 2 mg of absorbable copper and according to Boila *et al.* (1984) the Cu concentration in blood serum is indicative of a Cu deficiency and not necessarily the adequate of body stores of Cu. Administration of Cu orally or by injection has led to rapid improvement in breeding performance of the animals (Mahadevan and Zubery 1993).

Copper requirements

Net requirement in sheep and cattle

Muscle is the major contributor to the extra-hepatic pool of copper. On average, muscle contains only 0.8 and 1.2 mg Cu kg"1 fresh weight in cattle (Simpson *et al.* 1981) and sheep (Grace 1983), respectively, and declines with copper depletion (Simpson *et al.* 1981). Relatively large amounts of copper are deposited in the fleece, giving a significant requirement for wool growth in sheep (ARC 1980).

Copper requirements of poultry

NRC(1994) gives the following copper requirements (all mg kg"1 DM): starting chicks (0–8 weeks), 5; growing Leghorn-type chickens (8–18 weeks), 4; broilers (all ages), 8 ; starting turkey poults, 8; growing turkeys, 6; and breeding turkeys, 6. No requirements are given for laying or breeding hens, presumably because their previous low recommendations (3–4 mg kg"1 DM) have not been questioned.

Copper availability to animals from feeds and fodders

Copper availability to animal changes with the feeding of forage species and their method of preservation or storage. Curing or drying of forage may alter the chemical form of copper. Making it more available than copper in fresh green plants. Fresh grass has the lowest availability of copper. This availability is further reduced by small increase of molybdenum in the forage. In forages preserved as hay the antagonistic effects of molybdenum are detectable, but less than that of sulphur. In silages, molybdenum has small effect but sulphur reduces copper availability inalogarithmic manner. A copper molybdenum ratio 2:1 in feed is normal, and higher Mo tends to bind copper, making it unavailable for absorption (Multimore and Mason 1971)

Zinc

Zinc is an essential component of numerous enzyme systems and is an indispensable element for the growth and reproduction of male and female animal (Om and Chung 1996). Zn activates some enzymes and is a component of a large number of important metalloenzymes, such as carbonic anhydrase, carboxy peptidase, alcohol dehydrogenase, glutamic dehydrogenase, lactic dehydrogenase, alkaline phosphate and mannosidase. Zinc has beneficial effect in reproduction (Underwood 1977). Zn is also known to be essential element for proper sexual maturity, reproductive capacity and onset of estrus, repair and maintenance of the uterine lining following parturition and normal reproductive functions. It plays a major role in reproductive hormones (Cousins and Hempe 1990).

Deficiency

Dairy cattle

Lowered feed intake is one of the first changes observed in a zinc deficiency. The cattle grow slower due to a decreased feed intake and less efficient feed utilization. Other symptoms in a severe zinc deficiency are skin parakeratosis (usually most severe on the legs, neck and head), hair loss, unthrifty appearance, stiffness of joints, teeth gnashing, retarded testicular growth and excessive salivation. Reduced reproductive performance has been observed in both males and females fed zinc deficient diets (Pitts *et al.* 1966). Another effect of a zinc deficiency is a failure of wounds to heal normally. In most cases, when a zinc deficient animal is given zinc, there is a dramatic and quick recovery.

Sheep

Zinc deficiency in lambs results in a lack of appetite, reduced growth, slipping of wool, swelling around the eyes and hooves, excess salivation, general listlessness, impaired growth of testes and cessation of spermatogenesis. Loss of appetite is the first sign of a zinc deficiency in growing lambs. Recent studies have shown that lambs switch from meal eaters to nibblers (Droke and Spears 1993) as they become zinc deficient. The zinc deficiency caused a few deaths, a continuous loss in body weight during lactation and development of skin lesions and frothy saliva. A recent study showed that 7 of 30 ewes fed a low-zinc diet aborted, reabsorbed or delivered mummified and deformed lambs, while the other 23 ewes delivered lambs that were 20% smaller than the controls (Apgar and Fitzgerald 1985). Feeding a diet containing only 3 ppm zinc during pregnancy reduce survival of the newborn lambs and caused pregnancy toxemia in the ewes as a result of anorexia (Apgar *et al.* 1993).

Goats

Zinc deficiency in goats includes reduced feed intake, weight loss, parakeratosis (mange-like condition), stiffness of joints, excessive salivation, swelling of the feet and horny overgrowth, small testicles and low libido (Neathery *et al.* 1973).

Poultry

Zinc deficiency causes growth retardation and abnormal feather development in poultry. Feather fraying occurs near the end of the feather. The hock joint may become enlarged. The long bones of the legs and wings also become shortened and thickened with a zinc deficiency. Other symptoms include scaling of the skin, especially on the feet, loss of appetite, reduced efficiency of feed utilization, and mortality in severe cases. Zinc deficiency in the breeder diet reduces egg production and hatchability. Embryos produced in zinc deficient eggs show a wide variety of skeletal abnormalities in the head, limbs and vertebrae. The hatched chicks also may not stand, eat or drink (Jimenez 1982).

Requirement of zinc

Dairy cattle

The estimated zinc requirement for dairy cattle is 40 ppm in the diet (NRC 1989). There may be certain conditions or an interrelationship with other nutrients that might increase zinc needs. A review of the various experiments conducted on the zinc requirements of dairy cattle and the levels needed in each one indicate there is considerable variation in the requirements obtained.

Sheep

The requirement of zinc for sheep is 20-33 ppm in the diet. The maximum tolerable level in the diet is 750 ppm (NRC 1985). Excess zinc will cause a copper deficiency.

Goat

A level of 45 to 75 ppm zinc should be used in the total diet of goats until their zinc requirements are met.

Poultry

The 1984 NRC publication, Nutrient Requirements of Poultry, recommends a level of 40-75 ppm zinc in various poultry diets (NRC 1984). This publication indicates that the toxic level of zinc varied from 800 to 4,000 ppm in the diet.

Zinc sources

Supplemental zinc is usually added to animal diets in the form of zinc oxide or zinc sulfate. Recent comparisons of bioavailability in chicks suggest that feed grade zinc oxide has only 44-78% the availability of zinc sulfate when added to purify (Wedekind *et al.* 1992) or practical (Sandoval *et al.* 1997) diets. A recent comparison of availabilities of zinc sources for pigs in corn-soy diets ranked zinc sulfate > zinc methionine > zinc oxide > zinc lysine (Wedeking *et al.* 1994).

Iron

Iron has been recognized as an essential nutrient for over 100 years. Despite this fact, iron deficiency is still a major problem in several segments of the livestock industry. Approximately two-thirds of body iron is present in hemoglobin in red blood cells and myoglobin in muscle, 20% is in labile forms in liver, spleen and other tissues with the remainder in unavailable forms in tissues such as myosin and actomysin and in metalloenzymes. In hemoglobin, which contains 0.34% iron, an atom of ferrous iron in the center of a porphyrin ring connects heme, the prosthetic group, with globin, the protein. The iron in hemoglobin is essential for the proper function of every organ and tissue of the body. Iron also plays a role in other enzymes involved in oxygen transport and the oxidative process, including catalase, peroxidases, flavoprotein enzymes and cytochromes. Iron in blood plasma is bound in the ferric state (Fe^{+++}) to a specific protein called transferrin. Transferrin is the carrier of iron in the blood and is saturated normally only to 30-60% of it iron binding capacity (Church and Pond 1989).

Deficiency

Iron deficiency is rare in adult cattle but calves frequently meet with iron deficiency, especially if fed milk replacer containing no supplemental iron or whole milk for extended periods of time (Puls 1994). Iron deficiency usually occur in newborn animals whose source of Fe is the milk of the dam and Otto *et al.* (2000) reported that milk is a poor source of Fe. Lower levels of iron were reported in infertile heifer by *Sharmaetal* (1988). Hidiroglou (1979) found no differences in' circulatory iron levels in fertile and infertile animals. According to Mgongo *et al.* (1953) deficiency of iron causes stress due to malnutrition that results in normochromic anaemia (Sharma *et al.* 2001) in animals affecting response of ovarian receptors to estrogen hormones.

Supplementation

Iron deficiency anemia can be prevented by iron supplementation (Ullrey *et al. 1965).* Intramuscular injection of 200 mg iron (Green *et al. 1993)* and daily oral supplementation with 20-40 mg iron as ferrous sulphate is also effective in newborn calves (Kume and Tanabe 1994). The most inexpensive method of supplying iron is to use ferrous sulphate orally@2-4 g daily for 2 week to adult cattle with iron deficiency anemia (Sharma *et al.* 2001). The NRC (2000) recommends that milk replacers fed to herd replacements or dairy beef should contain 100 mg of Fe/kg of DM with an upper limit of 1000 mg of Fe/kg DM.

Dietary sources of iron

Most cereal grains contain 30–60 mg Fe kg"1 dry matter (DM) when sampled in the field and species differences appear to be small, although 10 and 20 mg kg"1 DM have been recorded for Egyptian-grown maize and barley, respectively (Abou-Hussein *et al.* 1970). Purchased cereals contain more iron and, in the UK, average values of 100, 100, 120 and 140 mg kg"1 DM have been reported for barley, maize, oats and wheat, respectively (MAFF 1990). The higher values recorded for cereal by-products – 480 mg kg"1 DM in maize gluten feed, 220 mg kg"1 DM in wheat feed and 2600 mg kg"1 DM in rice-bran meal – reflect both the uneven distribution of iron in the grain and contamination during processing. Leguminous seeds and oilseed meals commonly contain 100–200 mg Fe kg"1 DM.

Animal sources

Cow's milk in full lactation averages about 0.5 mg Fe kg"1 fresh weight (equivalent to only 4 mg kg"1 DM) and similar low iron concentrations are found in milks from other species, although sow's milk contains three times more, as does bovine colostrums. Most of the iron in secreted milks is present as the glycoprotein, lactoferrin.

Forages

Pasture iron shows marked seasonal fluctuations with peaks in spring and autumn, and iron values can range from 70–111 to 2300–3850 mg kg"1 DM in New Zealand (Campbell *et al.* 1974) and southern Australia. High values are attributable to soil contamination, which is most likely to occur on soils prone to water logging. Lower values of between 90 and 110 mg kg"1 DM have been reported for Egyptian green-cut clover and clover hay (Abou- Hussein *et al.* 1970).

Manganese

Manganese was first recognized as a necessary nutrient for animals in the early 1930s. Because manganese is found in many different feeds, a deficiency is less likely than with most of the other trace minerals. The need for manganese in livestock nutrition became evident when two diseases of poultry, perosis (slipped tendon) and nutritional chondrodystrophy, were prevented by feeding manganese supplements (Schaible *et al.* 1938) and breeder hens deprived of manganese produced chick embryos with malformed skeletons (Lyons and Insko 1937). Bone, kidney, liver, pancreas, and pituitary gland are the sites of highest manganese concentration. Relative concentration is quite low compared to the other trace minerals. Manganese is essential for chondroitin sulfate synthesis, which is critical to the organic matrix of bone. Many enzymes required for the synthesis of polysaccharides and glycoproteins require manganese to be active. Manganese is a key component of the metalloenzyme, pyruvate carboxylase, a critical enzyme in carbohydrate metabolism. The body has only a limited storage of mobilizable manganese reserves. The absorption of manganese in all domestic livestock is poor (Underwood 1981).

Deficiency

Dairy cattle

General symptoms of manganese deficiency include impaired growth, skeletal abnormalities, disturbed or impaired reproduction, and abnormalities of the newborn. Deformed calves at birth have weak legs and pasterns, enlarged joints, stiffness, twisted legs, general weakness, and reduced bone strength. Low manganese diets cause heifers and cows to be slower in exhibiting estrus and more likely to have "silent heats," and a lower conception rate. Abscessed livers and practically no bile in the gall bladder have also been reported (NRC 1989).

Poultry

Manganese deficiency in the diet of growing chicks and poults results in perosis, or slipped tendon. Manganese-deficient chicks have less proteoglycan in the cartilage of the tibial growth plate than manganese-repleted chicks (Liu *et al* 1994). This contributes to the perosis condition which involves a twisting and bending of the tibia, and slipping of the gastrocnemius tendon from its condyles. With increasing severity chicks are reluctant to move, squat on their hocks and soon die. In laying and breeding birds, a manganese deficiency results in lowered egg production and hatchability and reduced eggshell strength. In many cases, embryos that die as a result of manganese deficiency exhibit chondrodystrophy, a condition characterized by a parrot-like beak, wiry down and shortening of the long bones (NRC 1984).

Sheep

Depressed or delayed estrus and poor conception rates occur in ewes with a manganese deficiency. More services per conception have also been reported (Underwood 1981). One report indicates that manganese may have a possible role in the functioning of the corpus luteum (Hideroglou 1975). Early weaned lambs on a purified diet containing less than 1 ppm manganese exhibited bone changes similar to those seen in other manganese deficient animals (NRC 1985). This indicates that sheep probably need manganese for skeletal development, as various other species do. The levels of manganese in wool appear to be quite sensitive to changes in the manganese status of the lamb (NRC 1985).

Goats

Symptoms of manganese deficiency include reluctance to walk, deformity of the forelegs, and reduced reproductive efficiency (NRC 1981). The growth of female goats fed 20 ppm manganese for the first year of life and 6 ppm the following year was not affected, but the low manganese level delayed the onset of estrus and required more inseminations per conception (NRC 1985). In mature goats the manganese content of the hair is a better indicator of manganese status than manganese in any other part of the body (NRC 1985). Goats also show tarsal joint excrescences, leg deformities, and ataxia due to a manganese deficiency (Underwood 1981). In same study a decrease of 25% in the body fat content of kids born from goats on manganese deficient diets has been reported.

Requirement

Dairy cattle

In cattle the manganese requirement for growth is lower than that required for reproduction and birth of normal calves. Ohio researchers used regression techniques to estimate the intake required to meet the metabolic fecal manganese requirement (Food and Drug Administration 1997). The corresponding dietary concentrations, assuming dry matter intakes of 21 and 12 kg/day for lactating and dry cows, respectively, were 28 and 49 ppm in the diet on a dry matter basis. These concentrations are approximately 1.6 and 2.7 times higher than those needed to meet the manganese requirements for lactating and dry cows, respectively, as calculated using the 2001 National Research Council dairy requirements model.

Poultry

Manganese requirement of poultry is 25-90 ppm, It varies considerably with the different classes of poultry. Ducks normally require 30-50 ppm (Dean 2006). Manganese source also influences requirement. Manganese sulfate has the highest bioavailability, while manganese oxide and manganese carbonate are only about 30 and 55% as available as the sulfate (Henry 1995).

Sheep

The exact requirements for manganese are not known, but the National Research Council committee on sheep recommends a level of 20 to 40 ppm in the diet (NRC 1985). High intakes of calcium and iron may increase manganese needs. A level of 45 to 50 ppm is suggested for sheep until more definitive data become available (NRC 1981).

Dietary sources of manganese

Seeds and grains

Seeds and seed products vary widely in manganese, due partly to inherent species differences Barley commonly contains 15–28 mg Mn kg”1 dry matter (DM), maize grain far less and wheat or oats far more. Species differences amongst lupin seeds are striking: the white lupin (*Lupinus albus*) can contain 817–3397 mg Mn kg”1 DM, some 10–15 times more than other common lupin species growing on the same sites (Gladstones and Drover 1962, White *et al.* 1981) while *Lupinus angustifolius* may contain as little as 6 mg Mn kg”1 DM (White *et al.* 1981).

Forages

Pastures vary markedly in manganese about a high mean value of 86 mg kg”1 DM (Grace 1973). Soils usually contain 300–1100 mg Mn kg”1 DM (Suttle *et al.* 2003), much more than the pasture they support. Contamination can also occur during sample processing if mills with steel blades are used. Increases in soil pH markedly decrease manganese uptake by plants, whether they are cereals, grasses or legumes (Mitchell 1957). Low manganese levels have been reported for maize silages from Canada (approximately 22 mg kg”1 DM) (Buchanan-Smith *et al.* 1974).

Iodine

Around 1900, scientists first recognized that iodine was required for the proper functioning of the thyroid gland and that an iodine deficiency caused goiter. Shortly thereafter, iodized salt became widely accepted as a means of preventing goiter in man and animals.

Function

Iodine has only one known but vital function as a constituent of thyroid hormones. Structural studies of the thyroxine molecule have revealed that its four atoms of iodine are attached in two similar positions on linked ‘outer’ and ‘inner’ tyrosine rings (OR and IR). Thyroxine therefore became known as 3,3',5,5'-tetraiodothyronine (T4), a relatively inactive prohormone requiring removal of one atom by two OR-deiodinase enzymes (5’ORD types I and II) to generate the active hormone, 3,3',5-triiodothyronine (T3). Both T4 and T3 can be inactivated by an IR deiodinase (type III), and physiological activity of thyroid hormones is regulated peripherally by the three deiodinases rather than by the thyroid gland (Todini 2007).

Deficiency and requirement

Dairy cattle

Lactating dairy cattle require more iodine than beef cattle because approximately 10% of the iodine intake is normally excreted in milk, and this percentage may increase as milk production increases (NRC 1989). To ensure that the iodine requirements of the high-producing cow are met, 0.6 ppm iodine is recommended.

Sheep

In newborn lambs, the more common symptom of iodine deficiency is goiter. If the condition is not advanced, the lambs may survive. Other symptoms are lambs born weak, dead, or without wool. Mature sheep seldom show a change in their appearance. But an iodine deficiency may result in reduced yield of wool and reduced conception rate (NRC 1985).

Goats

Kids with goiter, as well as weak and dead kids, have been reported at birth. Very little experimental information is available with the goat. Therefore, extrapolated data from sheep studies must be used until additional experimental data become available from studies with goats (NRC 1981).

Poultry

Goiter occurs in the thyroid gland, causing the thyroid gland to grow to many times its normal size. Histological examination of the thyroid shows hyperplasia and an absence of colloid. Iodine deficiency in breeding hens results in reduced egg iodine levels, reduced egg production, decreased hatchability, prolonged hatching time, and thyroid enlargement in the embryos (NRC 1984).

Source of Iodine

Iodine occurs in trace in most foods and is present mainly as inorganic iodine in which form it is absorbed from the digestive tract. The richest source of this element are foods of marine origin, and values as high as 6g/kg DM have been reported. Fish meal is also a rich source of iodine.

Selenium

Selenium was recognized as a potentially toxic mineral many years before it was identified as an essential nutrient. It was not until 1957 that the role of selenium in preventing liver necrosis in the rat and oxidative diathesis in chicks was recognized. Selenium is present in all cells of the body, but the concentration is normally less than 1 ppm. Toxic concentrations in liver and kidney are normally between 5 and 10 ppm (Church and Pond 1989). Selenium is an important part of the enzyme glutathione peroxidase. This enzyme destroys peroxides before they can damage body tissues. Vitamin E is also effective as an antioxidant. Therefore, both selenium and vitamin E prevent peroxide damage to body cells. Rancidity in feeds causes formation of peroxides that destroy nutrients. Vitamin E for example, is easily destroyed by rancidity. Selenium spares vitamin E by its antioxidant effect as a constituent of glutathione peroxidase Selenium plays a critical role in increasing the immune response in animals. Selenium is important in sulfur amino acid synthesis. Sulfur amino acids protect animals against several diseases associated with low intakes of selenium and vitamin E. This protection is believed to be due to the antioxidant activity of selenium and vitamin E. Therefore, the sulfur amino acids, methionine and cystine, can spare vitamin E and selenium through their antioxidant role.

Deficiency

Dairy cattle

A major symptom of selenium deficiency is white muscle disease, which usually occurs in young calves. This causes chalky white striations, degeneration, and necrosis in cardiac and skeletal muscles. Heart failure, paralysis (usually of the hind legs), a dystrophic tongue and elevated SGOT (serum glutamic oxaloacetic transaminase) values may also be evident.

Poultry

The main symptom of selenium deficiency is exudative diathesis. This disease characterized by edema of the breast, wing and neck regions. This edema is caused by an abnormally high permeability of the capillary walls that allows fluid to accumulate between the muscle and skin. In broilers fed low-selenium grains, chicks between 3 and 6 weeks of age, begin to show signs of weight loss, leg weakness and eventual death. With severe selenium deficiency, the growth rate is reduced and mortality increased even in the presence of adequate vitamin E. Pancreatic fibrosis and a reduction in pancreatic output of lipase, trypsinogen, and chymotrypsinogen are associated with selenium deficiency. Pancreatic lesions can occur as early as 6 days of age and usually return to normal within two weeks of selenium supplementation. In laying hens egg hatchability is the most sensitive criteria of selenium deficiency. Selenium is also required to prevent myopathies of the gizzard and heart in turkeys and ducks.

Sheep

Selenium deficiency has serious effects on lamb production. The manifestations are reduced growth and white muscle disease, which affects lambs at two to eight weeks of age (NRC 1985).

Dietary sources of selenium

Forage selenium can range widely both within and between types on a given farm, making it difficult to assess total selenium intake in ruminants. In the UK, the average selenium concentration in fresh grass is 0.05 mg kg”1 DM (MAFF 1990), but in Australia and New Zealand, where selenium-responsive disorders have occurred more frequently, forage selenium may be 50% lower (Grace and Knowles 2002). Legumes tend to contain less selenium than grasses, but the difference diminishes as soil selenium status declines (Minson 1990) and lucerne is an exception to the rule.

Cereals and legume seeds

Cereal grains and other seeds vary widely in selenium content between and within countries. The average values for wheat grown in the UK, the USA and Canada have been calculated at 0.03, 0.37–0.46 and 0.76 mg kg”1 DM, respectively (Adams *et al.* 2002). Linseed meal is generally rich in selenium compared with extracted soybean meal (SBM) and rapeseed meal (MAFF 1990), Cereal by-products such as maize gluten feed, rice-bran meal and distillers' dark grains are good sources of selenium (MAFF 1990).

Molybdenum

The first indication of an essential rolc for molybdenum came from the discovery by de Renzo *et al.* (1953) that activity of the flavoprotein enzyme, xanthine oxidase (XO), depended upon its molybdenum content. Molybdenum is an essential nutrient because it is a constituent of the enzyme xanthine oxidase and other enzymes. Molybdenum has been shown to be essential for lambs, chicks, and turkey poults fed highly purified diets.

Sources

Soils vary widely in molybdenum content with sandy soils at the low extreme and those of marine origin at the high extreme and in the extent to which they yield the element to growing crops and pastures. Approximately 10% of soil molybdenum is normally extractable, but the proportion rises with soil pH and with it the concentration in crops and pastures. Soils can become contaminated with molybdenum by the burning of industrial wastes (Raisbeck *et al.* 2006). Cereals rarely contain >1 mg Mo kg”1 DM, but pasture levels vary widely depending on soil conditions.

Deficiency

A deficiency of mo, by itself, has not been described in ruminants. However low dietary Mo (< 0.2 ppm), insufficient to prevent adverse effects in sheep of Cu levels otherwise not excessive might be considered deficient

Newer trace elements

Nickle

Nickle (Ni) was first shown to be essential for chicks fed on a highly purified diet under strict environmental control conditions (Nielson and Sauberlich 1970). Later on, experimental Ni deficiency was produced under laboratory conditions in various species of animals, maintained in specially clean environments and fed diets very low in Ni (Nielson 1987). Depigmentation of the skin, thickened legs, swollen hocks, growth retardation and anemia were reported in chicks given a basal diet providing only 2-15 ug Ni/kg and an additional 50 ug Ni/kg alleviated all abnormalities (Spears *et al.* 1978).

Nickel was also found to improve growth and fertility in rats (Nielson and ollerich, 1974). The symptoms of deficiency have recently been confirmed in the first-generation off spring as growth retardation, impaired reproduction and anemia (Stangi and Kirchgessner 1996).

Goats fed a low Ni diet (100ppb) had higher abortion rate, decreased conception rate, along with a decreased viability of female goats and their offspring (Anke *et al.* 1983). Urease produced by bacteria has been found to be Ni dependent (spears *et al.* 1977, Nielson 1980). Supplementation of diet containing 0.26-0.85 mg Ni/kg DM with 5 mg Ni ($NiCl_3$) has increased ruminal urease, growth rate and feed conversion efficiency of lambs (Spears 1984). Supplementation of a similar basal diet has increased urease activity without affecting growth (Spears *et al.* 1986), urea utilization and N retention in lambs (Milne *et al.* 1990). Rumen microorganisms involved with methanogenises and sulphate reduction may require Ni (Nielson 1996). Ni supplementation has also frequently altered patterns of VFA production in the rumen, but the changes have been inconsistent (Milne *et al.* 1990).

The Ni requirement of ruminants may be less then 500 ppb, but since normal feeding situation usually provide Ni in excess of this amount, a primary deficiency of Ni is unlikely to be seen under practical conditions (Anke *et al.* 1983).

Optimum amount of Ni in the diet is 25 mg/kg which has a positive influence on bone strength characteristics and performance of broilers (Wilson *et al.* 2002).

Boron

Boron (B) was recognized as an essential element for plant in early part of 20th Century, but its role in animal has recently been advocated (Nielson 1997). Beneficial responses from giving boron to animals were first noted in vitamin-D deficient chicks in 1981 and have recently been confirmed (Hunt *et al.* 1994). Studies on chicks were suggestive of its role in ameliorating signs of vitamin D deficiency (Hunt and Nielson 1981), with a possible interrelationship involving Ca and Mg (Hunt *et al.* 1983). Boron seems to play an essential role in metabolism of Ca, P, Mg and vitamin D through regulation of paratharmone hormone action, as boron supplementation in the diet of men and women has increased values for markers of copper, vitamin D and has decreased bone turnover (Nielson 1994). Effect of dietary B supplementation as orthoboric acid in laying hens revealed a significant improvement in serum Ca levels and damaged egg ratio. (Kurtoglu *et al.* 2003). Boron is a well tolerated element, but loss of appetite and body weight occurred in heifers given 150-300 mg B/liter drinking water (Green and Weeth 1977), where as more than 300 mg B/kg DM being required to produce the first signs of toxicity in poultry (Puls 1994). The quantity of B in meat and bone tissue samples increased with increasing dietary B. It was suggested that it might be possible to add another marketing dimension to poultry products by using B as an additive for elderly people needing to combat brittle bones (Wilson and Ruszler 1995). Egg production, feed consumption and body weight were suppressed at a dietary concentration of 400 mg B/kg. (Wilson and Ruszler 1996). Ameliorative effects of B were studied in sheep given fluorine rich diet. Supplementation of 0.5mgB/mg F improved the total N balance, retention of F decreased (P,0.01) due to its increased (P,0.01) fecal excretion (Yadav *et al.* 1996). Boron has been found to be beneficial in broiler chickens, which were deficient in vitamin D3 (Kurtoglu *et al.* 2001). Boron addition has a positive effect on Ca, P and alkaline phosphatase activity of blood serum of chicks. Boron provided significant improvement in serum Ca levels and damaged egg ratio (Kurtoglu *et al.* 2002). Boron supplementation (15mg/kg diet) to pigs increased the production of cytokines following a stress, which indicates a role of B in the immune system.

Silicon

Silicon (Si) provides physical support to plants and may reduce their susceptibility to pests. However, Si may have negative effects on the feed digestibility and contribute to urinary calculi in animals (Mayland and Shewmaker 2001). When silicon was lacking in the diet, Ca and Mg concentration of the femur were decreased, this was reversed by feeding germanium (Ge) and /or Si (Seaboun and Neilson 1994). Si has been shown essential for rats and chicks by Carlisle (1970) with a requirement between 50 to 500 ppm (Carlisle 1974). Siliocn is involved in glycoamino glycon formation in cartilage and connective tissue.

Silicon deficient chicks had bone abnormalities due to reduced mucopolysaccharides, glucose aminoglycoine and collagen synthesis with pathological changes (Mc Dowell 1992).

Silicon was demonstrated to play a significant role for the activity of propyl hydroxylase, a measure of the rate of collagen biosynthesis (Carlisle *et al.* 1981). Growth rate and development of long and skull bone was reduced in chicks given a diet containing 1 mg Si/ kg as compared to those given 250 mg/kg diet. Sheep given graminaceous or leguminous diets excreted 99% of the ingested silicon in their faeces and the absorbed silicon was mostly recovered in urine (Jones and Handreck 1965). Silicon is believed to be absorbed and transported in the blood stream as monosilicic acid and readily enters tissues and cross membranes in this free form (Carlisle 1986).

Silicon supplementation @ 300mg/kg diet of broilers demonstrated to have positive impact on bone mineralization (Tekeli and Zohouri 1998).

Vanadium

Vanadium (V) is widely distributed in plants and animals (Underwood 1977). Although V is capable of causing toxicosis at relatively low dietary concentration, it is also an essential element recommended to be in the diet at very low levels (NRC 1994).

Female goats given a diet containing, 10 ug V/kg DM showed a higher abortion rate when compared with control goats given 2 mg V/kg DM (Anke *et al.* 1989). Vanadium depleted goats produced less milk and there was a higher mortality among their offspring than in those of control goats. Addition of V to rations at dose of 5 and 10 ppm induced a significant improvement in milk production, milk fat and total milk protein in goats (Bonomi 2003).

Vanadium added to the ration (5, 10 and 15 ppm) improved the weight, feed utilization and reproductive efficiency in heifers (Bonomi 2003). Vanadium added to the rations of pigs at the doses of 5 and 10 ppm improved the weight gain (8.5 and 12.5 percent), feed utilization (6.0 and 8.0 percent), meat yield (15.0 and 20.0 percent), respectively (Bonomi 2004).

Tin

Much work has not been carried out on the tin supplementation in the diet of livestock. The importance of tin for rats was demonstrated by Schwartz (1970) who supplemented the purified diet with 2-ppm tin as stannous sulphate and observed better growth rate. This has been found to play some role in leaf pigmentation (Milne *et al.* 1990). It has also been reported to play a significant role in tertiary strengthening of protein and oxidation, reduction reactions at physiological pH.

Lithium

Lithium (Li) is now regarded as an essential dietary constituent for some species but deficiencies are unlikely to occur naturally. Goats deprived of Li become unhealthy, they exhibited growth retardation, impaired fertility, low birth weights and reduced longevity in a diet containing 3.3 ug Li/kg DM, when compared with supplemented goats given 24 ug Li/kg DM (Anke *et al.* 1983). Conservation of relatively higher amount of Li in endocrine organs in the face of depletion may indicate and endocrine function of Li (Patt *et al.* 1978).

References

Abou-Hussein ERM, Raafat MA, Abou-Raya, AK and Shalaby, AS 1970 Manganese, iron and cobalt content in common Egyptian feedstuffs. United Arab Republic Journal of Animal Production 10: 245–254.

AFRC 1993 Energy and Protein Requirements of Ruminants. An advisory manual prepared by the AFRC Technical Committee on Responses to Nutrients. Commonwealth Agricultural Bureaux, Farnham Royal, UK.

Anke M, Groppel B, Nordmann, S and Kronemann H 1983. In: Fourth Spurenelement Symposium. Abteilung Wissenchaftliche Publikationen, Friedrich Schiller Universistat, Jena, Germany, p 19.

Anke M, Schneider, HJ and Bruckner C. 1980. In: Third Spurenelement-Symposuim Nickel, Abeteilung Wissenchaftliche Publikationen, Friedrich Schiller Universistat, Jena, Germany, P 375.

Annekov B N, 1981. Mineral feeding of cattle. Mineral Nutrition ofAnimals. Georgievskii V I, Annekov B Nand Samokhin VT (Eds). Butterworth, London.

Apgar, J. and JA Fitzgerald. 1985. Effect on the Ewe and Lamb of Low Zinc Intake

Apgar, J, G.A. Everett and G.A. Fitzgerald, JA 1993. Dietary zinc deprivation effects on parturition and outcome of pregnancy in the ewe. Nutrition Research 13:319.

ARC 1980 The Nutrient Requirements of Ruminant Livestock. Commonwealth Agricultural Bureaux, Farnham Royal, UK, pp. 166–168.

Armstrong, TA, Spears JW and Lloyd, KE 2001. J. Anim. Sci. 79(6):1549-56.

Beaton, JD, Tisdale, SL and Platou, J, (1971) Crop responses to sulfur in North America. Technical Bulletin 18. The Sulfur Institute, Washington, DC, pp. 1–10.

Bennetts HW and Chapman FE 1937 Copper deficiency in sheep in Western Australia: a preliminary account of the aetiology of enzootic ataxia of lambs and on anaemia of ewes. Australian Veterinary Journal 13: 138–149.

Bhattacharyya B N, Talukdar S C, Baruah R N, Baruah K K Sr., Baruah KKJr. and Baruah A. 1995. Studies on circulatory levels of trace minerals at different reproductive status in goat. Indian ·Journal of Animal Reproduction 16(2): 96-98.

Blaxter K L, Rook JA F and McDonald A M. 1954. Experimental Mg deficiency in calves. 1. Clinical and pathological observations. Journal of Comparative Pathology 64: 157-75.

Blumer C C, Madden F J and Walker D 1. 1939. Hypocalcaemia, grass tetany or grass staggers in sheep. Australian Veterinary Journal 15: 2-27.

Boila R J, Devlin T J, Drysdale RA and Lillie L E. 1984. Injectable copper complexes as supplementary copper for grazing cattle. Canadian Journal of Animal Science 64: 365-78.

Bonomi, A 2003. Nutr. Abst. Rev. 73(4):1824.

Bonomi A 2004. Nutr Abst Rev. 74(2): 871.

Braak A E, Vande Klooster A T Van, T and Malestein A. 1987. Influence of a deficient supply of magnesium during the dry period on the rate of calcium mobilization by dairy cows. Research Veterinary Science 42: 101-08.

Brown TF McCormick ME, Morris DR and Zerinque LK 1989. Nutr. Res. 9:503-512 Bureaux, Farnham Royal, England.

Buchanan-Smith JG, Evans, E and Poluch SO 1974 Mineral analyses of corn silage produced in Ontario. Canadian Journal of Animal Science 54: 253–256.

Call J W, Butcher J E, Shupe J L, Lamb R C, Boman R Land Olsen A E. i 986. Clinical effects of low dietary phosphorus concentration in feed given to lactating cows. Journal of Veterinary Research 48: 133-136.

Campbell AG, Coup MR, Bishop WH and Wright DE 1974 Effect of elevated iron intake on the copper status of grazing cattle. New Zealand Journal of Agricultural Research 17: 393–399.

Carlisle, E.M. 1970. Science 167:279.

Carlisle, E.M. 1974. Fed. Proc. 33: 1758

Carlisle EM 1980 J Nutr 10 : 352

Carlisle EM 1986. In: Trace Elements in Human and Animal Nutrition, 5th edn, Vol. 1, Academic Press, New York, P. 373.

Carlisle, E.M. and Curran, K.J. 1979. Fed. Proc. 38: 553

Carlisle EM Berger JW and Alpenfels WF 1981. Fed. Am. Soc. Exp. Biol. 40: 866.

Church DC and WG Pond 1989. Basic Animal Nutrition and Feeding. 3rd ed. John Wiley and Sons, New York.

Cook R H and Fadalla B 1987. Seasonal variations in plasma phosphorus level of the transruminant sheep in kordofan Sudan. Tropical Animal Health Production 19: 57-62.

Cornforth IS, Stephen RC, Barry TN and Baird GA 1978 Mineral content of swedes, turnips and kale. New Zealand Journal of Experimental Agriculture 6: 151–156.

Cousins R J and Hempe J M. 1990. Zinc. Present knowledge in Nutrition. (Ed.) M L Brown. International Life Science Institute, Nutrition, Foundation, Washington, DC. pp. 251-60.

DeanW, 2006. Duck Nutrition. International Duck Research Cooperative. http://www.duckhealth.com/ducknutr.html

Dickert G Konhessar, V Piechulla, KG and Thauer RK 1981. J. Bacteriology 148: 459-464.

Droke EA and JW Spears. 1993. In vitro and in vivo immunological measurements in growing lambs fed diets deficient, marginal and adequate in zinc. J. of Nutrition. 123

Dutta A, Baruah B, Sharma B C, Baruah K K and Goswami RN. 2001. Serum macro mineral profiles in cyclic and anoestrus local heifers in lower Brahmaputra valley of Assam. Indian Journal of Animal Research 35(1): 44-46.

Friedrich B, Heine, E, Fink, A and Friedrich, C 1981. J. Bacteriology 145: 1144-1149.

Gladstones JS and Drover, DP 1962 The mineral composition of lupins. 1. A survey of the copper, molybdenum and manganese contents of lupins in the south west of Western Australia. Australian Journal of Experimental Agriculture and Animal Husbandry 2:46-53.

Grace, N.D. (1973) Effect of high dietary Mn levels on the growth rate and the level of mineral elements in the plasma and soft tissues of sheep. New Zealand Journal of Agricultural Research 16:177-180.

Grace, ND 1983 Amounts and distribution of mineral elements associated with the fleece-free empty body weight gains of the grazing sheep. New Zealand Journal of Agricultural Research 26:59-70.

Grace GD and Knowles SO 2002 A reference curve using blood selenium concentration to diagnose selenium deficiency and predict growth responses in lambs. New Zealand Veterinary Journal 50:163-165.

Grace ND, Knowles SO and West D 2006 Dose–response effects of long-acting injectable vitamin B12 plus selenium (Se) on the vitamin B12 and Se status of ewes and their lambs. New Zealand Veterinary Journal 54:67-72.

Green GH and Weeth HJ 1977. J. Anim. Sci. 46: 812-818.

Green LE, Berriatua Eand Morgan K L. 1993. Anaemia in housed lambs. Research on Veterinary Science 54: 306-11.

Henry PR 1995. Manganese bioavailability. In: Ammerman, CB, DH Baker, and A.J.
Herrick J B. 1977. A paper contributed to 1st All India Symposia on Animal Reproduction. PAD, Ludhiana.
Hibbs J Wand Pounden WD. 1955. Studies on milk fever in dairy cows. 4. Prevention by short time, prepartum feeding of massive doses of vitamin D. Journal of Dairy Science 38: 65-72.
Hideroglou M 1975. Manganese Uptake by the Ovaries and Reproductive Tract of Cycling and Anestrous Ewes. Canadian J. Physiol. & Pharmacology. 53:969.
Hidiroglou M 1979 Trace element deficiencies and fertility in ruminants: a review. J. Dairy Sci. 62:1195.
Hidiroglou Mand Knipfel J E. 1981. Maternal fetal relationship of Cu, Mn and S in ruminants. Areview Journal of Dairy Science. 64: 1637-47.
Hove K. 1986. Cyclic changes in plasma Ca and the Ca homeostatic endocrine system of the postparturient dairy cow. Journal of Dairy Science 69: 2072-82.
Humphries W R, Phillippo M, Young B W and Bremner 1. 1983. The influence of dietary iron and molybdenum on copper ,metabolism in calves. British Journal of Nutrition 49: 77-86.
Hunt CD and Neilsen FH 1981. In: Trace Element Metabolism in Man and Animals., Proc. Eedinse Fourth International Symposia, p.597.
Hunt CD, Herbel, JL and Idso JP 1994. J. Bone and Mineral Res. 9 : 171-81.
Hunt CD, Shuler TR and Nielson FH 1983. In: Spurneyelement Symposium. Abteilung Wirsenschaftlichs Publikationen, Fredrishc Schiller Universitat, Jena, Fermany, p. 149.
Jamieson S and Allcroft R 1947 Copper pine of calves. British Journal of Nutrition 4: 16-31.
Johnson JH Crokshank HR and Smalley HE 1980. Vety. Human Toxicol. 22: 248-251.
Jones LHP and Handreck KA 1965. J. Agri. Sci. (Cambridge) 65 : 129-134.
Judson GJ, Brown TH, Kempe, BR and Turnbull RK 1988 Trace element and vitamin B12 status of sheep given an oral dose of one, two or four soluble glass pellets containing copper, selenium and cobalt. Australian Journal of Experimental Agriculture 28: 299-305.
Judson GJ, McFarlane JD, Mitsioulis, A and Zviedrans, P 1997 Vitamin B12 responses to cobalt pellets in beef cows. Australian Veterinary Journal 75:660-662.
Kennedy DG, Blanchflower, WJ, Scott, JM, Weir, DG, Molloy, AM, Kennedy, S and Young, PB (1992) Cobalt–vitamin B12 deficiency decreases methioninc synthase activity and phospholipid methylation in sheep. Journal of Nutrition 122:1384-1390.
King, N, Odom, TW Sampson, HW and Pardue, SL 1993. Nutr. Res. 13:77-85.
Kokorev VA, Fedin, AS and Matrenin AP 1990. Sel'skokho Zyaistvennaya Biologiya 6: 96-105.
Kokorev, VA, Fedin, AS and Yokorlev, VV 1991. Zootekhniya 11:46-49.
Kume S E and Tanabe S. 1994. Effect of twinning and supplemental iron-saturated lactoferrin on iron status of newborn calves. Journal of Dairy Science 77: 3118-23.
Kurtoglu V, Kurtoglu F and Coskun, B 2001. Res. Vet. Sci. 71(3) :183-187.
Kurtoglu V, Kurtoglu F, Coskun B Seker, E Balevi J, Cntingul IS 2003. Nutr. Abst. Rev. 73(5) : 2302.
Lewis (eds.) Bioavailability of Nutrients for Animals. Chapter 11. Academic Press, New York, p. 239.
Lines EW 1935 The effect of the ingestion of minute quantities of cobalt by sheep affected with 'coast disease': a preliminary note. Journal of the Council for Scientific and Industrial Research, Australia 8:117–119.
Liu AC, Heinrichs BS and Leach RM Jr. 1994. Influence of manganese deficiency on the characteristics of proteoglycans of avian epiphyseal growth plate cartilage. Poultry Sci. 73:663
Lyons, M and Insko, WM Jr 1937 Chondrodystrophy in the Chick Embryo Produced by Manganese Deficiency in the Diet of the Hen. Bulletin No. 371, Kentucky Agricultural Experiment Station.

Mahadevan V and Zubairy A W. 1993. The influence of copper sulphate supplement feeding on cows for early reproductions and reducing inter calving period. Indian Journal of Animal Research 11: 23-28.

MAFF 1990 UK Tables of the Nutritive Value and Chemical Composition of Foodstuffs. In: Given, DI (ed.) Rowett Research Services, Aberdeen, UK.

Marston T. 1999. Compendium on trace mineral supplementation in beef cattle. Part 1. Nutrition Management. Kansas State University, Manhattan, Kansas.

Marston HR 1935 Problems associated with 'coast disease' in South Australia. Journal of the Council for Scientific and Industrial and Industrial Research, Australia 8: 111–116.

Mayland HF and Shewmaker GE 2001. J. Range Management 54 (4): 441-46.

McChowell J 1968. Tlte effect of experimental copper deficiency on growth, reproduction and haemopoiesis in the sheep. Veterinary Record 83: 226.

McCollum EV 1957 A History of Nutrition. Houghton Mifflin, Boston, Massachusetts.

McDowell L R, Conrad J H and Ellis G L. 1986. Mineral deficiencies and toxicities of tropical forages for grazing livestock. International Range land Development. A symposium, Orlando, FI.

McDowell, ER 1992. Minerals in Animal and Human Nutrition, Academic Press, London.

Mgongo F 0, Gombe K and Ogaa S 1. 1953. Proceedings of 4th International Congress on Animal Reproduction 5: 16.

Miller C P, Winter W H, Coates D B and Kerridge P C. 1999. Phosphorus and beef production in northern Australia. 10. Strategies for phosphorus use. Tropical Grassland 24: 239-49.

Miller JK and MillerWJ. 1962. Experimental Zn deficiency and recovery of calves. Journal of Nutrition 76: 467

Milne, JS, Whitelaw, FG, Price, J and Shand, WJ 1990. Anim. Prod. 50,507-512.

Minson, DJ (1990) Forages in Ruminant Nutrition. Academic Press, San Diego, California, pp. 208–229.

Mosdol Gand Waage S 1981. Hypocalcaemia in the ewe. Nordisk Veterinary Medicine 33: 310-26.

Neal, WM, Becker, RB and Shealy, A.L. (1931) A natural copper deficiency in cattle rations. Science 74: 418-419.

Neathery, MW, WJ Miller, DM Blackmore, FM Pate and RP Gentry. 1973. Effects of Long-Term Zinc Deficiency on Feed Utilization, Reproductive Characteristics and Hair Growth in Mature Male Goat. J. Dairy Sci. 56:98.

Nielsen, FH 1987. In : Trace Elements in Human and Animal Nutrition. Academic Press. New York, p. 245.

Nielson, FH 1980. J. Nutr. 110: 965-73.

Nielson FH 1994. In : Trace Elements in Human and Animal Nutrition, 5th edn., Vol. 2, Academic Press, New York, p. 415.

Nielson FH 1996. In: Present Knowledge in Nutrition, Filer, I.J. and Ziegler, E.e. (eds), 7th edn, International Life sciences Institute, Nutrition Foundation, Washington, D.C.

Nielson FH 1997. In Handbook of Nutritionally Essential Mineral Elements, Marcel Decker, Inc. New York, pp. 1-12.

Nielson FH and Oilerich, DA 1974. Fed. Proc. 33:1767-72.

Nielson FH and Sauberlich, HE 1970. Proc. Soc. Exper. Biol. Med. 134 : 845-849.

NRC 1994 Nutrient Requirements of Poultry, 9th edn. National Academy of Sciences, Washington, DC.

NRC 1981 Nutrient Requirements of Goats. National Academy Press. Washington, DC.

NRC 1984. Nutrient Requirements of Poultry. 8th revised ed. National Academy Press.

NRC 1985. Nutrient requirements of sheep. National Academy Press. Washington, D.C.

NRC 1989 Nutrient requirements of Dairy Cattle. National Academy Press. Washington D.C.

NRC 1994, Nutrients Requirements of Poultry, 9th edn, National Academy of Sciences, Washington, D.C.

NRC 2000. Nutrient Requirements of Dairy Cattle. National Academic Science, Washington, DC.

NRC 2001. Nutrient Requirements of Dairy Cattle. 7th rev. edn. National Academic Science, Washington, DC.

O'Dell BL 1960. Magnesium requirement and its relation to other dietary constituents. Federation, Proceedings 19: 648.

Om AE Son and Chung Kyung Won 1996. Dietary zinc deficiency alters 5a-reduction and aromatization of testosterone and androgen and estrogen receptors in rat liver. Biochemical and meolecular roles of nutrients. Journal of Nutrition 12:842-48.

Otto M Radostits C G, Blood D and Hinchcliff W K. 2000. A textbook of the diseases of animal cattle, sheep, pigs, goats and horses. Veterinary Medicine. 9th edn. WB Saunders.

Patt EL, Packet, EE and O'Dell BL 1978. Bioinorganic Chemistry 9:299-310.

Phillippo M, Humphries W R, Atkinson T, Henderson G D and Garthwaite P H. 1987. The effect of dietary Mo and Fe on eu status, puberty, fertility and oestrus cycles in cattle. Journal of Agricultural Science 109: 321-36.

Pitts WJ, Miller WJ, Fosgate T, Morson J D and Clifton C M. 1966. Effect of Zn deficiency and restricted feeding from two to five month of age on reproduction in Holstein bulls. Journal of Dairy Science 49: 995-1000.

Poston JM and Stadman TC 1975 Cobamides as cofactors: methyl cobamides and the synthesis of methionine, methane and acetate. In: Babior, B.M. (ed.) Cobalamin: Biochemistry and Pathophysiology. John Wiley & Sons, New York, pp. 111-140.

Prochazkova J, Jeman L Kracmar, S and Pipalova, S 2004. Nutr Abst Rev.74 (6):3054.

Prohaska JR 2006 Copper. In: Filer, LJ and Ziegler, EE (eds) Present Knowledge in Nutrition, 7th edn. International Life Science Institute–Nutrition Foundation, Washington, DC.

Puls R. 1994. Mineral levels in Animal Health. Diagnostic Data. 2nd edn. Sherpa International 9: 96.

Qi K Owens FN and Lu CD 1994 Effects of sulfur deficiency on performance of fiber-producing sheep and goats: a review. Small Ruminant Research 14:115-126.

Quirk MF and Norton, BW 1987 The relationship between the cobalt nutrition of ewes and the vitamin B12 status of ewes and their lambs. Australian Journal of Agricultural Research 38:1071-1082.

Quirk MF and Norton BW 1988 Detection of cobalt deficiency in lactating heifers and their calves. Journal of Agricultural Science, Cambridge 110: 465-470.

Raisbeck M Siemion RS and Smith MA 2006. Modest copper supplementation blocks molybdenosis in cattle. Journal of Veterinary Diagnostic Investigation 18: 566-572.

Rossi AF Miles RD, Damron BL and Flunker LK 1993. Poult. Sci. 72: 2124.

Sandoval M, PR Henry CB Ammerman RD Miles and RC Littell. 1997. Relative

Schwartz K 1970. Biochemical and Biophysical Research Communiation 40: 22-28.

Schaible PJ, Bandemer SL and Davidson JA 1938 The Manganese Content of Feedstuffs and its Relation to Poultry Nutrition. Technical Bulletin No. 159, Michigan Agricultural Experiment Station.

Seaboun CD and Nielson FH 1994. Biological Trace Element Research, 42:151-164.

ShannaV K, Vadoderia V P, Kevaru K S and Sidiquie GM. 1988 'Indian Veterinary Journal 65: 332.

Sharma M C, SarkarTK, Joshi C and Dwivedi P. 2001. Prevalence of macro-mineral deficiency in bovine of tarai region and its effect on haemato-biochemical profile. Indian Journal of Pathology 25(1&2): 53-56.

Sharma M C, Joshi C, Pathak N Nand Kaur H. 2005a. Copper status and enzyme, hormone, vitamin and immune function in heifers. Research in Veterinary Sciences 79(2): 113-23.

Sharma MC, Raju S, Joshi C, Kaur Hand Varshney V P. 2003. Studies on serum micro mineral, hormone and vitamin profile of buffaloes, its effect on production and therapeutic management in Haryana state of India. Asian-Australasian Journal of Animal Science 16(4): 519-28.

Sharma MC, Joshi C and Sarkar T.K. 2002a. Therapeutic efficacy of minerals supplement in macro minerals deficient buffaloes and its effect on haematobiochemical profile and production. Asian-Australasian Journal of Animal Science 15(9):1278- 87.

Simpson AM, Mills CF and McDonald I. 1981. Tissue copper retention or loss in young growing cattle. In: Howell, J.McC., Gawthorne, J.M. and White, C.L. (eds) Proceedings of the Fourth International Symposium on Trace Element Metabolism in Man and Animals. Australian Academy of Sciences,

Sjollema B 1933 Kupfermangel als Ursache von Krankheiten bei Pflanzen und Tieren. Biochemische Zeitschrift 267:151-156.

Smith, E.L. 1948 Presence of cobalt in the anti-pernicious anaemia factor. Nature, UK 162, 144-145.

Smith GS and Urquart NS 1975. J. Anim. Sci. 41:882-890.

Souri M, Galbraith H and Scaife JR 1998 Comparisons of the effects of genotype and protected methionine supplementation on growth, digestive characteristics and fibre yield in cashmere-yielding and angora goats. Animal Science 66: 217-223.

Spears JW 1984. J. Nutr. 59 : 823-835.

Spears JW, Harvey RW and Samsell LJ 1986. J. Nutr. 116 :1873-1882.

Spears JW, Hatfield EE, Ferbes RM and Koening SE 1978. J. Nutr. 108:313-320.

Spears JW, Smith CJ and Hatfield EE 1977. J. Dairy Sci. 60 : 1073-76.

Stangl GI and Kirchgessner, M 1996. Anim. Physiol. Anim. Nutr. 75:164-174.

Suttle N F 1987. The nutritional requirements for Cu in animals and man. Copper in Animals and Man. Vol. 1. (Eds) Howell Mc C and Gawthome J M. CRC Press, Boca Roton. Florida.

Suttle NF 1991. Mineral supplementation of low quality roughages. In: Proceedings of Symposium on Isotope and Related Techniques in Animal Production and Health. International Atomic Energy Agency, Vienna, pp. 101-114.

Suttle NF, Bell J, Thornton I. and Agyriaki A. 2003 Predicting the risk of cobalt deprivation in grazing livestock from soil composition data. Environmental Geochemistry and Health 25: 33-39.

Szilagyl M, anke M, Balogh i, Regius-Mocsenyji A, Euri A. 1989. Sixth International Trace Element Symposium 4: 1101-1105.

Tekeli SK and Zohouri A. 1998. Turk Veterinerlik ve Hayvancilik Dergisi, 22 :111-117. throughout Pregnancy. J. An. Sci. 60:1530.

Todd, JR 1972. Copper, molybdenum and sulfur contents of oats and barley in relation to chronic copper poisoning in housed sheep. Journal of Agricultural Science, Cambridge 79:191-195.

Todini L, Malfatti A, Valbonesi A, Trabalza-Marinucci, M. and Debenedetti, A 2007. Plasma total T3 and T4 concentrations in goats at different physiological stages as affected by energy intake. Small Ruminant Research 68:285-290.

Tohami MMEl, Mamam AM and Al UA 1997. Some studies on Cu element with special reference to the effect of Cu deficiency and treatment on reproductive organs of female rabbits. Egyptian Journal of Comparative Pathology and Clinical Pathology 10(1):53-63.

Ullrey P S, Braly P S, Whetter P A, Ku P K and Mage WT. 1965. Selenium supplementation of diet for sheep and beef cattle. Journal of Animal Science 46: 559-65.

Underwood E J and Suttle N F. 1999. The Mineral Nutrition of Livestock. 3rd edn. CABI Publishing (UK).

Underwood EJ 1977. Trace Elements in Human and Animal Nutrition, 4th edn., Acadmic Press, New York, pp. 449-451.

Underwood EJ and Filmer, JF 1935 The determination of the biologically potent element (cobalt) in limonite. Australian Veterinary Journal 11: 84-92.

Van-Niekerk FE and Van-Niekerk CH 1989b. Effect of high levels Of dietary molybdenum and sulphur on SA mutton Merino sheep 11. Certain aspects of the estrus cycle and pregnancy.

South African Journal of Animal Science 19(3): 114-0.

Wedekind KJ, AE Hortin AE and Baker, DH . 1992. Methodology for assessing zinc bioavailability: efficacy of estimates for zinc-methionine, zinc sulphate and zinc oxide. J. Anim. Sci. 70:178.

Wedeking KJ, Lewis, AJ , Giesemann, MA and Miller PS 1994 Bioavailability of zinc from inorganic and organic sources for pigs fed corn-soybean meal diets. J. Anim. Sci. 72:2681.

Weeth HJ, Speth CF and Hanks DR 1981. Am. J. Vet. Res. 42:474-477.

Weiss WP and Socha, MT 2005. Dietary manganese for dry and lactating Holstein cows. J. Dairy Sci. 88:2517.

White C L, Chandler B S and Peter D W 1981. Zinc supplementation of lactating ewes and weaned lambs grazing improved mediterranean Pastures. Australian Journal of Experimental J. Agriculture 31: 183-89.

Wilson JH, Wilson EJ and Ruszler PL 2002. Transaction of the ASAE 45 (4):1071-1074.

Wilson JH and Ruszler PL 1995. Transaction of the ASAE 38:167-170.

Wilson JH and Ruszler PL 1996. Br. Poult. Scie. 37(4):723-729.

Yadav PS, Kapoor V and Mondal AB 1996. Indian J. Anim. Sci. 66:1055-59.

9

Heat Stress and Livestock: Impact and Adaptive Mechanisms

V Sejian, V P Maurya and Gyanendra Singh

Introduction

A huge increase in the demand of animal production is expected in the next decades. Food and water security will be one of the other priorities for humankind in the 21st century. Over the same period the World will experience a change in the global climate that will cause shifts in local climate that will impact on local and global agriculture (Nardone *et al.* 2010). Global climate change is expected to alter temperature, precipitation, atmospheric carbon dioxide levels and water availability in ways that will affect the productivity of crop and livestock systems (Hatfield *et al.* 2008). For livestock systems, climate change could affect the costs and returns of production by altering the thermal environment of animals thereby affecting animal health, reproduction, and the efficiency by which livestock convert feed into retained products (especially meat and milk).

Animals live in complex environment in which they are constantly confronted with short and long term changes due to a wide range of factors such as environmental temperature, photoperiod, geographical location, nutrition and socio-sexual signals (Kleeman and Walker 2005, Ali and Hayder 2008). Homeostasis, the state of relative physiological stability in an organism, is a prerequisite to survive. Despite changes in environmental conditions, many living species have the ability to maintain their homeostasis within fixed limits by means of a set of specific innate repertoire of counter regulatory behavioral and physiological mechanisms. When the individual innate and acquired repertoire of counter regulatory mechanisms is overridden by environmental or internal perturbations a state of stress is reached and the 'stress responsive systems' are activated (Johnson *et al.* 1992, Karman 2003). The 'stress system' consists of neuroanatomical and functional structures that produce the behavioral, physiological and biochemical changes directed towards maintaining homeostasis when it is threatened (Karman 2003).

Heat stress seems to be one of the several intriguing factors making animal production of many world areas difficult. In fact the animals can adapt to the hot climates, nevertheless the response mechanisms are helpful for survival but are detrimental to performance. In farm animals exposure to different environmental stressors elicit various physiological changes. Some of these changes are under emergency involving activation of adrenomedullary system that results in the release of catecholamines. The other change is the general adaptation response, which involves activation of pituitary–adrenal axis resulting in release of corticosteroids that controls the metabolic activity. Among the glucocorticoids, cortisol is the major stress relieving hormones in ruminant species, which plays a major role in adapting to the extreme environments. Well adapted animals are characterized by minimum production loss (McManus *et al.* 2009). Environmental factors and thermoregulatory mechanisms have a significant impact on the energy exchange between the animal and the environment (Srikandakumar *et al.* 2003).

There are substantial evidences which suggests that neuroendocrine responses varies with the type of stressor and are specific and graded, rather than 'all or none'. While acute responses bring about survival; chronic responses may result in morbidity and mortality. Both of these responses are integrated via a network of mutual interplay between immune system, central nervous system and the endocrine system. Infact, it's a network that exist between nervous and endocrine system which coordinate this stress response. An important component of this network is the hypothalamo-pituitary- adrenal (HPA) axis it consists of 3 components: corticotrophin releasing hormone (CRH) neurons in the hypothalamus, corticotrophs in the anterior pituitary and the adrenal cortex (Bernabucci *et al.* 2010). The HPA axis is a critical part of this mesh which is activated by the release of several neurotransmitters and hormones. Further, understanding the cellular dynamics behind the short and long term adaptation in the tropical animals is useful in developing mitigatory measures for improving the productivity (Collier *et al.* 2006).

Heat Stress Impact on Livestock Production

Environmental stresses reduce the productivity and health of livestock resulting in significant economic losses. Livestock productivity is thought to be affected by many factors including, biomass productivity, photoperiod, geographical location, age, breed, nutrient availability, water availability, management practices, environmental conditions so on and so forth (Khalifa 2003). Of all these factors, environmental condition influencing livestock productivity is of major concern (Shelton 2000, Koubkova *et al.* 2002). Among the environmental variables, heat stress seems to be the most detrimental factor affecting livestock production (Rivington *et al.* 2009). Heat stress can cause a significant financial burden to livestock producers by decreasing milk and meat production, decreasing reproductive efficiency, and adversely affecting livestock health. Heat stress is a significant issue for livestock grazing in the tropics and sub tropics and the effects of heat stress may aggravate with prospects of global warming caused by the accelerating emission of GHGs. Heat stress affects animal performance and productivity at all stages of the lifecycle. The adverse effect of heat stress in livestock may be attributed to repartitioning of energy necessary for maintenance of homeothermy. The impact encompasses decreased growth, reduced milk yield, decreased reproduction, increased susceptibility to diseases, and delayed initiation of lactation. Heat stress in cattle reduces, feed intake and growth, and in extreme cases can cause death, resulting in substantial revenue loss to producers. Heat stress also negatively affects reproductive function (Maurya *et al.* 2004, Sejian *et al.* 2011a). Infact reproductive inefficiency is one of the most costly production-limiting problems facing the livestock industry. Heat stress causes infertility in farm animals and this represents a major source of economic loss to the farmers. Reproductive processes in both the male and female are sensitive to environment. As a general rule, increased temperature decreases ovulation rates, shortens duration of estrus, decreases fertility of males, and increases rate of embryonic mortality. Given the associated economic losses through heat stress in several livestock species, additional research is needed on the interactions among heat stress, nutritional requirements, immunological status,

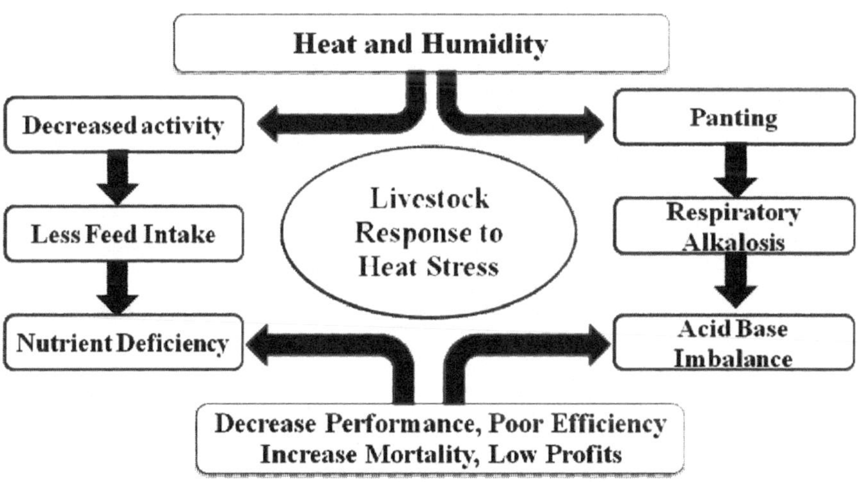

Fig. 1: The effects of elevated temperature on livestock

and the overall livestock performance. Figure 1 describes the mechanisms by which heat stress affects production in livestock.

Consequences of Heat Stress to Livestock

The livestock sector is socially, culturally and politically significant both nationally and globally. It accounts for 40% of the world's agricultural Gross Domestic Product (GDP). It employs 1.3 billion people, and creates livelihoods for one billion of the world's population living in poverty (Gaughan *et al.* 2010). Global demand for livestock products is expected to double during the first half of this century (FAO 2009), as a result of the growing human population, and its growing affluence. Hence it's of utmost importance to concentrate on improving the productivity of livestock to meet the growing needs of human population.

Although most livestock possess well developed mechanisms of thermoregulation, they may not be able to adequately maintain homeothermy under heat stress, especially the high producing animals (Gaughan *et al.* 2010). Indeed, the hyperthermia negatively affects any form of productivity, regardless of breed, and stage of adaptation (Marai *et al.* 1999). An understanding of the control of body temperature in livestock under heat loads, and the relationship of this to productivity must come from an approach in which the animal is viewed in relation to both its thermal and nutritive environments (Sejian *et al.* 2010a). Exposure to elevated ambient temperature evokes a series of drastic changes in

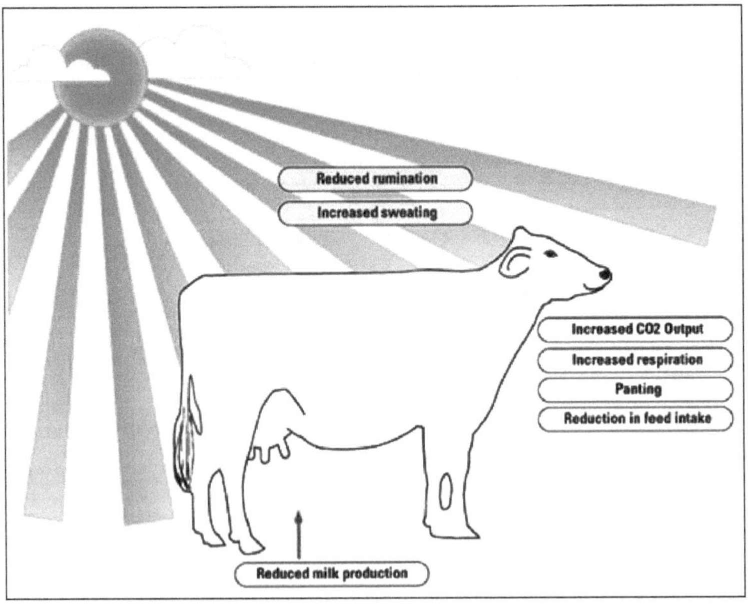

Fig. 2: Effect of heat stress on cow and milk production

the livestock biological functions that include depression in feed intake efficiency and utilization; disturbances in metabolism of water and protein; and alteration in energy, and mineral balances, enzymatic reactions, hormonal secretions and blood metabolites. Such physiological changes lead to a low live body weight and impaired reproduction, i.e. depression in age at puberty, reproductive activity and fertility (Marai *et al.* 2007). As a result the production potential of the livestock species are directly under threat leading to severe economic losses. Figure 2 describes the effects of heat stress on cow and its consequences on milk production.

Heat Stress Severity Measurement for Livestock

The severity of heat stress on livestock production can be measured using certain weather parameters, which measure the intensity of heat stress. The most noteworthy weather parameters included in such heat stress severity measurements were the ambient temperature (including dry and wet bulb temperature) and relative humidity. Few formulas developed by various researchers for measuring intensity of heat stress during summer season have been successfully used under hot semi-arid condition of Rajasthan (Naqvi and Hooda 1990). Certain mathematical formulas have been designed by various researchers and research institutes for measuring the severity of heat stress. All these formulae calculate the heat stress severity in terms of score on a 100 scale and finally these scores are compared on a standard 'Temperature Humidity Index' (THI) chart to assess the stress imposed by the impending climatic condition on livestock over a period of time. The THI is used as a guide to measure heat stress and it combines the effects of temperature and humidity into one value. There are three stress categories (temperatures given in Fahrenheit): Livestock alert is 75-78 degrees, Livestock danger is 79-83 degrees, and Livestock emergency is 84+ degrees. The higher the humidity the lower is the temperature for the Livestock alerts; for danger and emergency levels to occur. Table 1 describes the different THI categories and the corresponding responses of cattle.

One of the major limitations of THI is that it does not account for wind velocity and solar radiation. Devising a single index encompassing all the components of environment and use of such index in predicting the climatic stress would be desirable. Several attempts have been made over the past to develop a satisfactory and practical index to describe the climatic stress but most have fallen short of expectations. The simple reason for this drawback is the difficulty to measure the combined influence of temperature, humidity, radiation, wind velocity and their diurnal variation on various species of livestock because of experimental complexity.

Table 1: Description of different categories of THI and its corresponding response and effects in cattle

THI	Stress Level	Effect and symptoms in cattle
<72	None	
-72-79	Mild	• Dairy cows will adjust by seeking shade. • Respiration rate and dilation of the blood vessels will increase. • The effect of milk production will be increase.
80-89	Moderate	• Both saliva production and respiration rate will increase. • Feed intake will decrease and water consumption will increase. • There will be increase in body temperature. • Milk production and reproduction will be decreased.
90-98	Severe	• Cows will become very uncomfortable • Rapid respiration (Panting) and excessive saliva production. • Marked decrease in milk production and reproduction
>98	Danger	• Potential cow deaths can occur.

Heat Stress and Livestock Reproduction

Heat stressed cows have reduced duration and manifestations of estrus, reduced LH pulse and amplitude and increased incidence of anestrus or silent ovulation. Gonadotropin secretion is markedly reduced by high ambient temperatures (Gilad *et al.* 1993). Besides producing less GnRH, the sensitivity of the pituitary to GnRH is also reduced by heat stress (Gilad *et al.* 1993). In vitro studies using isolated follicles from Holstein heifers showed that heat stress decreases gonadotropin-induced estradiol synthesis and increases progesterone synthesis (Sejian *et al.* 2011, Sejian *et al.* 2012). Similar findings have been observed invivo where acute heat stress reduces estradiol production by granulose cells (Sejian *et al.* 2011). This altered balance between estradiol and progesterone causes premature luteinization of follicles and decreases ovulation and fertility (Sejian *et al.* 2012). In males, the process of spermatogenesis is highly temperature sensitive. Normal spermatogenesis requires 4-5^{0}C less than the core body temperature. Hence in heat stressed bulls, the percentage of viable and motile spermatozoa is reduced and the number of sperm abnormalities is increase (Rahman *et al.* 2011). Thus, heat stress reduces conception rate because of poor quality semen.

Furthermore, heat stress alters the uterine environment making it unfavorable for successful implantation. Heat stress during early gestation reduces uterine blood flow and embryonic attachment to the uterine wall thus favoring embryonic loss. Also, chronic heat stress decreases oxygen consumption by the fetal placenta with retardation in placental growth resulting in fetal growth restriction. Similar to high environmental temperatures, long term exposure to low temperatures can decrease gonadotropins as well. This could be attributed to shifting of resources to energy conservation, increased thermogenesis at low temperatures, decreased fat deposition, reduced leptin secretion and decreased reproductive functions.

Mechanisms of heat stress affecting reproduction

- The principal biological mechanism by which heat stress impacts on livestock reproduction is partly explained by reduced feed intake; but also includes altered endocrine status, reduction in rumination and nutrient absorption, and increased maintenance requirements resulting in a net decrease in nutrient/energy availability for reproduction
- Stress (heat, nutritional, pH, immunological, physiological, *etc.*) prevents the basic cell function by causing improper folding of proteins. To cope with this, expressions of HSPs like HSP70 are stimulated/ upregulated. The basic function of HSPs is to help in 'proper folding' of proteins so that the three dimensional structure of the protein is not compromised and thus maintaining its normal function. Further, the cell diverts its entire transcriptional and translational machinery to synthesize proteins that are required for maintaining 'house keeping functions' like cellular respiration, ATP synthesis, excretion of protons (H+ ions) at the expense of other functions of the cell, *e.g.* testosterone production. To combat stress, the Adrenocorticotropic hormone (ACTH) - Cortisol axis is activated and cortisol and other glucocorticoids are produced to enhance stress tolerance.
- Glucocorticoids are anti-inflammatory. They prevent the expression of cyclo-oxygenase-2 (cox2) and lipoxygenase which gives rise to prostaglandins and leukotrienes, respectively. Luteinizing hormone (LH) surge induces cox2 expression and PGE2 production. PGE2 is essential for ovulation but may be inhibited by glucocorticoids.
- The growing foetus is under stress due to space and nutrient constraint. This fetal stress activates ACTH-Cortisol axis. The fetal cortisol enters maternal circulation and activates the expression of 17 alpha hydroxylase enzymes in the placenta. This enzyme converts progesterone into estrogen. Declining progesterone levels with concomitant increase in estrogen levels induces parturition. Injection of glucocorticoids (dexa, beta, prednisone, triamcenalone, etc.) any time after first month of pregnancy induces abortion in sheep and goat. Glucocorticoids prevent the 'maturation of placentome'. Therefore, if parturition is induced with steroids, retained fetal membranes are common sequel.

- Cortisol 'stabilizes' the lysosomal membrane and thereby prevents the release of vasoactive substances (like histamine, serotonin, heparin, substance P, proteolytic enzymes) and subsequently prevents inflammatory response. Most of the common reproductive events like ovulation, luteolysis, implantation, parturition and involution of the uterus are 'physiological inflammation'. You may expect an adverse effect on these events. Reproduction is basically a 'luxurious phenomenon' and appropriate when the animal is in just right homeostasis. In case of severe stress, reproduction is typically the first physiological event to let go by the body.
- In female animals, stress can inhibit 3â HSD and minimize progesterone secretion from luteal cells. Aromatase, an enzyme present in the granulosa cells, may be inhibited to induce follicular atresia and anestrus. In male animals, stress adversely affects spermatogenesis perhaps by inhibiting the proliferation of spermatocytes. At cellular level, testosterone is converted into dihydrotestosterone (bioactive form of testosterone) by 5 alpha reductase and stress may adversely affect the expression of this enzyme.

Livestock Adaptation Mechanisms

High ambient temperatures, solar radiation, and humidity are environmental stressing factors that affect animals. Even when ruminants possess well developed mechanisms of thermoregulation, they could not able to maintain homeothermy under heat stress, especially the high producing animals. The evidences show that hyperthermia negatively affects any form of productivity, regardless of breed, and stage of adaptation. An understanding of the control of body temperature in cattle under heat loads, and the relationship of this to productivity must come from an approach in which the animal is viewed in relation to both its thermal and nutritive environments. One of the most important effects of increased body temperature is the depression of the metabolic rate as a result of reduced appetite and the decreased reproductive efficiency. However, animal response should not be attributed solely to lower intake or to forage quality. There is a direct physiological effect, mostly mediated by the effect high environmental temperatures exert on the concentration of different hormones.

When environmental conditions change, an animal's ability to cope (or adapt) to the new conditions is determined by its ability to maintain performance and oxidative metabolism (Pörtner and Knust 2007). The stress response is influenced by a number of factors including: species, breed, previous exposure to the stressor, health status, performance, body condition, mental state and age. Insufficient acclimatisation or adaptation would determine what an animal experiences as stressful. Subsequent acclimatisation or adaptation may alleviate the stress response (Kassahn *et al.* 2009), but performance may not return to pre-stress levels. Animal response to temperature stress (which could be considered as part of the adaptation process) usually results in a loss of performance (e.g. growth or reproduction) before cellular and molecular stress responses are activated (Kassahn *et al.* 2009). This suggests that the use of biological stress markers as an aid in selection may be limited. And indeed little progress has been made in this area to date. Typically the first noticeable livestock response to a heat challenge is a decrease in feed intake. A reduction in feed intake has a direct impact on performance, for example reductions in milk yield, milk quality, meat and egg production. In addition, the cellular and molecular stress responses will have an effect on mammary gland metabolism, energy partitioning, immune status and reproductive success. The effects on reproduction efficiency may be long term e.g. months rather than days (Santolaria *et al.* 2010). All of these changes lead to economic losses in livestock systems (St Pierre *et al.* 2003). Biological effects on livestock due to global warming are not predictable simply in terms of a response to increased heat load (Marcogliese 2001). Among other things, increasing temperature may also increase exposure, and susceptibility of animals to parasites and disease (Marcogliese 2001), especially vector-borne diseases (Tabachnick 2010). However, little effort has been dedicated to understanding the potential impact of climate change on parasite populations and subsequent effects on animal production (Tabachnick 2010). Table 2 describes the different type of heat stress responses exhibited by livestock to counter heat stress.

Table 2: Different types of adaptation responses exhibited by livestock after subjecting to heat stress

Type of Target organ	Action elicited	Type of response
Hypothalamus	Decrease feed intake, increase water intake, shade seeking, changing posture, wallowing	Behavioral response
HPA Axis	Increase CRH, increase ACTH, increase cortisol, decrease TSH, increase arginine vasopressin	Neuroendocrine response
Cell	Increase HSF-1, increase HSP, increase thermal tolerance gene expression	Cellular response
Heart, Lung, Liver, Skin	Increase respiration rate, increase pulse rate, increase rectal temperature, increase sweating rate,	Physiological response
Thyroid gland	Decrease T_3, decrease T_4	Metabolic response
Blood	Increase PCV, increase Hb, increase glucose, decrease protein, decrease cholesterol	Biochemical response

1. Processes of thermal adaptation

Adaptation is a progressive reduction in physiological strain in an animal when the animal is repeatedly exposed to stress (Taylor 2006). Within a limited range of body temperatures, animals have some capacity to modify their behavioural, physiological, or morphological characteristics in response to a thermal challenge (Angilletta 2009). Thermo tolerance is a cellular adaptation that is a result of a single nonlethal heat exposure that allows an animal to survive subsequent insults that would otherwise be lethal (Moseley 1997). Indeed, the precise physiological and biochemical mechanisms that define the upper and lower thermal tolerance limits in animals are largely unknown (Hofmann and Todgham 2009). The response mechanisms employed by an animal when exposed to stressors are helpful for survival but are usually detrimental to productive performance which means that animals that are actually adapting to specific changes may not be selected because they are doing so at the expense of production.

2. Genetic adaptability

Genetic improvement is an evolutionary action; evolution should be defined as a continuous process of adaptation of the populations of organisms to the ever-changing geological, biological and climatic conditions. Because of the almost infinite number of combinations of environmental factors organisms must have a great variety of genetic types that can deal with a range of climatic, nutritional, or other conditions. In a word, any population must be genetically heterogeneous - i.e., with a great genetic diversity - in order to be able to survive under the challenge of the changing environment. This is the basis for the livestock genetic improvement. Genetic selection has been a traditional method to reduce effects of environment on livestock by development of animals that are genetically adapted to hot climates. There are additional possibilities for meeting this goal. Identification of specific genes that control traits related to thermo tolerance makes it possible to select for thermal resistance without inadvertently selecting against milk yield. Possibly, there are also genes controlling cellular resistance to heat shock. Integration of marker-assisted selection into animal breeding systems should make selection for traits conferring thermotolerance more rapid.

a. *Behavioural adaptations*

Behaviour is one of the most effective adaptive mechanisms, at least in the short term. Behavioral change allows animals to avoid or reduce the impact of stressors. However not all behavioral changes have a positive impact. For example, there may be increased aggressive behaviour when cattle, with limited shade or water access, are exposed to high heat load i.e. the dominant animals will push less dominant ones from shade, or they may block water trough access. However, most behavioral changes would be considered to be attempts to reduce the impact of a given stressor. These changes include: seeking shelter (e.g. shade or wind breaks), changing posture (e.g. standing or altering orientation to the sun to reduce gain of heat from solar radiation), wallowing, reducing feed intake (when exposed to hot conditions) or increasing feed intake (when exposed to very cold conditions), and increasing or decreasing water intake again based on ambient conditions. Other behavioural changes such as bunching, by heat stressed cattle, may be a psychological response to the stress and is not considered to be a positive behaviour. It is largely thought that bunching will decrease heat loss from already heat stressed cattle. On the other hand, huddling in pigs and puppies is a positive behavioural adaptation to cold stress. Behavioral changes are an innate component of the animal. These behaviors may be enhanced by human selection within a breed or species. However, as with many other adaptive responses, there may be subsequent production losses. Understanding the 'normal' behaviors of animals under non stressful conditions is paramount for our understanding of behavioral changes when animals are stressed.

b. *Neuro-endocrine mechanism of adaptation*

Neuroendocrine responses to stress play an integral role in the maintenance of homeostasis in livestock. In general, activation of the stress circuitry inhibits functions such as growth and reproduction. Substantial evidence suggests that neuroendocrine responses varies with the type of stressor and are specific and graded, rather than 'all or none'. While acute responses have important adaptive functions and are vital to coping and survival, chronic stressors elicit endocrine responses that may actually contribute to morbidity and mortality. Integration of these responses is possible through the network of mutual interactions that exist between the immune system, the central nervous system and the endocrine system. A crucial component of this network is the stress axis. Activation of the stress axis is accomplished through the release of several neurotransmitters and hormones. The stress axis or the hypothalamo-pituitary- adrenal (HPA) axis consists of 3 components: corticotrophin releasing hormone (CRH) neurons in the hypothalamus, corticotrophs in the anterior pituitary and the adrenal cortex. A variety of molecular mediators have been implicated in the stimulation of CRH neurons ranging from neurotransmitters such as catecholamines to pro-inflammatory cytokines. CRH is an obligatory and primary stimulus for adrenocorticotropin hormone (ACTH) secretion by the pituitary gland. Subsequently, ACTH stimulates glucocorticoid synthesis from the cells of adrenal cortex. Glucocorticoid is the final activation product in the HPA axis and the primary effector molecule of this neuroendocrine circuit. The secretion of ACTH, which is very crucial in this neuroendocrine response, seems to be regulated by a variety of peptides, but principally by CRH and vasopressin (VP; arginine vasopressin in most mammalian farm animal species, lysine vasopressin in pigs). CRH seems to be active mainly in the acute phase of stress while VP is proposed to maintain HPA axis activity after repeated stimulation. In addition to the more traditional regulation of pituitary corticotrope function by hypothalamic peptides, pro-inflammatory cytokines are now recognized to play an important regulatory role in the HPA axis. Currently, interleukin-1, interleukin-6, and tumor necrosis factor are implicated strongly as stimulators of the stress axis. Cytokines are also capable of stimulating the secretion of the hormone leptin from adipocytes. Leptin is now recognized as an inhibitor of stress axis activity. Therefore, both leptin and glucocorticoids complete the negative feedback circuit to suppress stress axis activity and maintain homeostasis. We will explore the relationship between these hormonal mediators, immune cytokines, brain neurochemicals and stress axis activation and the implications for livestock production.

c. *Physiological adaptability*

To improve production, apart from infusion of high producing genes, the agro-climatic conditions have an important role to play by influencing the physiological mechanism(s). Physiological responses in homeotherms are important indicators of homeostasis. Normal fluctuations in physiological responses, i.e. respiration rate, pulse rate, rectal temperature and sweating rate vary with the changes in season in an effort to maintain normal body temperature independent of the fluctuation in environmental temperature. Hence these are considered important indices for comparative adaptability of different genotypes. Increased respiration rate is the first reaction when animals are exposed to environmental temperatures above the thermoneutral zone (Maurya *et al.* 2007). This response ensures direct heat stimulation of the peripheral receptors, which transmit nervous impulses to the heat centre in the hypothalamus. With the increasing environmental temperature, the respiration rate continues to rise linearly until a certain temperature afterwards rate of respiration slows down and depth of breathing increases. At this level, the respiratory muscle reach a threshold of maximum activity and thus the respiration rate cannot increase further. Respiration rate can be an indicator of heat stress (Habeeb *et al.* 1992).

Ambient temperature is known to influence the body temperature and cardio respiratory responses of livestock. Native had lower rectal temperature and pulse than exotic breeds, while crossbred had intermediate values. The magnitude of physiological responses evoked by thermal stress and time required for their subsequent return to normalcy after removing the stress were considered useful indices for assessing the thermal ability of animals. The physiological response under hot semi-arid environment varied enormously during forenoon and afternoon. This is because the animals maintained under semi-arid condition are usually exposed to high ambient temperatures during hot summer months. The increase in ambient temperature through the morning and towards the afternoon may explain the reason for rise in the physiological responses in the afternoon when compared to morning values (Maurya *et al.* 2004). The type of nutrition consumed and body condition scoring (BCS) also seem to affect the respiration rate under heat stress conditions (Sejian *et al.* 2010c). Among the physiological response, respiration rate showed significant changes with respect to the different BCS than the pulse rate and rectal temperature. The reason for this could be that the homoeothermic animals initially react to stress by enhancing the thermoregulation mechanism, such as respiration rate, to avoid an increase in rectal temperature. These findings are in agreement to the previous study conducted in our laboratory by Maurya *et al.* (2003).

d. *Cellular and molecular mechanisms of livestock adaptation*

Heat tolerance at the cellular level is directly related to the ability of the cell to maintain elevated levels of heat shock proteins (HSP's). As stated by Horowitz and Assadi (2010) " A hallmark of the acclimation process is the enhancement of cytoprotective networks-that of the heat shock proteins, anti-oxidative and apoptotic-and the stabilization of the Hypoxia Inducible Factor- Iá, the master regulator of oxygen homeostasis". Thermal acclimation and thermal adaptation is associated with increased basal levels of HSPs (Maloyan *et al.* 1999). Thermal acclimation and cyclopentenone prostaglandins have been shown to increase the DNA-binding activity of HSF-1 leading to increased HSP gene transcription (Buckley and Hofmann 2002). Several investigators have shown that heat acclimation also provides cross-tolerance against other types of stress such as hypoxia, ischemia, acidosis and energy depletion, (Kregel *et al.* 2002) and that HSF-1 is involved in this process. Although there is little evidence for endocrine regulation of HSF-1 gene expression activity, there is substantial evidence that expression of heat shock proteins and other cytoprotective proteins are modulated by the endocrine system.

Although there is a good amount of knowledge about the physiological aspects, the effects of heat stress at the cellular and genetic level are still being unrevealed. Functional genomics research is providing new knowledge about the impact of heat stress on livestock production and reproduction. Using functional genomics to identify genes that are up- or down regulated during a stressful event can lead to the identification of animals that are genetically superior for coping with stress and toward the creation of therapeutic drugs and treatments that target affected genes. Given the complexity of the traits related to adaptation to tropical environments, the discovery of genes controlling these traits is a very difficult task. One obvious approach of identifying genes associated with acclimation to thermal stress is to utilize gene expression microarrays in models of thermal acclimation to identify changes in gene expression during acute and chronic thermal stress. Another approach will be to use consomic lines with single gene deletions exposed to a defined thermal environment. This permits the identification of those genes that are involved in key regulatory pathways for thermal resistance and thermal sensitivity. Finally, gene knockout models in single cells will also allow better delineation of the cellular metabolic machinery required to acclimate to thermal stress. Those genes identified as key to the process of thermal acclimation will then need to be mapped to their chromosomal location and the sequences of these genes will need to be determined in order to see if there are single nucleotide polymorphisms (SNPs) that are associated with changes in the coding for gene expression or protein function. Identification of SNP's that are associated with variation in animal resistance or sensitivity to thermal stress will permit screening of animals presence or absence of desirable or undesirable alleles. However, further research is needed to quantify the genetic antagonism between adaptation and production traits to evaluate the potential selection response. With the development of molecular biotechnologies, new opportunities are available to characterize gene expression and identify key cellular responses to heat stress. These new tools will enable to improve the accuracy and the efficiency of selection for heat tolerance. Epigenetic regulation of gene expression and thermal imprinting of the genome could also be an efficient method to improve thermal tolerance.

e. *Metabolic response*

The major exogenous regulator of thyroid gland activity is the environmental temperature (Dickson 1993). Heat stress in general is associated with significant depression in thyroid gland activity resulting in lowering of thyroid hormones level (Nazifi *et al.* 2003, Rasooli *et al.* 2004). In addition blood thyroid hormones are considered to be good indicators of the nutritional status of an animal (Riis and Madsen 1985). Reduced thyroid hormone level after exposure to heat stress is the adaptive response exhibited by livestock to counter the heat stress (Sejian *et al.* 2010a). Indirectly this reduced thyroid hormone levels acts as the metabolic signal for reduced feed intake. All these activities of the livestock work towards reducing the metabolic heat load which will further aggravate the heat stress effects on these animals.

f. *Blood biochemical response*

Mechanism of homeostasis in an animal body exposed to heat will involve metabolic reactions which in part will reflect in changes in blood constituents as well as in body fluid compartments. Environmental temperature influences the extracellular fluid volume, plasma volume, red cell volume and water intake in livestock (Sejian *et al.* 2008). Seasonal variations in the hemoglobin concentration and PCV in heat stressed livestock has been reported (Sejian *et al.* 2010a, Sejian *et al.* 2010b). Both Hb and PCV increases significantly in heat stressed animals. This could be attributed to the severe haemoconcentration that arise due to dehydration in these animals (Sejian *et al.* 2010a). However, during thermal stress when the animals has the *ad libitum* access to water then there is a decrease in the hemoglobin concentration (Maurya *et al.* 2007), which could be attributed to hemodilution and increase in the blood volume of these animals. The packed cell volume (hematocrit percentage) also decreases in heat stressed animals provided with *ad libitum* water either due to red cell destruction or hemodilution (Marai *et al.* 1991).

Several researchers have studied the effect of thermal stress on the blood glucose concentrations but the results reported by them were conflicting. The decrease in glucose level during heat exposure relates to decrease in concentration of insulin and thyroxin, which are closely associated to energy metabolism during heat exposure (Sejian *et al.* 2010a). Decrease in plasma glucose could also be due to the marked dilution of blood and body fluid as a whole or increase in the plasma glucose utilization to produce more energy for greater muscular expenditure required for high muscular activity (Sejian *et al.* 2010a). Other study showed that glucose concentration is increased during thermal exposure of animals because there was decrease in the glucose utilization, depression of the catabolic and anabolic enzyme secretions leading to subsequent reduction of metabolic rate, and also due to breakdown to glycogen into free glucose. Serum proteins, particularly albumin concentration, usually decrease in heat exposed cattle. A negative correlation between plasma protein and elevated environmental temperature has also been reported in some studies. The decrease in the plasma protein may be due to dilution of plasma proteins and decrease in protein synthesis as a result of depression of anabolic hormone secretion. In addition to this, there is also increase in glucocorticoids concentration. The reduced feed intake in heat stressed animals also leads to decrease in nitrogen concentration in the body, which leads to low protein level. The serum total lipid concentration was also decreased in the animals exposed to prolonged thermal stress, which might be due to increase in utilization of fatty acids for energy production as a consequence of decrease in glucose concentration (Kamal *et al.* 1989). Plasma free fatty acids, total lipids and cholesterol concentration were decreased during prolonged exposure of hot environment. Lower levels of ruminal ammonia and blood urea nitrogen were also reported in Merino sheep maintained at high temperature (35°C) than those of same animals when kept at low temperature (15°C).

In a study conducted on Iranian fat tailed sheep, reduced serum concentrations of total protein, glucose, cholesterol, total lipid, calcium, inorganic phosphorus, magnesium, creatine kinase, and lactate dehydrogenase were reported after thermal exposure (Nazifi *et al.* 2003). They also reported increased levels of serum aspartate aminotransferase and alanine aminotransferase in heat stressed animals. Packed cell volume, total plasma proteins RBC and haemoglobin concentration had high positive correlation between each other indicating the significance of these parameters for heat tolerance in Brazilian sheep (Mc Manus *et al.* 2009). Al-Haidary (2004) reported significant increase in PCV while there was no effect on haemoglobin concentration after exposed to heat stress in Naimey sheep. Srikandakumar *et al.* (2003) conducted an experiment on Omani and Merino sheep and observed varying degrees of changes in blood chemistry. They established significant reduction in plasma calcium, creatinine phosphokinase and aspartate aminotransferase after imposing both the breeds to heat stress. Heat stress also increased BUN in Omani while it was decreased in Merino sheep. Likewise, heat stress increased the plasma glucose in Merino while it was decreased in Omani breed. These results indicated that Omani sheep was more heat tolerant than Merino sheep.

Heat stress causes an increase in the activities of serum glutamic oxaloacetic transaminase (SGOT) and serum glutamic pyruvic transaminase (SGPT) due to increased stimulation of gluconeogenesis by corticoids (Sejian and Srivastava 2010). The significant reduction in alkaline phosphatase by heat stress which was attributed to the reduction in thyroid hormone that occurred under heat stress. However, acid phosphatase generally is not affected more in heat stressed animals. Lactic acid concentrations were increased at 35°C and 40°C exposure. Blood plasma vitamin A level was decreased as the ambient temperature increased while atmospheric pressure was decreased in rams but in ewes it did not show specific trend. Further the activity of plasma acid phosphates and LDH were comparatively higher during summer than spring season.

Conclusion

The stress response initiated by a particular stressor integrates a variety of homeostatic mechanisms ranging from physiological, immune to behavioral adaptations. Allostasis helps the animal to maintain these adaptations in the face of adverse, constantly fluctuating environmental demands. Both phenomena are tightly coordinated by the brain and involve the whole body. Adaptation to prolonged stressors will most likely be accompanied by a production loss, and input costs may also increase. Increasing or maintaining current production levels in an increasingly hostile environment is not sustainable. It may make better sense to look at using adapted animals, albeit with lower production levels (and also lower input costs) rather than try to infuse 'heat tolerance' genes into a non-adapted breed. As a result, effective stress management by the animal leaves no impact on livestock productivity. On the other hand, inability of the animal to meet the demands posed by chronic stressors produces a major biological consequence to the animal impairing growth, productivity, reproduction and survivability. It is therefore essential to use good management practices, promote the welfare of livestock to help them maintain optimal responses to stressors.

References

Al-Haidary AA 2004. Physiological responses of Naimey Sheep to heat stress challenge under semi-arid environments. International Journal of Agriculture and Biology 2: 307-309.

Ali A, Hayder M 2008. Seasonal variation of reproductive performance, foetal development and progesterone concentration of sheep in the subtropics. Rep Dom Anim 43: 730-734.

Angilletta MJ 2009. Thermal adaptation – a theoretical and empirical synthesis. Oxford University Press Inc., New York.

Bernabucci U, Lacetera N, Baumgard LH, Rhoads RP, Ronchi B, Nardone A 2010. Metabolic and hormonal adaptations to heat stress in domesticated ruminants. Animal 4:1167–1183.

Buckley BA, Hofmann GE 2002. Thermal acclimation changes DNA-binding activity of heat shock factor 1 (HSF1) in the goby Gillichthys mirabilis: implications for plasticity in the heat-shock response in natural populations. J Exper Biol 205:3231-3240.

Collier RJ, Stiening CM, Pollard BC, VanBaale MJ, Baumgard LH, Gentry PC, Coussens PM 2006. Use of gene expression microarrays for evaluating environmental stress tolerance at the cellular level in cattle. J Anim Sci 84(E Suppl):E1–E13.

Dickson WM 1993. Endocrine glands. In: Swenson MJ, Reece WO (eds) Dukes Physiology of domestic animal, 11th edition, Comstock publishers association, Ithaca, London, pp 629-664.

FAO 2009. State of Food and Agriculture - Livestock in the balance. Rome, FAO.

Gaughan JB, Mader TL, Holt SM, Sullivan ML, Hahn GL 2010. Assessing the heat tolerance of 17 beef cattle genotypes. Int J Biometeorol 54(6):617-27.

Gilad E, Meidan R, Berman A, Graber Y, Wolfenson D 1993. Effect of heat stress on tonic and GnRH-induced gonadotrophin secretion in relation to concentration of oestradiol in plasma of cyclic cows. J.Reprod. Fertil. 99: 315-321.

Habeeb AA, Marai IFM, Kamal TH 1992. Heat stress. In: Philips, C., Piggens, D. (Eds.), Farm Animals and the Environment. C.A.B. International, pp. 27–47.

Hatfield J, Boote K, Fay P, Hahn L, Izaurralde C, Kimball BA, Mader T, Morgan J, Ort D, Polley W, Thomson A, Wolfe D 2008. Agriculture. In: The effects of climatechange on agriculture, land resources, water resources, and biodiversity in the UnitedStates. A Report by the U.S. Climate Change Science Program and the Subcommittee on Global Change Research. Washington, DC., USA, 362 pp.

Hofmann GE, Todgham AE 2009. Living in the now: physiological mechanisms to tolerate a rapidly changing environment. Annu Rev Physiol., 72:22.1 – 22.19.

Horowitz M, Assadi H (2010) Heat acclimation-mediated cross-tolerance in cardioprotection: Do HSP70 and HIF-1a play a role? Ann N Y Acad Sci 1188:199-206.

Johnson EO, Kamilaris TC, Chrousos GP, Gold PW 1992. Mechanisms of stress: a dynamic overview of hormonal and behavioral homeostasis. Neurosci and Biobehav Rev 16(2):115-130.

Kamal TH, Habeeb AA, Abdel-Samee AM, Abdel-Razik MA 1989. Supplementation of heat-stressed Friesian cows with urea and mineral mixture and its effect on milk production, in subtropics. Proceedings of International Symposium on the Constraints and Possibilities of Ruminant Production in the Dry Subtropics. Cairo, Egypt. pp 183-185.

Karman AG 2003. Neroendocrine adaptation to stress in pigs.CRH and vasopressin in the paraventricular nucleaus. PhD Thesis, Department of animal Science, Wageningen University, The Netherlands.

Kassahn KS, Crozier RH, Portner HO, Caley MJ 2009. Animal performance and stress: responses and tolerance limits at different levels of biological organisation. Biol Rev 84:277–292.

Khalifa HH 2003. Bioclimatology and adaptation of farm animals in a changing climate. In: Interactions between climate and animal production, EAAP Technical Series 7: 15-29.

Kleemann DO, Walker SK 2005. Fertility in south Australian commercial Merino flocks:

relationships between reproductive traits and environmental cues. Theriogenology 63: 2416-1433.

Koubkova M, Knížková I, Kunc P, Hartlova H, Flusser J, Dolezal O 2002. Influence of high environmental temperatures and evaporative cooling on some physiological, hematological and biochemical parameters in high-yielding dairy cows. Czech J Anim Sci 47(8): 309–318.

Kregel KC 2002. Molecular Biology of Thermoregulation Invited Review: Heat schock proteins: modifying factors in physiological stress responses and acquired thermotolerance. J Appl Physiol 92:2177-2186.

Maloyan A, Palmon A, Horowitz M 1999. Heat acclimation increases the basal HSP72 level and alters its production dynamics during heat stress. Am J Physiol 276:1506–1515.

Marai IFM, El-Darawany AA, Fadiel A, Abdel-Hafez MAM 2007. Physiological traits as affected by heat stress in sheep – A review. Small Rumin Res 71: 1-12.

Marai IFM, Habeeb AAM, Farghaly HM 1999. Productive, Physiological and Biochemical Changes in Imported and Locally Born Friesian and Holstein Lactating Cows under Hot Summer Conditions of Egypt. Trop Anim Health Prod 31: 233-243.

Marai IFM, Abdel-Samee AM, El-Gafaary MN 1991. Criteria of response and adaptation to high temperature for reproductive and growth traits in rabbits. Options Mediterraneennes Series A 17:127-134.

Marcogliese DJ 2001. Implications of climate change for parasitism of animals in the aquatic environment. Can J Zool 79:1331 – 1352.

Maurya VP, Naqvi SMK, Joshi A, Mittal JP 2007. Effect of high temperature stress on physiological responses of Malpura sheep. Indian Journal of Animal Sciences 77: 1244-1247.

Maurya VP, Naqvi SMK, Mittal JP 2004. Effect of dietary energy level on physiological responses and reproductive performance in Malpura sheep in hot semi-arid region of India. Small Rumin Res 55:117-122.

Maurya, VP, Naqvi SMK, Mittal JP 2003. Physiological responses and growth of native (Malpura) and crossbred (Bharat Merino) female lambs born in autumn season under semi-arid ecology of India. Indian Journal of Animal Sciences 73: 916-919.

Mc Manus C, Paludo GR, Louvandini H, Gugel R, Sasaki LCB, Paiva SR 2009. Heat tolerance in Brazilian sheep: physiological and blood parameters. Trop Anim Health Pro 41: 95-101.

Moseley PL 1997. Heat shock proteins and heat adaptation of the whole organism. J App Physiol 83:1413 – 1417.

Nardone, A., Ronchi, B., Lacetera,N., Ranieri, M.S., Bernabucci, U (2010). Effects of climate changes on animal production and sustainability of livestock systems. Livestock Science 130 (2010) 57–69.

Nazifi S, Saeb M, Rowghani E, Kaveh K 2003. The influence of thermal stress on serum biochemical parameters of Iranian fat-tailed sheep and their correlation with triiodothyronine (T_3), thyroxine (T_4) and cortisol concentrations. Comp Clin Path 12: 135-139.

Parsons PA 1994. Habitats, stress, and evolutionary rates. J Evolution Biol 7:387– 397.

Pörtner HO, Knust R 2007. Climate change affects marine fish through the oxygen limitation of thermal tolerance. Science 315:95 – 97.

Rahman U, Faisal M, Safeer M, Tariq M, Ubaidullah, Rehman J ur, 2011. The effect of cigarette smoking onte smoking on testosterone in young male adults of Peshawar. J. Med. Sci., 19: 106-108.

Rasooli A, Nouri M, Khadjeh GH, Rasekh A 2004. The influence of seasonal variations on thyroid activity and some biochemical parameters of cattle. Iranian J Vet Res 5(2): 1383-1391.

Riis PM, Madsen A 1985. Thyroxine concentration and secretion rates in relation to pregnancy, lactation and energy balance in goats. J Endocrinol 107: 421–427.

Rivington M, Matthews KB, Buchan K, Miller D, Russell G 2009. Investigating climate change impacts and adaptation options using integrated assessment methods. Aspect Appl Biol 93: 85-92.

Santolaria P, Lopez-Gatius F, Garcia-Ispierto I, Bech-Sabat G, Angulo E, Carretero T, Sanchez JA, Yaniz J 2010. Effects of cumulative stressful and acute variation of farm climate conditions on late embryo/early fetalloss in high producing dairy cows. Int J Biometeorol 54:93 – 98.

Sejian V, Srivastava RS 2010. Pineal-adrenal-immune system relationship under thermal stress: effect on physiological, endocrine and non-specific immune response in goats. Journal of Physiology and Biochemistry, 66(4): 339-349.

Sejian V, Maurya VP, Naqvi SMK 2010a. Adaptive capability as indicated by endocrine and biochemical responses of Malpura ewes subjected to combined stresses (thermal and nutritional) under semi-arid tropical environment. International Journal of Biometeorology, 54:653-661.

Sejian V, Maurya VP, Naqvi SMK 2010b. Adaptability and growth of Malpura ewes subjected to thermal and nutritional stress. Tropical Animal Health and Production, 42:1763-1770.

Sejian V, Maurya VP, Naqvi SMK 2011. Effect of thermal, nutritional and combined (thermal and nutritional) stresses on growth and reproductive performance of Malpura ewes under semi-arid tropical environment. Journal of Animal Physiology and Animal Nutrition, 95:252-258.

Sejian V, Maurya VP, Kumar K, Naqvi SMK 2012. Effect of multiple stresses (thermal, nutritional and walking stress) on the reproductive performance of Malpura ewes. Veterinary Medicine International doi:10.1155/2012/471760.

Sejian V, Maurya VP, Naqvi SMK, Kumar D, Joshi A 2010c. Effect of induced body condition score differences on physiological response, productive and reproductive performance of Malpura ewes kept in a hot, semi-arid environment. Journal of Animal Physiology and Animal Nutrition, 94(2): 154-161.

Sejian V, Sanyal S, Das PK, Ghosh PR, Sivakumar B, Pandiyan GDV 2008. Effect of unilateral Adrenalectomy on the blood Biochemistry of Black Bengal Goats (Capra hircus). Turkish Journal of Veterinary and Animal Sciences, 32 (4): 249-254.

Shelton M 2000. Reproductive perfoprmance of sheep exposed to hot environments. In: Malik RC, Razaqque MA, al-Nasser AY (cds), Sheep production in hot and arid zones., published by the Kuwait Institute for Scientific Research, pp. 155-162.

Srikandakumar A, Johnson EH, Mahgoub O 2003. Effect of heat stress on respiratory rate, rectal temperature and blood chemistry in Omani and Australian Merino sheep. Small Ruminant Research, 49L 193-198.

St-Pierre NR, Cobanov B, Schnitkey G 2003. Economic losses from heat stress by US livestock industries. J Dairy Sci 86(E Suppl):E52 – E77.

Tabachnick WJ 2010. Challenges in predicting climate and environmental effects on vector-borne disease episystems in a changing world. J Exp Biol 213:946 – 954.

Taylor NAS 2006. Ethnic differences in thermoregulation: Genotypic versus phenotypic heat adaptation. J Therm Biol 31:90 – 104.

10

Examination of Sick Animals and Management of Diseases

Ajay Kumar Upadhyay

Deviation from normal well being may be for various kinds involving one or more organs and exhibiting array of symptoms. Examination of animals for diagnosis of disease is strenuous process because investigator has just got to know by several means of examination. The evidences produced by owner, animal house, feed being provided supported by prevailing environment and climate are immensely helpful in accurate diagnosis. Diagnosis, of production related diseases, is more difficult because of biasness in history revealed by the owner. At the same time diagnoses of poisonings are also not easy because owner's description about possible cause and symptoms exhibited by sick animals are vague. So careful quizzing of owner about each and every aspect of disease and animal obviously seems to be first step towards diagnosis, followed by examination of animal. Examination of environment in relation to history revealed by owner and observations drawn from the examination of animal can infer some conclusion about cause of disease. Support of laboratory examinations such as fecal test, blood test and urinalysis are important tools in disease investigation. Following steps are essential in disease diagnosis.

1. History recording

The patient in cannot reveal any information about his suffering, so, owners observations about his animal provides clue for further progress towards final diagnosis. Quizzing of owner in a way so that he reveals many things about ailments helps in arriving at any conclusion. Failure of proper history recording misguides you to waste your time, energy and expenses. Therefore, while taking history from owner, the veterinarian must be cautious and appreciate his concern towards animal to evolve client-clinician relationship. Moreover, the use of local language and non-technical terms are essential for better elucidations. At the time of history recording, clinician must consider that all the ailing animals suffer with reduced appetite and production, rumination decreases in case of bovines. Clinician must also consider that there are diseases, which are species-specific, like Hemorrhagic septicemia, Black leg occurs in bovines while distemper in canines, breed-specific, like milk fever, mastitis and ketosis occurs in high producing animals, age-specific such as Black Quarter in month to 2 years and rickets in young ones and similarly mastitis and metritis are sex specific diseases occurring in females only.

A. Present history

Present history deals with things related to present problem. Sequence of events related with major complain of animal health observed by owner must be recorded. Observations made by owner should be thoroughly cross-examined to arrive at any conclusion. All informations regarding present sickness, such as how long animal is sick, chronology of deviations from normally and treatment provided, if any, must be recorded. Single animal may be affected with present problem or many other animals may be affected. Moreover, owner always tries some home medicine so exact detail of prior treatments must also be taken as they help in diagnosis.

B. Past history

History about previous ailments, specially related to present situations must also be taken in account while dealing with owner for best diagnostic possibilities. In case animal(s) has/have been affected with any disease, what was the cause? How long animal remained sick? What kind of treatment was given to the animal? How it recovered? What was the effect on performance, whether ailment occurred in isolation or too many animals were found to be affected with such disease? History about feeding, management, treatment and other related factors during previous illness must also be recorded to facilitate diagnosis.

2. General inspection

Inspection of animal from distance is of utmost importance because many observations can be made without disturbing the animal about his behavior, appearance, posture and gait. Attempt should be made to inspect the animal at his place because it not only provides the opportunity to assess the environment and management of animal house but also reduces the change of behavior of animal in a strange environment such as veterinary hospital, where completely different kind of persons and animals are present. General inspection may be of two types. One is related with change in appearance and expression and other is related with change in regions or system.

A. Behaviour of animal

Behaviour of animal is a reflection of his health. Diseased animals always behave differently to healthy animals. Most of the sick animals separate themselves from healthy animals; stand apart, either they become dull or excited, respond abnormally to the external stimuli such as sound, touch and movement. Appearance of animal in which it become indifferent to external stimuli and stand apart is called as "Dammy Syndrome'. This can be seen in acetonemia and listeriosis in cattle and encephalitis in horse and dogs. Extreme apathy or depression is termed as 'coma' and under this state animal become completely unconscious. Increased kicking during milking of a lactating cow and disobedience to master's order by dog also shows change of behaviour and indicates about disease conditions.

B. Feed intake

Change in appetite is first symptom. Appetite may be reduced or increased as per the disease condition. Besides quantitative change in appetite, there may be abnormality in prehension, mastication, swallowing, rumination and regurgitation. These abnormalities are indicative of several diseases related with nutrition, muscular weakness and nervous system or of infectious origin.

C. Defecation

Deviation in amount of defecation is indicative of digestive disturbances of various origins such as infection, parasitism, nutritional imbalance and nervous manifestation. Diarrhoea and constipation are the commonly seen symptoms. Sometimes, scant faecus in mature animal is confused with constipation and at the same time paralysis of anal sphincter get confused with diarrhoea. Besides, amount and frequency of defecation may also get changed in Vagus indigestion and paralysis of rectum.

D. Urination

Anomaly in urination is indicative of diseases pertaining to urinary system involving partial obstruction of urinary tract or inflammation of urinary tract. Frequent urination is related with painful conditions of urinary tract or estrous. Dribbling of urine occurs because of paralysis or obstruction of urethra or its sphincter.

E. Voice

Frequent attempt of voicing in animals are indicative of pain in some or other part of body, mostly abdomen. Soundless voicing occurs in throat paralysis or rabies. Frequent bellowing also occurs in cases of uterine torsions in advance pregnant cows. Change in voice pattern and frequency must be examined for nervous manifestation due to various etiologies.

F. Skin

Close examination is required to know about changes in various layers of skin. In cases of hyperkeratosis, hypertrophy, folding and lumping of skin may occur. There may be dermatitis, urticaria, scab or scale over the skin in various disease conditions. Normal shine of skin may be lost as it may become greasy in seborrhea or may be dry in chronic disease conditions. In certain diseases, abnormal sweating, enlarged lymph node and hemorrhage may also be noticed.

G. Posture

In disease of joints, ligaments, tendons, bones, muscle, nerves and skin, change of posture may be noticed. In osteo-dystrophia fibrosa, the horse continually shifts its weight in laminitis. During acute colic, arched back with extension of all the legs can be seen. During early stage of pregnancy toxemia, sheep keeps its head in elevated position with pricked (listening attitude) ear. Deviation of head toward flank with sternal recumbancy in recently parturited cow is indicative of milk fever. Extension of hind legs posteriorly in a frog-like attitude is indicative of bilateral hip dislocation in large animals. In case of tetanus, rigidity of ear, limb and tail can be observed.

H. Gait

Voluntary movement of animal can be examined in connection with rate, range, fever, and direction. Lameness, stiffness, extension and flexion of limbs and speed of movement can be examined in relation with diseases of skeleton, muscular and nervous system. High stepping gait with bounding progression is evident in louping ill of sheep. Range diminishes in arthritis, laminitis with expression of shuffling and stumbling gait. Circling is commonly founding in gid, listeriosis and ketosis. Dummy syndrome is characteristic of liver disease and encephalomyelitis. Sometimes abnormality in gait is associated with diseases of allergic origin or simply due to weakness. Examination regarding gait must be carried out at natural flooring where animal is kept.

I. Respiration

During respiratory movement, air from the outside is inhaled followed by exhalation and pause. Expiration is slightly, longer than inspiration. 'Biphasic expiration' is normal in equines with two peaks and should not be confused with alveolar emphysema. In case of dogs and pigs, 'interrupted breathing' or sniffing is a common phenomenon. Any form of excitement or exercise and change in environment alters the rate and depth of respiration, because of increase in oxygen demand due to stress. Deep respiration is called 'laboured respiration'. Dropped respiration causes respiratory arrest or 'apnoea' for 15-30 seconds followed by gradual resumption. Such condition is commonly found in advanced cardiac and renal diseases being known as 'cheyne-stokes respiration'. Pause in respiration followed by deep breathing that gradually becomes shallow until apnoea is known as 'syncoptic respiration' occurs in toxemia. Recurring series of rapid and shallow breaths with periods of 'apnoea' is called as Biol's respiration, which is seen in affections of medulla oblongata.

Table 1: Normal respiration rates (per minute) of domestic animals

Animal species	Respiration rate (per minute)
Cattle	10-30
Buffalo	10-24
Horse-	10-14
Sheep/Goat	15-30
Pig	10-20
Dog	20-40
Cat	30-40

3. Palpation

Examination of animal body parts through mere touching is called as palpation. This may be direct with help of fingers or mediated (indirect) with the help of some instrument like probe. Palpation is carried out to know the size, temperature, consistency and sensitivity of a lesion or organ. The lesion may be hot in acute inflammatory condition of cold in chronic lesions or in lesions of hard body parts. Changes may be 'resilient' when pressure is released, tissues regain its position (normal) or 'doughy', when pit formation remains after release of presser (edema). It may be 'firm' like liver, hard like bone, fluctuating when it is soft, elastic and undulates on presser like scrotum or normal udder. It may be tense, as in case of recent acute sever inflammation, feels like a viscous distended with gas or fluid under presser or emphysematous; puffy swelling filled with movable gas.

4. Percussion

In this technique, strikes are made over some part of body to vibrate the deeper part and records the sounds emitted by the tissue. It may be 'direct' where palm of one hand is put on the body part and strike is made by finger of another hand on the back of palm to cause the organ to emit audible sound to observe. It may also be 'indirect' where circular or oval, irony or hardwood plate termed as pleximeter and a firm rubber end hammer or plexor is used for the purpose. In case of small animal direct percussion more useful. The sound observed on percussion may be 'resonant', that is produced by striking a hallow organ containing gas as from lungs. 'Tympanic' should is produced by organs containing gas under pressure like in tympanitis. 'Dull' sound may be revealed by solid organs like bony part of body or liver.

5. Ballottement

Modified forms of percussion to detect floating organs are known as ballottement. In this first a firm push stroke is applied to an appropriate part of body to move away the underlying or mass causing it to rebound on fingertips to sense the presence, shape and size of organ or mass. Presences of fetus, abscess, large tumors and other such thing in abdominal cavity can be detected by ballottement.

6. Auscultation

Many organs during their functional activities produce some kind of sound. Listening of such sound directly through ear or with the help of any instrument is known as auscultation. Auscultation is being used to assess the normalcy of sounds produced by lungs, trachea, heart and alimentary tract. Stethoscope is being used extensively for the purpose instead of ear. While performing auscultation one should always compare diseased area with healthy to understand the abnormality better. Normal respiratory sound resembles with the sound produced alike letter 'V' is whispered softly in a drawn out manner and termed, as vesicular murmur is loud enough in cases of hyperpnoea, painting, laboured breathing and young animals. However, it is amplitude reduces with thick chest wall, early stages of pneumonia, congestion of lungs in pluritis. Vesicular murmur may be absent at the areas where lungs, bronchi, bronchioles are devoid of air. Hoarse sounds during respiration are present in bronchitis and shrill vesicular murmur is listened in pulmonary congestion. Jerking respiration sound or cogwheel respiration is heard when there is loss of elasticity of lungs tissues or in severe bronchitis.

7. Temperature recording

Recording of body temperature is useful in homoeothermic animals. Body temperature regulation against environment temperature fluctuations is achieved by action of peripheral thermo-receptors present in skin and mucous membranes and central thermo-detectors in anterior hypothalamus. In animals temperature is recorded by placing thermometer in rectum after bringing the mercury below the lowest point. Bulb of thermometer should be carefully inserted into the rectum touching the mucous membrane for two minutes. Moistening of bulb or application of lubricant will facilitate entry. Finger should be inserted along with thermometer to ensure the touch of bulb with mucous membrane. Average resting temperature and critical point above which animal is suffering with hyperthermia is given below:

Species	Normal		Critical point	
	°C	°F	°C	°F
Cattle	38.0 °C	(100.6°F)	39.0°C	(102.°F)
Buffalo	37.6°C	(100.0°F)	39.0°C	(102.°F)
Horse	37.2°C	(99.0°F)	38.0°C	(102.°F)
Sheep/Goat	38.6°C	(101.5°F)	39.8°C	(105.5°F)
Pig	37.8°C	(100.0°F)	39.0°C	(102.°F)
Dog	38.0°C	(100.6°F)	33.0°C	(102.°F)
Cat	37.8°C	(100.°F)	39.0°C	(102.°F)

8. Pulse recording

In case of cattle pulse may be recorded from middle coccygeal artery about 10 cm away from anus. In equine, commonly used site is external maxillary artery. In case of goat and sheep, femoral artery, high up in the inguinal region, is suitable. In pigs, recording of pulse is very difficult, so heart rate recording may be performed. Palpation of pulse is performed to record its rate, rhythm and quality to determine the status of health and disease. While recording the pulse one should consider the age, sex, environment and pregnancy etc. to interpret the results. Young animals have more pulse rate as compare to older animals.

Table 2: Normal pulse rates of domestic animals are herein

Animal species	Pulse rate/min
Cattle	42-60
Buffalo	42-62
Horse	33-41
Sheep	60-70
Goat	58-72
Pig	60-90
Dog	70-120
Cat	100-130

9. Ruminal movement

Normally, movement occurs at the rate of 1-3 cycles per minute. Functional status of rumen can be measured by palpating over the left flank. Palm or back of the closed fingers may be used to estimate the movement of rumen. Sufficient pressure should be applied to overcome the tone of abdominal muscle. Absence of ruminal movement may be encountered during simple indigestion, impaction and traumatic reticulo-peritonitis. Reduction in rate of ruminal movement may be observed in vagus indigestion, actinobacillosis, neoplasm, ulceration, grain engorement etc. On palpation of left flank, pain may also be felt along with impacted ruminal movement. In impacted rumen filled with food staff dull sound appears on percussion. Froth bloat resonant sound may be heard on percussion, where as in gaseous bloat, lymphatic sound evokes by percussion.

10. Examination of mucous membrane

Mucous membrane examination is mainly useful in diagnosis of inflammatory and deficiency diseases. Conjunctiva is most commonly examined mucous membrane. Both eyes should be examined to explicate any abnormality. Followed by conjunctival, buccal, vaginal and rectal mucous membranes are inspected for the purpose. Yellow coloured mucous membranes are indicative of jaundice, pale and watery mucosa are seen in anaemia, however, pale and dry mucosa may be observed in shock. Brick-red mucous membranes are encountered in arsenic poisoning and bluish mucosa reveals cyanosis of blood. Apart from these, presence of hypraemia and swelling may be seen in inflammatory conditions such as conjuntivitis, vulvovaginitis etc.

Viral Diseases

Foot and Mouth Disease (FMD)

(*Apthous fever, Contagious aptha, panzootic aptha, Vesicular aptha, Aphteuse, Aftaepizootica Fiebreaftosa*)

Foot and mouth disease is an acute febrile highly contagious disease of cloven footed animals. It is characterized by vesicular eruptions in the epithelium of buccal cavity, tungue, nares, muzzle, feet, taet and udder. Lesions may also appear in rumen pillars. In calves, there is focal disease imposed considerable economic stress on the cattle rearers due to reduction of milk meat and working capabilities of dairy and draught animals respectively.

The causal agent of the disease is a filterable virus. These filterable viruses belong to "picorna" virus group and have the size of 8-10 mμ. Disease causes morbidity up to 100%. Mortality is 20% in exotic animal and 20% in young animals. Cattle are more susceptible. A wide range of hosts like buffalo, sheep, goat, deer, hedge hog, reindeer, pig and camel are susceptible to this infection. Dog, cat and man may occasionally be infected after close association or intimate contacts will the infected host. Guinea-pig, rabbit and chicks can artificially be infected. The disease is spread at an extremely rapid rate through direct contact with the infected animal. Virus may gain the entrance in a herd from infective materials, food stuff, feeding utensils, waterer, etc. Spread through airways like human influenza virus can also take place.

The foot and mouth disease Virus initially settle in the mucous membrane of the mouth cavity and invade the basal layer of the stratified epithelium of the tongue and produce primary vesicles causing pyknotic changes in the nucleus of the cells with leukocytic infiltration. Following penetration of the mucous membrane, virus multiply in the stratum spinosum where cytolysis takes place giving rise to small cavities in the epithelial layer. These in turn develop into characteristic F.M.D.-vesicles. From the primary lesions virus invade the lymph and enter the blood circulation and viraemia is produced. Again from the blood stream virus go to the site of predilictions i.e. epithelium of mouth, dental pad, feet and udder where typical vesicles are formed. If the lesion is not complicated by the secondary bacterial invader lesions used to heal up within a short Span. Heart muscles may show acute degeneration of the myocardial cell fibres in calf. There is heavy mortality in young calf due to cardiac damage with absence of appreciable vesicular eruption. Some indigenous cattle show superficial brown necrotic lesion of the epithelium with a tendency of rapid healing.

The disease is not usually fatal. Morbidity is 100% but mortality is less in indigenous cattle and comparatively more in pure and cross bred cattle. The incubation period of the disease is 2-8 days under natural condition but it may extend to as long as 2-3 weeks. Viraemic condition elevates the body temperature. Temperature may range from 104-106°F. This is accompanied with generalized malaise, depression, dejection and apathy to food and water. There is reduction of milk yield. Following peak temperature characteristic vesicles appear in the oral mucosa (vesicular stomatitis), interdigital space, udder etc. At this stage there is reduction of temperature, profuse salivation and lameness. Drooling of foamy, roapy salivation and protrution of the tongue are the features in this stage. Vesicles may rupture due to jaw movements leaving an angry looking ulcer. Eroded surface produces quivering and frequent smacking of lips. The animals refuse to eat due to pain in the oral mucosa. Simultaneous lesions in the feet appear in the junction of the hoof and skin. A large vesicle may appear in the digital skin which will rupture and produce severe pain and lameness. In severe case, there is shedding of the hooves. Mastitis is often noticed in dairy cows. Pregnant animal may abort and infertility may ensure following abortion. Young animals die without exhibiting clinical manifestations. Signs of gastro-enteritis are noted in some animals. Suckling calf usually die as a result of myocarditis and myocardial degeneration.Piglets may show high mortality rate showing the signs of severe gastroenteritis.

Diagnosis is based on history, clinical findings and laboratory tests. A febrile animal showing lameness and salivation may be suspected for FMD and such animal should be kept for close observation for more than 6 hours and these allowances of time will help for the appearance of foot and mouth lesions. Differences amongst malignant catarrhal fever, mucosal disease, rinderpest and vesicular stomatitis should be made based on laboratory tests and necropsies.

There is no specific treatment. Symptomatic treatments may be rendered depending on clinical manifestations. Antiseptic mouth wash with potassium permanganate, sodium carbonate, boric acid and glycerin may be applied over mouth lesions. Similarly antiseptic or antibiotics may be given on foot lesions and lesions on mammary tissues. The control measures vary from country to country. The premises is thoroughly cleaned and disinfected. By adopting this method and vaccination programme as well, it has become possible to eradicate this disease from many countries of the world. The vaccines at present available are not enough to guarantee the continuing protection of the high grade animals against constant risk of exposure to viruses which may be circulating in the indigenous stock Several types of vaccines have been tried to combat with FMD. Recovered animals are immune to the homologous virus only.

Peste des petits ruminants

(Pest of small ruminants, stomatitis-pneumoenteritis complex or syndrome pseudorinderpest of small ruminants and kata)

Peste des petits ruminants (ppr) is an acute or subacute viral disease of goats and sheep characterized by fever, erosive stomatitis, conjunctivitis, gastroenteritis, and pneumonia. Goats are usually more severely affected than sheep.

The causative agent virus belongs to family paramyxoviridae, genus morbillivirus. Antigenically close to rinderpest virus and resistance to physical and chemical action.

A Peste des petits ruminant is primarily a disease of goats and sheep. Cattle and pigs are susceptible to infection with PPRV, but they do not exhibit clinical signs. Such subclinical infections result in seroconversion, and cattle are protected from challenge with virulent RPV. It is transmitted through direct contact between animals. Tears, nasal discharge, coughed secretions, and all secretions and excretions of incubating and sick animals are source of infection. There is 90% morbidity rate and 50-80% mortality rate.

The virus has got affinity for epithelial mucosa. On entry they invade the retropharyngeal lymph nodes and mucosa. From there the mucous membrane of the alimentary and respiratory tracts are affected. There is proliferation of cells and damage of them. Inflammation of the alimentary mucosa set up cells and damage of them.

Incubation period is 3-10 days. Inflammation of the alimentary mucosa set up enteritis causing diarrhoea, dehydration and death in young animals. Inflammations of the respiratory cels lead to pneumonia and respiratory distress. Sudden rise in body temperature (40-41°C) with effects on the general state: restlessness, dull coat, dry muzzle, depression of appetite, serous nasal discharge becoming mucopurulent and resulting, at times, in a profuse catarrhal exudate which crusts over and occludes the nostrils. Small areas of necrosis are visible on the nasal mucous membrane. Congestion of conjunctiva, crusting on the medial canthus and sometimes profuse catarrhal conjunctivitis may be observed. The young goat may die due to anoxic anoxia. Bacteria like E. coli and Pasteurella and protozoa like coccidia have been found to be asociated with this problem causing aggravation of pneumonia and diarrhoea and death. Recovered animals may possess lifelong immunity.

Diagnosis is based on history and clinical findings such as necrotic stomatitis with halitosis, severe non-haemorrhagic diarrhea, bronchopneumonia evidenced by coughing and abortion.

There is no specific treatment. Antibiotics such as oxytetracycline and chlortetracycline may prevent secondary pulmonary infections. The tissue culture rinderpest vaccine protects goats for at least 12 months against ppr.

Canine distemper

It is also known as canine influenza or hard pad disease and is characterized by diphasic fever, leukopenia, bronchopneumonia, gastroenteritis and encephalitis. The disease is caused by paramyxovirus belonging to morbilli group and this virus is closely related to the measles and rinderpest viruses. This virus is stable at pH range of 4.5-9.0 as well as at low temperatures. It can survive at 45°C for 10 minutes; at 37°C for one hour; at 20°C for 3 hours; at 4°C for several days and at frozen for several years. It is one of the oldest viral infections recognized in dogs and it was first characterized in 1809 by Edmard Jenner. The infection is more common in dogs of foreign breeds while indigenous breeds are quite resistant. Though the susceptible dogs of all age groups may be infected but its incidence is more in younger animals of 3-6 months age. The transmission of virus occurs by inhalation either by droplet or aerosol. The disease may spread indirectly also through insects and vectors.

After gaining entry through aerosol droplet, the virus reaches in the macrophages of respiratory tract. Within 24 hr of entry into macrophages, the virus multiply and spread to tonsils, bronchial lymph nodes and other adjoining lymph nodes via local lymphatics and multiply there. By 2-4 days post infection, virus numbers increase in tonsils and retropharyngeal and bronchial lymph nodes. Some of the mononuclear cells infected by virus may be found in bone marrow, thymus and spleen also. In 4-6 days of infection, virus multiply in lymphoid follicles in spleen, lamina propria of stomach, small intestine, mesenteric lymph nodes and Kupffer cells of liver. After 8-9 days of infection, virus spreads to epithelial and central nervous tissues via hematogenous route as a cells associated and plasma-phase viremia. The clinical symptoms of disease vary depending upon the virulence of virus, environmental conditions and age and immune status of host. The disease may occur in mild or severe forms. In the mild form of illness, dogs reveal listlessness, decreased appetite, and mild fever and upper respiratory tract infections. Due to the infection of upper respiratory tract, bilateral serous coulee-nasal discharge, coughing and dyspnoea are noticed. In later stages, the oculo-nasal discharge becomes mucopurulent. Many mildly infected dogs develop clinical signs that are indistinguishable from those due to other causes of 'Kennel cough'. In severe illness, multisystemic signs are noticed and it can occur in dogs of any age group with poor immune status. However, it is more common in unvaccinated, exposed puppies of 12-16 weeks of age which have lost their maternal immunity or younger puppies which did not receive adequate amount of maternal antibodies. Initially there is rise in body temperature (104-106°F) lasting for 1-3 days. Later on temperature subsides and is again increased after 7-12 days lasting for a weak or more, so there is biphasic temperature. Afterwards mild serous to mucopurulent conjunctivitis develops and is followed by dry cough. It becomes moist and productive and on auscultation, lower respiratory sounds can be heard. Subsequently diarrhoea of varying consistency from fluid to frank blood and mucus develops. Due to this diarrhoea, there is fluid loss, severe dehydration and emaciation. Besides, there may be Cutaneous or skin form in which there is appearance of rash, vesicles and pustules in this form. The vesicles or eruptions commonly occur on lower abdomen or belly and at inner side of thigh. In some cases, the skin of footpads and nose becomes hard due to hyperkeratitis and this condition is usually referred as "hard pad". Due to thickening of food pad, animals feel difficulty in walking on hard floor and they may become lame. There may also be nervous form, which occurs in dogs showing hyperkeratosis and symptoms in it may vary depending upon the part of central nervous system involved. They may show restlessness, excitement, chewing movements, excessive salivation and convulsions. The dog may fall and show epileptic fits. Leukopenia is a common finding in these dogs but it may not occur with delayed encephalitis. The adult dogs usually suffer from 'Old dog encephalitis' where they suffer with mental disorders and unable to make judgements and are not able to follow the commands of owner or handler. Such dogs usually pass urine and faeces involuntary.

For diagnosis haematological examination of dogs reveals absolute lymphopenia, thrombo- cytopenia (30,000 cells/ul) and regenerative anaemia. Inclusions can be noticed in stained blood smears in low numbers in circulating lymphocytes and even less frequently in neutrophils and erythrocytes. Changes in cerebrospinal fluid particularly in dogs showing neurological signs have also been reported. There is increased intracranial pressure due to inflammation resulting in rapid

CSF flow. The CSF has high protein content (25mg/dl) and high cell count. The impression smears prepared from conjunction or vagina has intracytoplasmic, eosinophilic inclusion bodies. These inclusions persist longer than viral antigen in cell and are pathognomonic. Besides, history of unvaccinated puppies of 3-6 months age gives suspicion of the disease. Nervous signs with fever, respiratory symptoms, diarrhoea, ocular discharge and hyperkeratosis of footpad is indicative of CD also helps in diagnosis.

Since the disease is caused by a virus, its treatment requires specific antiviral drugs. However, the antiviral preparations available today, are not effective against canine distemper virus invivo. Similarly intravenous live virus involution has been found ineffective, but attenuated vaccine virus may protect the dogs if given 4 days after exposure. The costly dogs may be protected by giving anticanine distemper serum (Homoserum) @ 1-5ml/kg b wt sc, im or iv but it should be given as early as possible. In severe cases, the serum may be given @ 5-10 ml/kg b wt.

Immunization by controlled vaccination is the only effective approach for preventing canine distemper. Active immunization with modified live virus vaccine induces long lasting immunity and it has helped in preventing the disease in dogs since last two decades. For control of the disease, infected dogs should be isolated as virus is shed in all the body excretions during the acute systemic disease.

Infectious canine hepatitis

It is also known as hepatitis contegiosa canis, contagious hepatitis, Rubarth's disease. The disease is caused by canine adenovirus-1 that is serologically related only to canine adenovirus-2 (causal agent of infectious canine laryngotracheitis or respiratory disease). The virus is resistant to environmental inactivation, disinfection with, acid, chloroform or formation or stable for 30 minutes at pH 3-9 at room temperature. It can survive for 3-11 days at room temperature on solid fomites, wet or dry hypodermic needles and for 6-8 years in glycerin or in vacuum. It can survive up to a month at 37°C but can be killed in 3-5 minutes at 60^0C. So for disinfection, steamcleaning is an effective means. The disease was recognized as clinical entity in dog for first time in 1947 by Rubarth since prior to it, the clinical cases were thought to be of canine distemper or toxic hepatitis. The disease can occur in any age group, but dogs up to one year of age are more susceptible. The main route of infection is through ingestion of contaminated food or water, which may be contaminated by urine, faeces or saliva of infected animal. Recovered dogs shed virus up to 6 months in their urine. The disease can also spread by direct contact. After gaining entry into the body of host, virus initially localizes in tonsilar crypts and Peyers' patches and then spreads to regional lymph nodes and lymphatics. Through thoracic duct, the virus reach into blood and viremia develops. Afterwards, virus is rapidly disseminated to other tissues and body secretions including saliva, urine and faeces. Hepatic parenchymal cells and vascular endothelia cells of many tissues are prime targets of viral localization and injury. The initial cellular injury to the liver, kidney and eye occurs due to cytotoxic effects of virus. Further damage and development of symptoms are based on host immune response. The ocular lesions develop when virus spreads to uveal tract after 4-6 days of parenteral inoculation and there is mild uveitis. Virus enters the aqueous humor from blood and spreads to corneal endothelial cells.

The disease is commonly seen in the dogs of less than one-year age though unvaccinated dog of any age group may suffer. The disease occurs in three forms- peracute, acute and subacute. In peracute form, the dogs die within few hours after the onset of clinical signs. Sometimes the dog apparently looking normal in night may be found dead in the next morning. Dogs that survive viremic period reveal vomition, abdominal pain and diarrhoea with or without evidence of heamorrhage. Temperature becomes high and haemorrhages may be noticed in eye, intestine and liver with rise in body temperature (104-106°F), enlargement of tonsils and red colouration of buccal mucosa. In acute form, apathy, anorexia and high rise of body temperature (105-106°F) followed by vomition or diarrhea occur. Faeces are often blood tinged and there is abdominal pain. The first rise of temperature usually falls after 24-48 hr without reaching to normal and then again rises to remains elevated for 5-7 days and it gives 'saddle curve'. Pulse and respiration rates are increased. Enlargement of tonsil, pharyngitis, laryngitis, coughing and hoarse lower respiratory sounds on auscultation and pneumonia may be noticed. The dogs show intense thirst, haemorrhagic buccal mucosa and abdominal tenderness. Deep palpation of xiphoid region or at liver yields pain due to enlargement of liver and dogs may show 'tucked up' condition of abdomen. Small haemorrhagic spots may be noticed over the abdominal skin. If there is venipuncture, severe bleeding occurs due to disturbance in clotting mechanism. The virus damage the endothelial cells due to which widespread clotting mechanism activation occurs and already damaged liver cannot remove activated clotting factors. In subacute form, there is more common in dogs of above 1 year of age and they show mild or subclinical attack. There is slight rise in body temperature (103-104°F), mild photophobia and enlargement of tonsils. The animals recover easily but regaining of weight loss is very slow in affected dogs.

The disease is diagnosed by recording of history of unvaccinated dogs along with age, may give suspicion of disease. Clinical signs like widespread haemorrhage, digestive system involvement and respiratory signs with pattern of fever are indicative of ICH. Analysis of blood and CSF may help in identification of the disease.

Clinical management of dogs developing ICH is primarily symptomatic and supportive. The treatment should begin with the intravenous administration of polyionic fluid like Ringer's solution. It helps in restoring ionic losses caused due to vomition and diarrhoea as well as assists in lowering body temperature. During intravenous infection, haemorrhage should be avoided since in affected dogs, blood coagulation is delayed. Since the dogs may develop hypoglycemia as a result of hepatic encephalopathy, 50% glucose solution should be given @ 0.5 ml/ kg b wt over a 5 minutes period. It should be continued to check the recurrence of hypoglycemia. To prevent secondary bacterial contamination, broad-spectrum antibiotics should be given. Use of nonabsorbable antibiotics like neomycin may be preferred and given orally.

For prevention of disease, now days, modified live CAV-1 vaccines are in use since these are commercially available. Dogs vaccinated with CAV-1 vaccine were also found to have cross immunity against CAV-2 infection as it protected dogs against CAV-2 aerosol challenge.

Canine parvovirus infection

It is also known as haemorrhagic enteritis and characterized by high fever, diarrhoea and myocarditis. The disease is caused by canine parvovirus type-2 (CPV-2) belonging to family parvoviridae. The virus is closely related antigenically to feline panleukopenic virus, mink enteritis virus and raccoon parvovirus. It can survive up to 15 min at 80°, 1 hr at 60°, 24 hr at 56°, 2 weeks at 37°, 3 months at 20° and 6 months at 4°C. It remains alive in faeces in environment for many years. However, it can be inactivated by 0.5% formalin, 1:32 dilution of hypochlorite solution, by solutions of propiolactone, hydroxylamine, oxidizing agents and UV irradiation. The CPV-1 was isolated in 1967 from faeces of clinically normal. Under natural conditions, only members of family canidae are infected. Faeces are the primary mode of spread of infective virus particles and these particles may remain in faeces for long periods. During severe illness, virus may be shed in saliva and vomitus also. Peak faecal shedding occurs during 4-7 days post-infection. Neonatal transmission of disease from infected mother or litter is also possible.

Severity and type of illness primarily depend on the age of pups and the virus grows and multiplies only in actively dividing cells. Myocardium and intestinal epithelium are the main target tissues for CPV-2. During first two weeks of life of pup, there is rapid cardiac myocyte replication while intestinal epithelium turnover rate is slow. After 8 weeks of life, the intestinal crypt epithelium replicates actively. Therefore infection of neonatal pups results in myocarditis while infection of older pups and dogs causes enteritis.

In infected dogs enteric form or myocardial form may be seen.

a) *Enteric form*

After an incubation period of 4-7 days, there is sudden vomition, depression, anorexia and pyrexia (104-106°F). Vomiting is often severe and prolonged and vomitus can vary from clear to bile coloured or blood stained. After 6-24 hr of vomiting, diarrhoea develops with patchy to fluid in consistency, greyish or yellow grey in colour and often blood streaked faeces. With the progression of disease, frank haemorrhage and increased mucus may appear in faeces. Dehydration and weight loss occurs within 24-48 hr. Secondary bacterial complications and die within 48-72 hr after onset of signs.

b) *Myocardial form*

This form occurs in neonatal puppies or pups up to 8 weeks of age. Most of the pups of infected litter die by 8 weeks of age while remaining die afterwards as a result of subacute heart failure. Before death pups reveal dyspnoea, crying and retching along with terminal convulsions.

Disease can be diagnosed by obverving leukopenia, which is proportional to the severity of clinical illness, but is not a consistent finding. Lymphopenia and neutropenia are also observed but lymphopenia is more characteristic. In chronic illness, some puppies reveal heart enlargement prior to electrocardiographic changes, which can be easily detected by radiological examination of cardiac area. In myocardial form, severe pulmonary edema, congestion of liver, ascitis and hydrothorax can be seen. The heart wall may be flaccid while cardiac chambers are dilated.

Since myocardial form occurs in younger pups and difficult to manage, usually the treatment is done for enteric form which may be managed if treated early. There is fluid loss through diarrhoea and vomiting, so fluid therapy is done immediately. For it, lactated Ringer's solution is used as it can take care of metabolic acidosis due to persistent diarrhoea. Now attenuated live CPV-2 vaccine is available which provides protection against disease as well as infection for long period. However, pregnant bitches should not be vaccinated with it. Combined vaccination with other canine products is well tolerated.

Rabies

It is also known as hydrophobia, mad dog, madness or Rabera. Though all warm-blooded animals are susceptible but it is recognised a natural disease of dogs, cats, bats and wild carnivores. The disease is caused by rabies virus belonging to the genus Lyssa-virus in the family Rhabdroviridae. The virus is destroyed by various concentrations of formalin, phenols, halogens, mercurials, mineral acids, various antiseptics, soap and detergents and high temperatures. Boiling inactivates the virus immediately while at 55°C or above, it is killed within few minutes. It is highly pathogenic, has short incubation period of 4 to 6 days only and does not invade the salivary gland.

After entry of the virus through bite, it replicates in myocytes and then through peripheral nerves, virus reaches into the central nervous system (CNS). Nerve endings of neuromuscular and neurotendinal spindles near the bite site are the probable spots for virus entry into the nervous system. The clinical signs appear with the replication of virus in brain and spinal cord and centrifugal movement of infection occurs in the peripheral organs. The virus reaches into salivary glands and saliva via peripheral nerves. This dog when bites to other animals or man that suffers to rabies.

Rabies in dogs manifests either in furious (hyper excitable) or in dumb (with paralysis predominant) form, although a variety of gradations between these two are seen. Contrary to the common belief, dumb rabies is more prevalent in our country. Probably the earliest recognisable signs in affected dogs are changes in the usual behaviour with or without the changes in the tone of the bark. The animal may refuse its normal feed and instead try to eat or bite abnormal objects like wood, stone, rags etc. There is increased nervousness, elevated body temperature, partial paralysis of the jaw, restlessness and an increased tendency to bite when objects are placed near the mouth of dog. After this initial furious stage which may last for short period, there is incoordination, increasing paralysis of limbs, a far-off look to the eyes and possible further changes in temperature. Excessive salivation, drooping of the lower jaw and paralysis of hind limbs are characteristic features of rabies in dogs through these appear in late stage of disease. The paralysis in progressive and terminally there is complete limb paralysis, stupor and death. Disease is diagnosed by a history of bite. Besides, fluorescent antibody technique is virtually 100% effective in diagnosing the disease.

The management of wound is very important and this can be done by washing the wound with plenty of soap and water. After removal of soap, any quaternary ammonium compound may be applied as the antiseptic along with antirabies serum. Cauterisation or application of oil, turmeric etc. on wound should be avoided. The wound must not be stitched at least within 24-48 hr of injury. There are post-exposure as well as pre-exposure immunoprophylaxis methods available for prevention of disease. Tissue culture antirabies vaccine is given on day 0, 3, 7, 14, 30 and 90 post-exposure. The dogs that are not severely exposed, the vaccination on day 90 is not necessary. For pre-exposure immunization, the initial vaccination is done in the pups at the age of 12-14 weeks. It is followed by a booster vaccination one year later and then it is repeated after every year.

Equine influenza

The disease is also known as pink eye, shipping fever, grippe or epizooly and is an acute highly contagious disease of horses characterized by explosive non-productive cough, respiratory distress and mild fever. The disease is caused by equine influenza virus, which belongs to myxovirus type A influenza viruses. There are two serologically distinct viruses IA/E1 and IA/E2 that produce disease in horses. These outbreaks occurred mainly due to international movement of horses. Now the disease is present in most of the countries throughout world. The disease occurs in equines of all age groups including newly born foals. Most of the cases are noticed in younger horses or animals up to two years of age. However, foals of 2-6 months of age are at greater risk. Through disease can occur at any time but most of the cases are noticed during winter months in our country. The morbidity rates may vary from 30-100% but mortality in uncomplicated cases is rare. The cases that are complicated with viral pneumonitis or bacterial pneumonia may die of the disease.

After gaining entry through inhalation, the virus enters and multiplies in epithelial cells of mucosae of upper respiratory tract. Due to this erythema, edema and focal erosions develop and secondary bacterial agents may invade the damaged mucosae. Despite persisting coughing, the affected horses do not remain infective after one week. After an incubation period of 1-3 days, there is rapid development of fever and body temperature becomes 105-106^{0}F within short period. Chills lasting 15-30 min may develop. The skin becomes rough and dry while hairs remain more or less erect. Such horses stand immobile with their head hanging. Respiration is slightly accelerated and may become 25-30 per min. There is loss of appetite with excessive thirst while visible mucous membranes become deep saffron or violet red in colour due to which it is commonly named as pink eye. There is unproductive hatching coughing which is pronounced by slight exercise or palpation of pharyngeal region. Sometimes watery discharge from eyes and nose may come out. If such animals are put to rest in comfortable stables, they recover quickly. However if they are put to work or other stress, the secondary bacterial infection may set in causing severe illness. The foals may suffer from high rise of temperature (105-106°F), severe dyspnoea, bronchopneumonia and increase in respiratory rate (80-90/min). Such foals usually die in 4-5 days due to secondary bacterial contamination with streptococci or *E. coli.*

The disease is diagnosed by taking history and observing clinical signs. Blood examination reveals leukopenia. The virus can be isolated from nasal swabs and can be grown on developing chicken embryo as well as established tissue culture. The disease can be confirmed by serum neutralization, complement fixation, haemagglutination inhibition and ELISA tests. On post-mortem serous to seromucoid nasal discharge, fluid accumulation in bronchioles and bronchiolitis are the lesions found.

There is no specific treatment available against the disease but symptomatic treatment is of value. The animals should be given broad-spectrum antibiotic for 3-5 days. The foals may be given hyperimmune sera to save the life. All the affected animals must be kept in hygienic stables and be provided good diet and fresh drinking water along with comfortable bedding. For the prevention of disease inactivated adjuvanted vaccines containing both IA/E1 and IA/E2 viruses are available. For controlling the disease, infected and suspected animals should be quickly isolated from healthy animals and should be kept in good and properly ventilated houses.

Equine infectious anaemia

It is also known as swamp fever, American surra, malarial fever, slow fever, mountain fever, typhoid fever of horse or plains paralysis and charactezised by long relapsing fever and anaemia. The disease is caused by equine infectious anaemia virus belonging to the subfamily Lentiviridae in family Retroviridae. It has been detected from almost all the continents and has global distribution. From India, the cases were noticed for the first time in 1986 at the Bangalore Turf Club. The disease occurs in horses, mules and donkeys of all age groups. Most of the cases occur in low lying areas and in rainy season since the vector population is high during the period. Transmission of the disease is mainly insect-borne and tabanid flies play significant role than stable fly.

After entry through insect bite, virus localizes in spleen, liver, kidney and lymph nodes. Then there is viraemia and virus disappears from tissues during periods between attacks. Virus multiplies mostly in the tissue macrophages and blood monocytes. There is involvement of reticuloendothelial cells, damage to the intima of small blood vessels and excessive destruction of erythrocytes due to invasion by virus. Inflammatory reaction in hepatic parenchyma occurs. Vascular changes are also noticed in nervous tissue resulting in encephalomyelitis, ataxia and spinal leptomeningitis. Anaemia develops in first fever phase due to enhanced erythrocytolysis and decreased production of erythrocytes. after the incubation period of 2-4 weeks, disease occurs in acute, moderate or chronic form. The disease starts with the recurrent fever (105°F) that may be continuous also in some horses. There is marked weakness, anorexia, muscle tremors and visible haemorrhage on mucous membrane. Petechial haemorrhage is prominent in sub-lingual and nasal mucosae. Irregular recurrent fever is the chief clinical manifestation. Inflamed icteric mucosae, anaemia and edema of ventral abdomen, prepuce and legs are also noticed. Respiration is accelerated and mainly is of abdominal type. Animal is dejected, head hangs low, leg weakness is marked, body weight is shifted from one leg to another and hind feet are frequently placed well forward under the body. Frequent urination and in severe cases, diarrhoea may develop. There is an increase in rate and intensity of heart sounds.

History of the disease indicates high vector population and presence of intermittent fever. There is prominent decrease in erythrocyte, leukocyte and platelets count. There is normocytic normochromic anaemia. Sideroleukocytosis and presence of Heinz bodies in large number of erythrocytes are seen. Total serum protein and albumin-globulin ratio are also reduced. One of the earliest tests to detect the disease is Coggin's test. Post-mortem examination reveals jaundice, subcutaneous edema and petechial or ecchymotic haemorrhages in subserosa along with enlargement of liver, spleen and local lymph nodes. Chronic cases reveal emaciation and paleness of tissues.

There is no specific treatment therefore cases are managed by symptomatic and supportive therapy. For preventing secondary bacterial contamination, broad-spectrum antibiotics should be given for 3-4 days. In moderately ill horses, haematinic drugs may be used. Oral haematinic mixture containing copper, cobalt and iron salts may also be supplemented for early recovery. In severely ill horses, blood transfusion may be done to save the life as there is severe anaemia in such cases. The animals should never be given corticosteroids or other immunosuppressive drugs as such drugs may cause immunosuppression and condition of horse may further aggravate. There is no vaccine available against the disease so animal husbandry practices may be adopted for control of disease.

Swine fever

It is also known as hog cholera or classical swine fever and characterized by piling of affected animals, discolouration of abdominal skin and nervous sings. The disease is caused by swine fever virus (SFV) which is a pestivirus belonging to the family Togaviridae. Initially an outbreak swine fever was noticed in 1830s in USA and since then disease has been recorded from almost all the countries of world. Among domestic animals, the virus naturally infects only pigs. Pigs of all age groups are equally susceptible but the severity of illness depends on the virulence of SFV. Spread of the disease can occur through ingestion or inhalation. Transplacental spread of virus from infected pregnant sows to foetus is also possible. The disease usually occurs in the form of severe outbreaks causing 100% morbidity and almost 100% mortality when virulent strain is involved. When disease is caused by moderately virulent strains, the morbidity and mortality rates are low.

The viruses enter through oronasal route and reach in the tonsil which is site for its primary replication. Thereafter, virus reaches in blood capillaries through lymphatic vessels and cause initial viraemia. The virus is trapped in spleen, peripheral and visceral lymph nodes, bone marrow and Peyers' patches. Then there is secondary viraemia and virus invades parenchymatous organs. Most of the lesions are produced by hydropic degeneration and proliferation of vascular endothelium that occurs due to the occlusion of blood vessels. As a result of this, congestion, haemorrhages and infarction in arterioles, venules and capillaries and thrombosis in small and medium-sized arteries occur. These changes are severe in spleen, lymph nodes, kidneys and gastrointestinal tract. There is atrophy of thymus, inflammatory reaction in spleen, depletion of lymphocytes and germinal follicles in peripheral lymphoid tissue and disturbance in fibrinogen synthesis. Clinical signs are noticed after an incubation period of 2-6 days. The young pigs may die suddenly without showing symptoms. Other animals reveal rise in body temperature, reduced appetite and dullness and these are reluctant to move. Within 24-48 hr, the body temperature becomes 106-107^{0}F and leukopenia develops. Conjunctivitis, marked ocular discharge, mucous to purulent nasal discharge and constipation followed by diarrhoea may also be seen. Affected pigs shiver and pile upon each other. In some cases convulsions may also be seen while in terminal stage, typical weaving, staggering gait and posterior paresis are seen. A purplish discolouration extending over abdomen, snout, ears, and medial sides of legs may also observed. Circling, incoordination and muscle tremors are also seen. Small areas of necrosis may be found on the edges of the ears, on tail and lips of vulva. About 90-95% of affected pigs die between 10-20 days post-infection. In persistent infection, pigs survive for a period of over 30 days, in such cases the affected pigs reveal anorexia, fever, diarrhoea and alopecia followed by dermatitis, blotching of ears and deep purple colouration of abdominal skin. It may lead to the development of runt pigs due to severe growth retardation and such pigs have relatively large head and small trunk. These pigs frequently stand with arched backs and hind legs placed under the body. Most piglets infected inutero and born alive, die shortly after birth. In such piglets, haemorrhages over skin are commonly seen. In congenital infections abortions, foetal mummification, stillbirths and neonatal deaths are usually noticed. The born piglets usually remain weak and reveal tremors. Other piglets may look healthy but carry persistent SFV infection and may show signs of disease. Such animals die after 2-11 months of birth.

History of outbreaks and clinical sings may give some idea about the disease. In affected pigs leukopenia is a consistent finding and leukocyte count may vary from 4000 to 9000/ul. It is accompanied with thrombocytopenia. However in terminal stage, leukocytosis occurs due to secondary bacterial invasion. Antibodies can be detected by virus neutralization, fluorescent antibody, agar gel diffusion, passive haemagglutination, complement fixation and indirect ELISA tests.

There is no effective treatment available against the disease. The acute cases can be saved only by giving hyperimmune sera as early as possible. It can be given at the dose rate of 50-150 ml and can be repeated after 24 hr if necessary. For controlling the disease, infected or suspected animals should be isolated from healthy animals. All the healthy animals must be vaccinated while infected or suspected cases should be given hyperimmune sera @ 20-75 ml depending on the size of pig.

Swine influenza

It is highly contagious disease characterized by fever, catarrhal inflammation of respiratory tract epithelium, laboured jerky breathing and coughing. The disease is caused by type A influenza virus which is pleomorphic having 80-120 nm diameter and belonging to the family Orthomyxoviridae. The first outbreak of swine influenza was observed in 1918 in USA. Pigs of all age groups are susceptible but most of the cases are seen in young pigs. The disease occurs in cold months of the year. Disease generally appears with the introduction of new pigs into a herd so the outbreaks are mostly related to the movement of animals from infected to susceptible herd.

After gaining entry through intranasal route, the swine influenza virus attaches to the cilia and adsorbs to the membranes of nasal, tracheal and bronchial mucosae. Virus replicates in epithelial cells and spreads throughout the respiratory tract within 24-72 hr. There is drainage of copious exudate from bronchi which contains cellular debris and mononuclear cells. Thickening of alveolar septae, interstitial pneumonia and distinct hyperplasia of bronchial epithelium may be seen. The symptoms of disease appear suddenly after an incubation period of 2-7 days in the form of high rise of body temperature (107-108^{0}F), anorexia, prostration and apathy. Generally there is muscular weakness, stiffness and pain accompanied by malaise and due to this, pigs are disinclined to move or rise and huddle together. Sick animals have a paroxysmal cough, laboured jerky breathing, sneezing and rise in respiration rate with abdominal respiration. On auscultation of chest, strong bronchial breathing noise during inspiration and expiration can be noticed. Rhinitis, nasal discharge, conjunctivitis and weight loss occur. The animals recover quickly after 3-6 days of illness.

History of involvement of large number of pigs and symptoms of upper respiratory tract may indicate about the disease. A moderate leukopenia in early stage of disease along with lymphopenia, neutrophilia, shift to left and increase in fibrinogen and reduction in protein fibrinogen ratio are noticed. The disease can be diagnosed by virus neutralization, haemagglutination inhibition and single radial haemolysis tests. The viral antigen can be demonstrated in nasal epithelium by indirect immunofluorescence test using monoclonal antibodies.

There is no specific treatment of the disease. To prevent secondary bacterial contamination, broad-spectrum antibiotics should be given for 3-4 days. The animals must be kept in dust-free, well-bedded and comfortable pens. They should be given fresh and clean drinking water and easily digestible soft food.

Application of turpentine liniment over chest area and use of expectorants may help in relieving pain and coughing. There is no effective vaccine available against the disease though an inactivated, formolized, avianized vaccine with adjuvant has been developed. But the vaccination is not successful due to great viral antigenic diversity. As the disease occurs rapidly, prevention of its spread in a herd is difficult.

Bacterial diseases

Haemorrhagic septicemia

(Pasteurellosis, Shipping fever, Stockyardis disease, Stockyardis pneumonia, Barbone disease)

It is acute septicemia occurring most often in cattle, buffalo, camel, sheep, goat and pig, usually following some form of stress. The disease is characterized by high-rise of temperature, depression, pneumonia and pericarditis. The name pasteurellosis was attributed for the disease because of extensive work done by Pasteur on fowl cholera caused by same causative agent.

Pasteurella multocida type-I or B causes the disease. It is gram-negative coccoid rod and filamentous organism that stains more distinctly at either pole (bipolar). The organism is susceptible to heat, light and phenol. Organism does not produce exotoxin. The organism may be present in the respiratory tract as commensally. Predisposing factors like fatigue, transportation, over exertion, starvation, close confinement may precipitate the disease.

The disease is wide spread in almost all tropical and subtropical countries. Boolinger, investigated the disease in cattle in 1978. The disease occurs in Europe, Africa and Asia. In India, the disease is enzootic in nature. Following monsoon with the rise of relative humidity most of the outbreaks are seen. Cattle and buffaloes of 6 months to 2 years are most often affected. Sheep, goats and pigs also suffer, but horse may acquire the disease to lesser extent. Dogs are resistance to disease. The disease is transmitted through contact with the infected animals. However, the pneumonic pasteurellosis occur through inhalation of contaminated droplets. Saliva contains abundant of bacteria to spread the disease. Disease occur in out break farm with heavy mortality.

The passteurella organisms present as commensal or entered from environment in the terminal boncheol and alveoli in presence of predisposing factors bring about changes in the lung. The organisms are efficient in destroying bovine mononuclear leukocytes of blood and macrophages of lungs. From the death macrophages pharmacologically active amines like histamine and prostaglandin and some fibroblastic elements are released. These substances are responsible for septicemia changes in the body and inflammatory changes in the lung parenchyma. The other secondary invaders like para influenza III virus, bovine herpes virus and bacteria produce significant changes in the lungs. The changes thus produced in the lungs are cause of pneumonia.

In cattle septicemia flares up as an outbreaks during the environmental stress. The incubation period is 2-5 days. There is high-rise of temperature (104-107^{0}F) with concurrent shivering followed by profuse salivation, lachrymation and nasal discharge. There is sign of conjunctivitis and the visible mucous membrane become deep red in colour. There is sharp drop in milk yield. There will be swelling at neck and rapid respiration rate. The affected cases may also exhibit signs of abdominal pain, severe diarrhoea or dysentery. Auscultation of lung reveals increased vesicular rales and moist rales. Due to the progression of the disease lung sound may become inaudible as a result of consolidation. In almost all the cases sububmucosal potechiation are evident. Death occurs due to hypoxia and toxemia. In less acute cases, there is rise in body temperature and localization of edema in the subcutaneous pocket of the head, dewlap and brisket region. The edematous swellings are hot and painful. The edema produces severe dyspnea due to obstruction of respiratory passage. Death usually occurs within 20-24 hours. Dog and cat usually remain as symptom less carrier.

Apart from history and symptoms diagnosis can also be made by identification of the organism through blood smears examination from heart blood, liver, lungs, spleen and intestinal content by stained with leishman stain to observe bipolalr short ovoid rods. Attempt should be made for culture of the organism from nasal swab or washing and the exudates may be used for culture from living animals. The samples may be cultured on blood agar and brilliant green agar plates on which non-hemolytic dewdrop like colonies are seen. Animal inoculation in rabbit and mice is the most suitable to inoculated through subcutaneous or intra-peritoneal route. Susceptible animals usually die within 24-48 hours following inoculation.

Intravenous administration of Sulphamethazine or Sodium sulphadimindine @ 150 mg/kg body weight for 3 days may be given. Oxytetracycline @ 5-10 mg/kg body weight, Chloramphenicol @ 10 mg. Kg. body weight and Ampicillin @ 10 mg/kg body weight may also be used in absence of sulphonamides. Long acting Oxytetracycline may be used to reduce the treatment cost. Many other antibiotics are effective in treatment of the disease. Symptomatic treatment with anti-inflammatory drugs e.g., Betamethazone or dexamethazone @ 1 mg/5 kg body weight may be required. Treatment should be continued for 3-4 days.

Various types of vaccines may be used for prophylaxis. Broth vaccines are used for quite a long period. @ 5 ml for animals upto 300 kg body weight and 10 ml for animals above 300 kg body weight having immunity for about 90 to 129 days.

Anthrax

(Spleenic fever, Carbuncles, Wool sorter's disease, Hide porter's disease, Malignant pustular dermatitis)

It is per cute disease characterized by septicemia and sudden death accompanied by the exudation of tarry blood from the body orifices of the cadaver. Failure of the blood to clot, absence of rigor mortis and the presence of spleenomegali or the most important necropsy finding. Muscle tremors, dyspnoea, congestions of the mucosa are seen during course of disease.

The diseases is caused by bacteria know as *Bacillus anthrasis*. Robert Koch discovered the etiology first in 1873. Bacilli are Gram positive, spore forming, large rod shaped non-motile capsulated, aerobic, spore bearing organism. The organism remains in chain. Spores are formed while the organisms are exposed to air (oxygen). The capsules are formed in the body of infected animals. Spores are very much resistant to chemical disinfectants and drying. Spore may survive in the soil for a long time and for ten yearn in the culture. The organism produces 'Medusa head' appearances on nutrient agar. The spore remains resistant to 100c for five-minute.

Anthrax is worldwide in distribution. In tropical and subtropical countries with high annual rain fall infection persists in soil. Anthrax mostly occurs as sporadic barring few places in India. Death rate is high in per acutely and acutely infected cases, while subacute and chronic cases recover very slowly. The infective agent of anthrax is commonly present in the soil. Under the influence of climate spontaneous out break take place out of spores present in the soil. Therefore, carcasses should not be opened. Hygiene is the biggest single factor in the prevention of spread of the disease. Infection gains entry through ingestion, inhalation or through the skin penetration. Ingestion of contaminated food water and injury of oral mucosa facilitate invasion by the organism, while inhalation root is of minor importance in animals. The disease occurs most commonly in cattle and sheep followed by in goats and horses. Dogs and cats are resistant for the anthrax.

Upon ingestion of the spores enter through defects in the epithelium around erupting teeth or through scratches from tough caused by fibrous food materials. After entry the bacteria are moved to the local lymph nodes by motile phagocytes after proliferation in this site the bacilli pass via the lymphatic vessels into the bloodstream and septicemia takes place followed by massive invasion of all body tissues. During multiplication it produces a lethal toxin that causes edema and tissue damage. Death occurs due to shock and acute renal failure, and terminal anoxia mediated by the central nervous system.

The incubation period ranges from 2 to 10 days. The disease appears mostly in all the four, per - acute, acute, sub-acute and chronic. Per acute cases reveal death without any premonitory symptom but in acute cases body temperature becomes high up to 42°C (107°F), heart rate increases and respiration becomes rapid as well as deep. Mucosae are congested and hemorrhagic. Milk may be blood -stained or deep yellow in color. Alimentary tract involvement is usual and is characterized by diarrhea and dysentery. Local edema of the dependent parts can also be observed. However, in sub acute and chronic farm edema, loss of appetite and blood in milk persists for weeks together.

Diagnosis is based on findings as all natural orifices usually exude dark tarry blood that does not clot. Putrefaction and bloating are rapid. Although necropsy should not be performed but presence of ecchymosed hemorrhages throughout the body tissues and presence of blood stained serous fluid in the body cavities severe enteritis and gross enlargement of the spleen with softening and liquefaction of its structure are most common findings. Ascoli's test which is an thermo precipitine test can be performed in suspected cases. Other serological tests like ELISA, aglutination tests are helpful in diagnosis of the disease.

Anthrax bacilli are sensitive to penicillin and other broad-spectrum antibiotics. Penicillin 10,000 units per kg body weight twice daily through parenteral route or Streptomycin @ 8 to 10 gm.

In countries where the anthrax is enzootic Anti anthrax spore vaccine @ ml sub cutaneously should be given. In heavily contaminated surroundings two doses of vaccines may be given at 15 to 30 days intervals. Vaccination should be carried out during hot summer season in India.

Black quarter

(Felon Rush brand, Black leg, Quarter ill, Symptomatic anthrax, emphysematous gangrene)

This is an acute infectious but not contagious disease of cattle, goat and sheep. The disease is characterized by development of focal gangrenous and emphysematous myositis gives rise to crepitating and sero hemorrhagic swelling in the heavy muscles like gluteal muscles. The disease produces severe toxemia with a very rapid course and high mortality. Black quarter is a predominant disease of cattle but traumatic black quarter may be seen in other animals.

The disease is wide spread in most of the tropical countries of the world. The disease spreads rapidly following heavy rainfall in summer season. The disease is most commonly seen in areas where well-developed cattle and sheep rearing programmes are in existence. Bollinger first recognized the disease. In India, the disease is sporadic in nature. Disease has been recorded from Madras, Bombay, Mysore and Hyderabad and from many other parts of the country.

The disease appears in almost all the States of the country during rainy season.

Black quarter is caused by *Clostridium chauvoei*, a gram-positive rod shaped, spore-forming toxin producing anaerobe. Cattle are the most susceptible host but the infection may spread to other animals like in sheep, buffalo and goat, due to trauma of the muscles. All breeds of cattle are susceptible but the incidence is more in 4 to 24 months of age group with good body condition. In the laboratory animals, guinea pigs and rabbits are most susceptible. The disease spreads from contaminated soils that occur due to infected carcasses. The organisms enter in host through ingestion of infected feeds or possibly through contamination of wounds.

Ingested organisms are carried from the intestine via circulation to the skeletal muscles. The spores from the alimentary tract penetrate tissue from the places of breach of alimentary mucosa due to trauma. Sometimes muscles of gluteal region, loin, and shoulders are get affected. The infection may affect the muscles and intermuscular tissues. There is necrosis of the muscle and blood capillaries. Hemorrhages from ruptured capillaries may occur. The muscles of the tongue, jaw and heart may be affected at a later stage due to dissemination of infection. Toxins produced from the organisms causes systemic reaction characterized by toxemia and local reaction characterized by narcotizing myositis. Severe toxaemia may lead to death. Following death there is rapid tissue decomposition and spores are formed in the dehydrated tissues. Recovered animals are immune.

After the incubation period of 2 to 5 days toxin produced by multiplying bacteria degenerate the affected muscles and gases are evolved. The toxic products are absorbed by the body fluids causing systemic disturbances and a decrease in the animal's vitality. First symptom is a rise in body temperature, which may be as high as 106° C or 108° C, but sometimes, there is hardly any sign of fever. Most commonly the lesions are located on the thigh, buttocks, shoulder, neck and lumber region and more rarely in the interamandibular space or in the tongue. Swellings are hot and painful in the early stage and become cold and painless latter. The muscle tissues are swollen, dark in colour and turn dry. Skin over the swelling is discolored and dry. There is laboured breathing and accelerated pulse rate (100-120/minutes). Sometimes there is distinct abdominal pain. Finally, the temperature drops and the patient die within 12 to 48 hours after manifestation of clinical signs.

In the filed outbreak, a tentative diagnosis is made from the history, clinical observation and postmortem findings. Apart from this disease may be diagnosed by the following methods.

By examination of smear made from the affected tissues or fluids of the swellings, gram (+ ve) rods with sub terminal spores will be seen.

Anaerobic media provides growth within 24 hours. Materials to be used for cultural examinations are heart blood, peritoneal fluids, and affected muscles either from the carcasses of natural host or from the inoculated laboratory animals.

The antibiotic like Penicillin, Chlorumphenicol, Aureomycin and Oxytetracycline may be injected into the affected muscles. Penicillin @ 10,000 units per kg body weight per day may be used. Crystalline penicillin may be given through intravenous route followed by procaine penicillin through interamuscular route.

Passive immunity is induced by the inoculation of immune serum. Prophylactic dose is 15 ml for cattle and 10 ml for sheep. Apart from this a multicomponent vaccine having *Cl. chauvoei, Cl.septium* and *P. multocida* may be given. Black quarter inactivated suspension vaccine of *Cl. Chauvoei* in formaldehyde solution with Aluminum hydroxide gel adjuvant for double immunity. This vaccine should be given @2.5 ml through subcutaneous route before onset of monsoon annually in cattle and buffalo.

Tetanus

(Lockjaw, Shaw horse disease)

Tetanus is non-contagious, non-febrile, highly fatal and infectious disease of all species of Domestic animals caused by the toxin of *Clostridium tetani*. It is characterized clinically by hyperesthesia, tetany and convulsions.

The disease is caused by *Clostridium tetani*. The organism is long rod shaped anaerobe with rounded ends measuring 2µm × 0.5µm. The organism develops terminal spores that give it a drumstick appearance. Organism is gram positive, motile and form long filament in young culture. The organism can be grown under strict anaerobic condition. Organisms remain in the intestine of the herbivorous animals as normal habitat. Soil may have good concentration of organisms if it is fertilized with animal manure. Three strains are distinguished as H, A and O antigens and produces three types of toxins. The toxins are tetanospasmin (neurotoxin), tetanolysin (haemolysin) and fibriolysin. The toxins are released under anaerobic condition especially form contused and abraded wounds.

It occurs in all farm animals mainly as individual, sporadic cases, although outbreaks are occasionally observed in young cattle, young pigs and lambs. The mortality in young ruminants is over 80%, but the recovery rate is high in adult cattle. *Cl. tetani* organisms are commonly present in the feces of animals, especially horses, and in the contaminated soil. Survival of the organism in soil varies very widely. Horse and mules are most susceptible followed by sheep, goat, pigs and cattle. All age group are equally susceptible. The organisms are very much resistant and therefore remain in the environment especially in the dust, garden soil and animal manured soil in large number for a considerable period. Organisms may continue to live in the faeces for a long period of time and thus remain as a potential source of infection to animals.

Organisms on entry through wound may not elicit the clinical signs of tetanus. The affected tissues must have low O_2 tension. Appreciable tissue damage from whatever cause will help the *C. tetani* spores to germinate, grow and toxin production. Localized impairment of blood supply may results in severe damage to the tissues. For the production of the disease there are three conditions that are required. The spores must have access to the animal's body tissues. There must be a lowered oxidation-reduction potential within the tissues where the spores are introduced and the organism must enter to an animal which is susceptible to the action of tetanus toxin. Thus, when a favorable condition is reached the bacteria proliferate and remain localized in the wound. Then tetanospasmin or neurotoxin is released by cellular autolysis. Tetanospasmin thus produced is absorbed from the site and then disseminate in the blood stream or exert neurotoxic effect through absorption by peripheral nerve endings. It may operate centrally in the spinal cord and brain stem or peripherally via the motor nerves. Spastic paralysis occurs due to distribution of toxin through blood stream and flaccid paralysis occurs due to local invasion of the nerves. Toxins may get attached with a substance known as 'protagon' in the nerve tissues.

The incubation period varies from few days to several weeks (1-2 week). The toxin causes hyperirritability responsible to induce titanic spasms. Animals those recover form tetanus may not develop immunity and therefore remain susceptible to recurring attack. Death usually occurs due to asphyxiation as a result of fixation of respiratory muscles. Horse is most susceptible to tetanospasmin 1ml of culture filtrate is enough to kill a horse.

Besides symptoms, there are no gross or histological findings by which a diagnosis can be confirmed. Search should be made for the site of infection, and culture of the organism attempted.

The principles of treatment are (a) destruction of tetanus organisms (b) neutralization of circulating toxin and (c) relaxation of muscles to prevent asphyxiation. Penicillin may be infiltrated locally to minimize further multiplication of the bacteria and antitoxin has to be injected to neutralize toxin depending on the size of the animal at 12 hours interval. Muscle relaxants like chlorpromazine @ 500 mg for an adult horse or 0.2 mg per kg. body weight may be given to other animals. A sterile 10% solution of magnesium sulphate (30 to 60 ml) may be given through subcutaneous route.

Positive immunity can be achieved by injection of antitoxin, Dose-1500 to 3000 IU. in horse @ 10 ml and 1-2 ml for cattle. Duration of immunity is about one year.

Colibacillosis

(Neonatal calf diarrhhoea)

The disease is characterized by acute profuse watery diarrhoea; prolong dehydration, acidosis and death in a few days.

E.coli having K-88$^{=}$,K-99^{+}and K-987^{+} antigen are responsible for the disease. The bacteria are Gram-ve rod, non-sporulated, non-capsulated bacteria. Smith and Little discovered that *E. coli* is causative agent of Neonatal calf diarrhoea.

Disease is found worldwide including India and western Canada where farm animals are kept. Morbidity is 75% if the disease is in intensive form and for dairy calf morbidity is 30%. Mortality ranges from 3-60%. Calves under 30 days of age are more susceptible and piglets in first week of their life are susceptible. Economic loss in cattle and swine heard are in greater importance as future livestock production are decreasing if the diseases is more intensified. Economic impact of this disease in animal is high mortality of animal and for owner it is very expensive treatment if animal get affected and loss in condition of surviving. Most common method of infection is through the alimentary canal by ingestion of contaminated food or through contaminated utensils or udder. Vaginal discharges are the infecting media. Infection through umbilical stump may occur. In foetus intrauterine infection is common.

The disease is found in several forms such as Septicemic colibacillosis, which is commonly found in animal deprived of colustrum. The enteroinvasive *E.coli* enters in intestinal lumen through ingestion, it invades extra intestinal tissue and multiply in body fluid. As a result it kills intra cellular part by phagocytosis and tissue damage occurs by release of cytotoxins. These toxins enters in blood and produces Septicemia resulting into hypothermia, lowering B.P., Tachycardia, decrease cardiac output, change in blood count, hyperglycemia followed by hypoglycemia, depletion of liver glycogen, Meningitis and Arthritis.

The enterotoxigenic *E.coli* enters through ingestion. It colonizes and proliferates in upper part of small intestine then bacteria adhere to intestinal epithelial cell where it multiplies and form colonies over surface of villi. Here it produces enterotoxins which stimulate Adenylocyclase which increases cyclic AMP resulting into increased net secretion of fluid, diarrhoea, dehydration, electrolyte imbalance, acidosis, hyperkalamia, shock and death. Colibacillosis occur in newborn animal which are aganoglobilimia been they have not ingested sufficient colustrum early enough and absorbed sufficient immunoglobulin thus rendering them highly susceptible.

Entericcolibacillosis on the other hand occurs in colustrum fed animal and in calf's by the colonization and proliferation by enteropathogenic *E.coli* which produces enterotoxin and cause varying degree of diarrhoea, dehydration while single infection occur commonly as in piglet diarrhhoea and what was previously described as enterotoxigenic colibacillosis in calf's, multiple infection with EPEC.

On the basis of history of outbreak of diarrhoea in heard, hypothermia, alaxia and weakness etc one can diagnose the disease. Faeces become voluminous and cardly to gelatinous in consistency in this disease.

Treatment

It is common practice to decrease milk intake of diarrheic hand fed dairy calf up to 24 hrs because lactose digestion is impaired and that resting the intestine for a few days would prevent an additional osmotic diarrhhoea caused by fermentation of undigested lactose in large intestine. Dehydration, acidosis and electrolyte imbalance are corrected by paventernal or oral use of simple or balanced electrolyte solution.

Vaccination of pregnant cattle with either purified *E.coli* i.e. $k99^{+}$ pile or whole cell preparation containing sufficient $k99^{+}$ antigen can significantly reduce the incidence by enterotoxigenic colibacillosis in calves. Oral vaccination is not effective in pregnant cows. Oral vaccination with $k99^{+}$ specific monoclonal antibody in calves during the first 12 hrs after birth may be an effective method of reducing the incidence of fatal entrotoxigenic colibacillosis

Mycobacterium Tuberculosis

(Pearly Disease)

Disease is clinically is characterized by the progressive development of tubucler in any of the organs in most species.

The disease is caused by *Mycobacterium tuberculosi.,* This name given because of a lipid substance mycolic acid. This substance makes by mycobactarium impermeable even for antibiotics.

All species, including humans, and age groups are suspendable to M. bovis, with cattle, goats and pigs most susceptible and sheep and horses showing a high natural resistance through out the world. In India it is found at Rajasthan and many state. Morbidity and mortality is very high for the disease. Infected animals losses 10-25% of their productive efficiency.

Mycobacteriums enter in feed-water and dust and through these agent reaction responsible digestive tract or touch wound infestation. Tubercular organism introduce to the tissue. Here local reaction is set up consist of edema, hyperemia and neutrophill infiltration - engulf the bacilli but are not able to kill them, they are probable helpful in limiting the movement of the organism in to tissue and neutrophills die and are infested by macrophage which are attracted to the area. Macrophages arrive at the site in response to the lipid content of the bacilli cells are derived from phagocytes of the reticulo end -othelium and from the monocytes of the blood. Macrophages ingest the neutrophill containing organism as well as the bacilli and bacilli continue Proliferate within the macrophage (8-10 days). Subsequantly more and more macrophage collected arround the earlier ones and nodule formation take place. Bacilli are killed and lipid substance gets dispersed into the cytoplasm of the macrophage covered by it into an "Epithelioid cell" organisms are spread by the lymph or the organism enters in the blood reach to different organ of the Animal.

Tuberculus granulosma may be found in any of the lymph nodes, but particularly in bacterial and mediashnal nodes, and many organ. In lungs, Miliary abcess may extend to cause a suppurative branchopneumanea. Characterstics of pus is cream to organs color and various in consistency from thick cream to crumbly cheese. Nodules are hard painful and calcified. On cut cheesy material comes out from the nodules. Tuborclosis tend to stimulate enveloping fibrous capsule but degree of encapsulation varies with the rate of development of lesion. Miliary tuberclosis with small transparent shot like lesion in many organ.

In advanced stages when much lung has been destroyed dyspnea with increased rate and depth of respiration become apparent. At this stage abnormalities may be detected by Auscultation and precusion of the chest areas with no breath sounds and dullnen. Enlrgement of the mediastinal lymph - node is commonly associated with recurrent and then persistant ruminal tympany. Tuberculous ulcers of the small intestine cause diarrhoea. Retropharyngeal lymph node Enlargement causes dysphagia and noisy breathing due to pharyngeal obstruction. By the Palpation: pharyngeal palpation with the aid of a speculum reveals a large firm rounded swelling in the dorsum of the pharynx.

Symptom as irregular grayish yellow/ white nodules of vary size, nodules are firm and surrounded by a fibers capsules. Diagnosis of the tuberculosis in living animal is base on the tuberculosis test, X-rays findings, and demonstrations of the organisms in executers or excretion. In times obtained surgically (biopsy) or at necropsy demonstrations of acid-fast organism within typical tuberclosis is sufficient to establish the diagnosis. However, bacterial isolation is necessary to establish its type. Acid-fast bacteria appear to infect many different species. It is important that in each case organism be recovered in culture for careful identification.

In animal claims have been made for the efficiency of long-term oral medication with isoniazid both-as treatment and as prophylaxis. Howevere, Raphempacine, Streptomycine, Ethembutol and Para amino salicylic Acid may be given @ 10mg /kg body weight daily for three months for the treatment of tuberculosis.

Actinomycosis

(Lumpy Jaw)

It is a chronic infectious disease, primarily of cattle and secondarily to pig, horse and man that is caused by a bacterium – *Actinomyces bovis.* The most common manifestation of this disease in cattle is a rarefying osteomyelitis of bones of the head, particularly the mandible and maxilla. On rare occasions it involves soft tissues, particularly the alimentary tract.

Actinomycosis is caused by *Actinomyces bovis*. It is a gram positive, rod-shaped anaerobe. It is pleomorphic, non-sporulating and non-motile organism. Actionmyces bovis is a strict obligatory parasite of man and animals being present in the mucosa of the mouth and pharynx.

The disease is common only in cattle where a tendency towards inherited susceptibility has been observed. Occasional cases occur in pigs and horses. A bovis is a common inhabitat of the bovine mouth and infection is presumed to occur through wounds to the buccal mucosa caused by sharp pieces of feed or foreign material. Infection may also occur through dental alveoli, and may account for the common occurrence of the disease in young cattle when the teeth are erupting. Infection of the alimentary tract wall is probably related to laceration by sharp foreign bodies.

The organisms are present in the buccal cavity and cause the disease after injury. There is invasion via oral mucosal lesions, erupting teeth sockets or laceration reticular, intestinal mucosa by sharp foreign body. Organism, a natural habitat of mouth is carried into tissue by sharp objects, may get entrance during change of teeth. Bite wounds at milking times especially in sows. The objects penetrating into cause necrosis, thereby initiating the anaerobic conditions for growth of the organism. The organism causes low grade inflammation reaction. Leucocytes invade the area that becomes walled off by connecting tissue growth. Throughout the mass, small abscess form containing organism growing in clubs sulphur granules. The lesion is chara cteristically granulomatous both in this site and where visceral involvement occurs. The effects on the animals are purely physical. Rarely the localization occurs in other organs, caused apparently by hematogenous spread from these primary lesions.

Actinomycosis of the jaw commences as a painless, bony swelling that appears on the mandible or maxilla, usually at the level of the central molar teeth. The more common, discrete lesions on the lateral surfaces of the bones are more readily observed. Some lesions enlarge rapidly within a few weeks, others slowly over a period of months. The discharge of pus is small in amount and consists of sticky, honey like fluid containing minute, hard yellow-white granules. Teeth embedded in the affected bone become malaligned and painful and cause difficult mastication with consequent loss of condition. In severe cases, spread to contiguous soft tissues may be extensive and involve the muscles and fascia of the throat. Excessive swelling of the maxilla may cause dyspnoea. Involvement of the local lymph nodes does not occur. Eventually the animal becomes so emaciated that destruction is necessary although the time required to reach this stage varies form several months to a year or more.

There is no satisfactory treatment against this malady. Early lesions get regressed satisfactorily with iodine and sulphur therapy. The lesion should be drained surgically and the sinus tract should be packed with lugol's iodine daily through intravenous route. Sodium iodide @ 15-30 gms, in 250-500 ml distilled water may be given through the intravenous route. The disease has been treated effectively with isoniazide at the dose level of 20mg/kg. body weight for 100 days. Streptomycin combined with iodide and surgical drainage may yield good response. There is no vaccine against this disease. Isolation of infected animals and treatment are to be rendered. Removal of contaminated materials and disposal of animals with discharging foci may be made.

Strangles

(Equine distemper)

It is contagious disease of equine characterized by fever, anorexia, depression and serous nasal discharge. There will be submandibular lymph nodes enlargement.

It is caused by *Streptococcus equi* subspecies *equi*. Is a Gram-positive coccobacillus that produces a B-hemolysin.

Strangles occurs in horses, donkeys and mules world widely. Strangles can affect horses of any age but morbidity rate is usually greater in younger horses. The morbidity usually remains 35% and case-fatality rate without treatment is about 100%, but with adequate early treatment may be as low as 1-2%. Nasal and abscess discharge from infected animals act as source of contamination of pasture, stalls, feed, water, grooming equipment and animal handlers which intern transmit the disease in susceptible horses.

After getting entry in the oral and nasopharyngeal mucosa bacteria get lodged in the pharyngeal and tonsillar lymphoid nodes to multiply. Subsequently inflammation, swelling and abscessation of lymph nodes and lymphatic vessels take place. Further infections may extend to cause bronchopneumonia. Shedding of *Strep. equi* from the nasal secretions usually begins 4-7 days after infection. Death usually occurs due to pneumonia caused by aspiration of infected material.

Disease occurs in acute form but may covert in chronic to in the form of purpura hemorrhagica. In acute cases after an incubation period of 1-3 weeks the disease develops suddenly with complete fever, anorexia, depression and serous nasal discharge. There will be submandibular lymph nodes enlargement and palpation elicits a painful response due to Lymphadenopathy. Obstruction due to swelling of the retropharyngeal lymph nodes may cause dyspnea and death because of asphyxiation. Sometimes complications occur in about 20% of cases in the form of fatal suppurative necrotic bronchopneumonia.

Although history helps in tentatively diagnosing the disease, however clinical manifestations are also helping. On hematology during the acute phase of the disease leucocytosis with a neutrophilia are observed. Hyperfibrinogenemia is characteristic of both the acute and chronic disease. Hyperproteinemia, hypoalbuminemia may also be present. Confirmation of strangles depends on the isolation of *Strep. Equi.* From nasopharyngeal swabs and discharges from abscesses. Various serological tests may be carried out for research and academic purposes.

For treatment of this disease penicillin either as procaine penicillin-G @ 22000 IU/kg intramuscularly at every 12 hours or potassium or sodium penicillin-G @ 22000 IU/kg intravenously at every 6 hours may be given. Tetracycline @ 6.6 mg/kg. body weight at every 12-24 hours or sulfonamide-trimethoprim combinations @ 15-30 mg/kg. orally or intravenously at every 12 hours may be used along with other supportive treatment and local dressing of lesions.

To control the disease transmission of *Strep. equi* should be prevented form affected horses. There is no suitable vaccine for the disease. Elimination of infected animals is a general practice to control the further outbreaks.

Glanders

(Farcy)

Glanders is a highly contagious disease of solipeds characterized by nodular lesions of the lungs and other organs Ulcerative lesions are also present on the skin and mucous membranes of the nasal cavity and respiratory passages. It has a progressive course and risk of zoonosis.

Glanders is caused by the bacteria *Pseudomonas mallei*. This pathogen was formerly known as *Loefflerella mallei*, *Pfeifferella mallei*, *Malleomyces mallei*, *Actinobacillus mallei*, *Corynebacterium mallei*, *Mycobacterium mallei*, and *Bacillus mallei*. The organism is closely related to *Ps. pseudomallei*, the cause of melioidosis, and is serologically related. It can be destroyed by direct sunlight and is sensitive to desiccation. It is readily killed by common disinfectants and may survive for up to 6 weeks in infected premises.

Glanders is currently occurs sporadically at Asia, Africa, the Middle East, and Asia. In Asia it is prevalent at India, Pakistan, Turkey, Syria, Iraq, Iran, Burma and China. Besides horses it also occur in donkeys and mules. Carnivores are susceptible to disease if they consume glandered meat. Humans also are susceptible to infection with glanders, which is an important occupational disease of veterinarians, farmers, and other animal workers. Swine and cattle are resistant to infection but goats may acquire infected. Ingestion of the pathogen, present in secretions from infected animals, constitutes the major route of infection in glanders. Inhalation of the organism is less likely to result in typical cases of the disease. However, invasion by way of skin lesions is possible.

Entry of bacteria occurs through the intestinal wall. There they are taken up by lymphocytes and brought to the lymphnodes. After multiplication septicemia take place in acute form and bacterimias in chronic form. Localization maximally occurs in the lungs followed by skin and nasal mucosa. Other viscera may become the site of the typical nodules. These nodules convert in abscess. In the respiratory passes bronco-pneumonia occurs and deaths in typical cases are caused by anoxic anoxia.

After infection fever develops in about 3 days and clinical signs within a week or months in chronic cases. Glanders may occur in cutaneous, nasal, and pulmonary forms, but in most outbreaks these forms are not clearly distinct and may occur simultaneously. Chronic infections with slow progression of an insidious course are more common than the acute disease. The acute form is more common in donkeys and mules than in horses. The nasal form of glanders is characterized by unilateral or bilateral nasal discharge. The yellowish-green exudates are highly infectious. The nasal mucosa has nodules and ulcers. These ulcers may coalesce to form large ulcerated areas, or they may heal as stellate scars. In some cases the perforation of septum may occur. Nasal lesions are accompanied by enlargement, indurations and suppuration of regional lymph nodes. In the cutaneous form of glanders, multiple nodules may develop in the skin of the legs or other parts of the body. These nodules may rupture, leaving ulcers that discharge a yellow exudate to the skin surface and heal slowly. Cutaneous lymphatic vessels of the region become distended and firm by being filled with a tenacious, purulent exudates. In the pulmonary form of glanders there will be formation of firm nodules. The nodules are either gray or white and firm. They are usually surrounded by a hemorrhagic zone, become caseous latter and then calcified. When horses, donkeys, and mules are concentrated, the morbidity can be high.

Ps. mallei are sensitive to many antimicrobial, but risk of spreading the infection to other animals or to humen forces to elimination of the affected animals. This policy has successfully eradicated glanders from most parts of the world. Sulfonamides have traditionally been used for the treatment of human infection.

In endemic areas, routine testing and destruction of positive animals have proven successful in the eradication of the disease. In endemic areas, communal feeding and watering sites should be avoided. During outbreaks it is important to bury or burn all contaminated bedding and foodstuffs to prevent infection of susceptible animals. Stalls and harness equipment should be thoroughly disinfected. Removal of susceptible species from contaminated premises for a period of months is advisable. Protective vaccines have not been developed.

Fungal diseases

Dermatophytosis

Ringworm

Dermatophytosis or ringworm is a clinical condition caused by fungal infection of the skin in humans, pets such as cats, and domesticated animals such as sheep and cattle. The fungi that cause parasitic infection (dermatophytes) feed on keratin, the material found in the outer layer of skin, hair, and nails. These fungi thrive on skin that is warm and moist, but may also survive directly on the outsides of hair shafts or in their interiors. It has been estimated that currently up to twenty percent of the population may be infected by ringworm or one of the other dermatophytoses.

Infections on the body may give rise to typical enlarging raised red rings of ringworm, infection on the skin of the feet may cause athlete's foot and in the groin jock itch. Involvement of the nails is termed onychomycosis, and they may thicken, discolour, and finally crumble and fall off. Fungi thrive in moist, warm areas, such as locker rooms, tanning beds, swimming pools and in skin folds. It can be spread by sharing sport goods, towels, and clothing.

Ringworm in pets may often be asymptomatic, resulting in a carrier condition which either infects other pets, or shows disease only when the companion animal develops an immunosuppressive condition. Circular bare patches on the skin suggest the diagnosis but no lesion is truly specific to the fungus (similar patches may result from allergies, sarcoptic mange, and other conditions). Three species of fungi cause 95% of dermatophytosis in pets, these are Microsporum canis, Microsporum gypseum, and Trichophyton mentagrophytes. Veterinarians have several tests to identify ringworm infection and identify the fungal species that cause it:

Woods Test: This is a black light (ultraviolet light) with a magnifying lens. Only 50% of Microsporum canis will show up as an apple-green fluorescence on hair shafts, under the black light. The other fungi do not show. The fluorescent material is not the fungus itself (which does not fluoresce) but rather an excretory product of the fungus which sticks to hairs. Infected skin does not fluoresce.

Microscopic test: The vet takes hairs from around the infected area and places them in a staining solution to view under the microscope. Fungal spores may be viewed directly on hair shafts. This technique identifies a fungal infection in about 40%–70% of the infections but cannot identify the species of dermatophyte.

Antifungal treatments include topical agents such as miconazole, terbinafine, clotrimazole, ketoconazole, or tolnaftate applied twice daily until symptoms resolve — usually within one or two weeks. Topical treatments should then be continued for a further 7 days after resolution of visible symptoms to prevent recurrence. The total duration of treatment is therefore generally two weeks, but may be as long as three.For prevention avoid sharing clothing, sports equipment, towels, or sheets. After being exposed to places where the potential of being infected is great, one should wash with an antibacterial and anti-fungal soap or one that contains tea tree oil, which contains terpinen-4-ol.

Aspergillosis

(Brooder pneumonia)

Aspergillosis has been observed in almost all birds and animals, including man. The disease is observed in one of two forms; acute outbreaks with high morbidity and high mortality in young birds, and a chronic condition affecting adult birds. It is more of a problem in turkeys than in chickens.

The condition is caused by Aspergillus fumigatus, a mold or fungus-type organism. Occasionally other types of molds are involved. These organisms are present in the environment of all poultry. They grow readily on many substances such as litter, feed, rotten wood and other similar materials.

The bird comes in contact with the organisms through contaminated feed, litter or premises. The disease is not contagious and does not spread from one bird to another. Most healthy birds can withstand repeated exposure to these organisms. Inhalation of large amounts of the infectious form of the mold or reduced resistance of the bird apparently results in infection. In adult turkeys, the disease more often affects the male.

In the acute form in young birds, main symptoms are gasping, sleepiness, loss of appetite and sometimes convulsions and death. Occasionally the organism invades the brain, causing paralysis or other forms of nervous symptoms. The more chronic form in older birds usually results in loss of appetite, gasping or coughing and a rapid loss of body weight. Mortality is usually low and only a few birds are affected at one time.

The disease produces hard nodular areas in the lungs and an infection of the air sacs. Sometimes the air sac lesions are similar to those produced by infectious sinusitis or CRD. In some birds, colonies of mold growth can be seen on the air sac membranes.

Diagnosis is usually made from history, symptoms and lesions. It may be necessary to base diagnosis on microscopic lesions.

The disease can usually be prevented by avoiding moldy litter, feed or premises. There is no treatment for the affected flock. Cleaning and disinfecting the equipment is often helpful.

Parasitic Diseases

Fasciolosis

(Liver fluke disease, liver rot, pipe stem liver)

It is characterized by both acute and chronic form of disease caused by large flukes parasitic in the bile ducts and intestines of mammals, elongated leaf like and animal shoe clinical manifestation loss of appetite, debility, jaundice, anemia, bottle jaw

Caused by *Fasciola hepatica* and *F. gigantica.* It is leaf shaped, broader anteriorily than posteriorly with aneteriorly cone shaped, broader anteriorly than posteriorly with anteriorily cone shape projection followed by pair of broad shoulders.

Cosmopolitan distribution but more where water lodging. *F. gigantica* is found in Asia, Africa, Australia and India and *F. hepatica is* found in European country and South and North America, high altitude of Asia and Africa. In India fasciola is mainly caused by *F. gigantica* but *F. hepatica* also present at high attitude and hilly area. A. P. W. Thomas (1883) discovered this trematode. Fluke occurs in the bile duct of sheep, goat, ox and other ruminants, pig, rabbit, elephants, horse, dog, cat, kangaroo and man. Diseases were common in cattle, buffalo and sheep. In horse and man, fluke may found in lungs. In goat, horse and pig incidence of disease is low. It is transmitted through snail of *Lymnea* spp. (L. *truneculata*, L. *aunculata*).

Life cycle of the parasite is indirect. Eggs come in faeces and hatch in the environment to release miracidium that penetrate inside snail and undergoes asexual reproduction to develop sporocyst and rediae. These leave the snail in 5-8 weeks from the time of infection in the form of cercarie. Cercarie immediately convert in metacercarie. Which may survive for 4-6 months depending upon humidity. Metacercarie are ingested by definitive host and excyst in small intestine. Immature larvae penetrate the lumen and come in peritoneal cavity than goes to liver capsule to damage liver parenchyma during movement, where it live for 6-8 weeks then reaches to bile duct. By the time it become mature and start laying eggs.

Time taken from sporocyst to metacercarie is 21 days in *F. hepatica* and 75 days in *F. gigantica* in dry weather. Prepatent period of *F. hepatica* is 9 weeks and *F. gigantica* is 13-14 weeks. Life span of fluke is 9 month.

Exclusively encounter in sheep and occurs after 5-6 weeks of ingestion. It occurs by liver parenchyma damage by moving immature fluke. It is essential traumatic hepatitis, liver very much enlarge; anorexia, distended abdomen with painful to touch, hypoprotenemia If number are excessive, rupture of liver capsule and hemorrhage into the peritoneal cavity. In acute cases the animal dies suddenly, blood stained froth appears at the nostrils and blood is discharged from the anus.

In cattle and buffalo there are digestive disturbances, constipation is marked, diarrhoea and in extreme cases, decrease milk yield, anaemia, weight loss, lack of vigor, paler mucus membrane, anorexia, edema of dependent part and edema in intermandibular space called as bottle-jaw may also be seen. Skin becomes dry and wool brittles and falling out in patches. Prostration occurs in severely affected animals.

The disease can be diagnose by faecal examination for detection of typical elliptical eggs, having less distinct embryonic mass, operculum and yellow in colour. Hematology will reveal hypercytic, normochromic anaemia and eosinophilia

Carbon tetrachloride, hexachloroethane, hexachlorophene, brotianide and nitroxynil are old drugs, presently not in use. Now-a-days following medicines are in common use.

1. Refoxanide @ 7.5 mg/kg body weight
2. Clioxanide @ 20-40 mg/kg body weight
3. Dianiphenenthial @ 100 mg/kg body weight
4. Oxyclozanide @ 10 mg/kg body weight
5. Albendazole @ 15 mg/kg body weight
6. Fendendazole @ 20 mg /kg body weight
7. Closental @ 10 mg/kg body weight
8. Triclabendazole @ 12 mg/kg body weight

Control of the disease can be achieved by regular prophylactic treatment at least two times in a year during months of February-March and September-October. Therapeutic treatment as and when required should be carried out to minimize the chances of spread of disease in other contact animals. Management practices in aspects of hygiene should be regularly monitored. If possible attempt should be made for disruption of miracidia, as it may be most suitable aspect of disease control. Control of snail by $CuSO_4$, molluscicide and by ducks or rearing that feed on snails. Chemical treatment of pasture by calcium cyanide, $CuSO_4$ etc may also be useful. Vaccination are being tried but still at experimental stage.

Paramphistomiasis

(Stomach fluke disease)

Paramphistomes are mainly parasite of the fore stomach of ruminants, although a few species occur in the intestine of the ruminants, pigs and horses and manifests with digestive disturbances.

Their shape is not typical of the trematodes but it is conical in place of flat. A water snail is necessary as an intermediate host.

Migrating paramhistome flukes cause intestinal paramphistomiasis. Cattle and sheep also open for infection. The paramphistomes are cosmopolitan but disease is most common in warmer regions, particularly Australia, Africa and India. Paramphistomiasis mainly depends for its continuous endemicity on permanent water masses, such as lakes and ponds, from which snails are dispersed. Paramphistome egg deposited by animal grazing in that area where snails loading, hatch and infect snails. Subsequent production of cercarae also helps in disease transmission. Most outbreaks of disease occur when pastures are heavy contaminated with encysted cercarae. Adult cattle's are less susceptible to disease due to development of immunity while sheep and goat are more susceptible.

The immature fluke excyste in the duodenum as they migrate, they attach firmly with to the mucosa and may penetrate with the muscularis mucosa. They are plug feeder and this result in severe erosion of duodenal mucosa. In heavy infection these cause enteritis, characterized by edema, haemorrhage and ulceration. Mature fluke in the rumen does not usually elicit any significant response but in massive infection papillae are short and red, becoming fused into aggregation with ruminal contents adhering firmly to the surface.

In heavy infections the most common sign is diarrhoea along with anorexia and intense thirst, something in cattle, there is rectal haemorrhage following a period of prolonged straining. Death may occur in 20 days, after the first sign appear. The mortality rate in acute outbreaks can be as high as 90%. Mature flukes in the fore stomach of the animals normally cause little harm although loss of weight, rough-dry.

The diagnosis is based on clinical sign and history of feeding fodder grown around snail habitats. Faecal examination is also helping in out breaks since the disease occurs during the prepatent period. Confirmation can be obtained by postmortem examination and recovery of the small flukes from the duodenum.

In paramhistomosis following drugs are commonly used.-

1. Oxyclozanide @ '10-15 mg/kg 2 days apart for immature and adult flukes.
2. Niclosamide @ 160 mg/kg as a single dose.
3. Resorantel effective for both immature and mature flukes.
4. Hexachlorophene @ 20 mg/kg as a single dose.
5. Fenbendazole @ 5-7.5 mg/kg body weight single dose.

Prevention of disease is very difficult because of practical inability in drainage of low-lying area and abundance of host snail. Molluskicides are usually ineffective in elimination of snails; moreover chances of toxicity are there due to accumulation of chemicals at some places. Animal should be kept away from the floodwater because metacercarae may persist during this period. Treatment at the time of peaks in the availability of metacercarae reduces the number of eggs in the feces.

Toxocariasis in rminants

(Round worm)

It is the largest intestinal parasite of cattle. It is clinically characterized by loss of appetite, poor growth rate and digestive disorders.

Toxocara vitulorum, a nematode parasite. The female are upto 30 cm and males 25cm long. It is a thick worm, pinkish when fresh and cuticle is rather transparent so that internal organs can be seen.

The disease is worldwide in presence. The most important feature is the reservoir of larvae in the tissues of cows, with subsequent milk born transmission ensuring that calves are exposed to infection from the first day of life. Morbidity is high but mortality is moderate. Mortality generally appears in neonates. Susceptible host ingests contaminated eggs having L2 larvae.

Parasite larvae (L2) enter in host intestine through ingestion. Larvae undergo somatic migration in tissues and create prenatal infection. Under natural condition mature parasite may be found in calves aged 10-42 days and even upto first 4 or 5 months of calf hood. In animals infected after birth, extensive migration of larvae occurs in organs like liver, lungs and kidneys. Life span of parasite is 11-12 months and prepatent period is 12-13 weeks.

Heavy infections are associated with poor thriving and intermittent diarrhoea, and in buffalo calf mortality may occur. Frequently the breath of animal has a butyric odour. Animal becomes dull.

History, clinical manifestation, and sub globular eggs, with thick-pitted shells are helping in diagnosis.

The adult worms are susceptible to a wide range of anthelmintics like piperazine, levamisole, and benzimidazole, Ivermectin etc, at a proper dose. Regular treatment of infected animals should be combined with good hygienic condition. The prevalence of infection can be reduced by treatment of calves at 3 and 6 weeks of age preventing developing worms reaching patency.

Jaundice

It is also referred as icterus and characterized by deposition of bilirubin leading to yellow colouration of plasma, visible mucous membranes and other tissues. Jaundice can be of three types:

1. Haemolytic jaundice
2. Hepatocellular jaundice
3. Obstructive jaundice

Haemolytic jaundice is also referred as pre-hepatic or overproduction jaundice. It can be caused by haemoprotozoan, viral and bacterial toxins, inorganic or organic toxins and plant toxins and immunological reactions. In anaplasmosis, babesiosis, leptospirosis, bacillary haemoglobinuria, infectious anaemia, chronic copper poisoning, selenium toxicity and due to excessive ingestion of brassica plants or berseem, there is occurrence of haemolytic jaundice.

Hepatocellular jaundice is also known as intrahepatic jaundice and occurs due to damage of hepatocytes by infectious or toxic agents. It can be caused by different infections and toxins causing hepatitis and deficiency of vitamin E or selenium.

Obstructive jaundice is also referred as extrahepatic or post-hepatic jaundice and occurs as a result of obstruction of bile duct. Biliary calculi or infection with nematodes or trematodes are common causes of bile duct obstruction. Mechanical stasis of biliary flow can also occur due to fibrosed tissue.

Under normal circumstances, there is destruction of erythrocytes when their life span is over. Due to destruction of these cells, haeme and globin moieties are separated and haeme is converted into iron molecules and bilirubin. This bilirubin is conjugated in the liver to form bilirubin diglucuronide. Some amount of bilirubin is passed into kidneys also. The conjugated bilirubin is converted into mesobilirubinogen and stercobilinogen, which after oxidation are converted to urobilin and stercobilin. Some amount of this urobilin and stercobilin are passed through urine and a faeces while some is reabsorbed.

In haemolytic jaundice, there is excessive destruction of erythrocytes and excessive bilirubin formation occurs. The bilirubin thus formed, is conjugated in the liver and excreted into the intestine. Liver is capable of excreting large amounts of bilirubin. However when limit of liver threshold is crossed by excess release of bilirubin, there is accumulation of unconjugated bilirubin in the circulation. In hepatocellular damage, the capacity of hepatic cells to conjugate free bilirubin is decreased. Due to it, there is diffusion of bilirubin from the biliary canaliculi to blood stream and free bilirubin is accumulated in excessive amount in the circulation. Intrahepatic cholestasis occurs due to swelling of hepatocytes and edema, which results in partial obstruction of biliary excretion. It results in excess bilirubin accumulation in circulation as well as its excess excretion in urine. There is rise in urobilinogen in urine also. When bile duct is obstructed, the conjugated bilirubin can not pass into intestine for bacterial oxidation. This results into the diffusion of conjugated bilirubin from bile duct into circulation and there is elevation of conjugated bilirubin in serum.

Irrespective of type of jaundice, there is yellowish discolouration of mucous membranes and skin. Urine becomes dark coloured. In infectious diseases, the degree of jaundice indicates the degree of haemolysis. In general, the degree of icterus has a direct relation to the severity of disease whether it is haemolytic, toxic or obstructive. The cases become anorectic and anaemic and reveal muscular weakness and mental depression. In terminal stage, they may suffer from recumbency and coma if treatment is not initiated. In haemolytic jaundice, haemoglobinuria also occurs while in hepatocellular jaundice, hepatic enlargement and pain on hepatic palpation are also noticed. The cases suffering from obstructive jaundice reveal light or clay coloured faeces. It is commonly seen in monogastric animals.

The disease can be diagnose by proper history recording and signs of yellow mucous membrane discolouration and deep yellow urine are sufficient to indicate the presence of disease.The blood can be examined for total bilirubin and ratio of free and conjugated bilirubin. Other tests like van den Bergh test and measurement of icteric index are also valuable in its detection. Examination of urine for bilirubin and urobilinogen are also helpful in its diagnosis.

For the management of jaundice it is essential that the primary cause be treated. The bacterial agents, protozoan infections or other factors causing excessive erythrocyte destruction should be treated. Similarly the agents associated with obstruction of bile duct, should be taken care of. The factors responsible for hepatocellular damage may be treated accordingly. However dietary management helps in early recovery of the cases. The affected animals should be given highly palatable and laxative diet containing excessive carbohydrates and rich in calcium salts. The diet should have minimum amount of fats. Intramuscular administration of liver tonics and vitamin B complex @ 5-10 ml once daily for 4-6 days also helps in early recovery.

Parturient paresis

(Milk fever, Post parturient hypocalcaemia)

It is a production diseases occurring most commonly in recently calved animals within 48 hours after parturition and characterized by general muscular weakness, recumbency, circulatory collapse and depression of consiousness.

Depression of levels of ionized calcium in tissue fluids is the major biochemical defect in parturient paresis. The onset of lactation results in a sudden large demand of calcium. Approximately 23 grams of calcium is drained in single milking in a cow giving about 10kg of colostrum. This calcium drain should be appropriately replaced by the intestinal absorption and bone resorption. Most of the recently calved animals adopt to this demand with the help of parathyroid hormone (PTH) and 1, 25- dihydroxy vitamin D. The cows which do not adopt to this develop milk fever.

It is most common in high yielding dairy animals in their 3-5th lactation. The Jersey and its crosses are more susceptible. Majority of the cases occur within 48 hours after calving but the danger period extends upto 10th day post partum. Generally the disease is sporadic but the incidence may be high on certain farms. Majority of the treated, have uneventful recovery but 15-25% cases get complicated. The case fatality rate is very low unless complicated with some other conditions.

Normal calcium level in bovines ranges between 8-12 mg/dl. The clinical signs are evident when the level falls below 5.5 mg/dl. The hypocalcaemia exerts its effect on skeletal, cardiac and smooth muscles. Effect on skeletal muscles is indicated by fatigue, muscular weakness, hypothermia and recumbency. Decrease in cardiac output, decreased stroke volume, decreased, amplitude of pulse, decreased tissue perfusion and hypoactivity of the vital organs, alimentary stasis, decreased amplitude and contractions of rumen are other effects of hypocalcaemia. There may be concurrent hypophosphataemia.

The disease is seen in three arbitrary stages.

First stage (Stage of excitement): This is a brief stage and many a times goes unnoticed. Hypersensitivity, muscle tremor of the head and limbs, grinding of the teeth, normal to slightly above normal rectal temperature, ataxia is noticed in this stage.

Second stage (Stage of Sternal recumbency): Majority of the cases are seen in this stage. The animal is seen in sternal recumbency with a typical posture of lateral kink in the neck and head resting over the flank. Subnormal temperature, cold extremities, dialated pupil unresponsive to light, relaxed anal spincture, increase in rate (80/min) but decreased intensity of heart sounds, weak pulse, ruminal stasis are important findings in this stage.

Third stage (Stage of lateral recumbency): The animal is almost comatose and lies in lateral recumbency. Heart sounds are almost inaudible and the rate is high (up to 120/min). It is almost impossible to raise the jugular pulse indicating venous collapse.

Spontaneous recovery is very rare and untreated cases deteriorate and die in about 12 to 24 hours.

The history of recent calving and classical clinical signs are indicative of the condition. Negative Sulcowitch test on urine is preliminary indication of hypocalcaemia. Hypocalcaemia (£ 5.5 mg/dl), elevated Creatinine phosphakinase in blood/serum are important biochemical changes indicative of the condition. Marginal hypermagnesaemia (> 3 mg/dl) and hypophosphataemia (1.3 - 3 mg/dl) can be seen in few cases. Eosinopenia, lymphopenia and neutrophilia are some haematological findings indicative of adrenocortical hyperactivity in the cases of parturient paresis

Standard treatment of milk fever comprises supplementation of parenteral calcium preparations. Adult dairy animals require about 8-12 gms of calcium as a total dose. Commercial prepartations usually contain 20-25% of calcium borogluconate. Boric acid in this type of preparations make calcium gluconate complex more stable. The requirment of such preparations in a dairy animal weighing 300-400 kgs is about 400-500ml. Underdosing is the main reason for incomplete recovery and relapse. 50% of the calculated dose is administered by intra-venous, route and remaining 50% is administered by subcutaneous route. The subcutaneous calcium acts as a depot for steady calcium level maintenance. Composite solutions containing calcium, magnesium, phosphorus and glucose are recommended to non-responsive and relapse cases with initial calcium therapy.

Response to Calcium therapy: Belching, muscle tremors over flanks, increase in the intensity of the heart sounds, sweating of muzzle and defecation are the sings indicative of good response. Unusual reaction includes increase in heart rate beyond 100-120/min, Grunting, sudden excitement which indicates overdose or rapid infusion rate. Calcium therapy should be stopped immediately to avoid further complications.

Miscellaneous Therapeutically regimen

Udder inflation technique is a good old miscellaneous treatment for milk fever. The udder is inflated to the moderate shape with the help of special hand pump and the teat orifice is sealed with cellophane tape. This prevents the secretion of milk in the acini there by preventing further calcium drain into the gland. Thorough aseptic measures are required to prevent the risk of introducing the infection.

The Downer Cow syndrome

The downer cow syndrome which usually occurs as a complication of hypocalcemic parturient paresis which is clinically characterized by prolonged recumbency even after two successive treatments with calcium.

Though the etiology of downer cow syndrome is not clear the available evidence and perusal of literature indicates that it is a complication of parturient paresis wherein the treatment is delayed. The other causes attributed are:

- Traumatic injuries to the hind legs during parturition due to slippery floor,
- Difficult parturition due to oversized calf leading to maternal obturator paralysis,
- Electrolyte imbalances during prolonged recumbency after parturition like hypophosphataemia, hypokalemia and low level hypomagnesemia.

The condition is most commonly recorded in first 2 to 3 days after calving in heavy milk producers. The exotic and cross bred cows in 3rd to 5th lactation are most susceptible.

Prolonged recumbency after parturition and delay in the treatment of such cows is the most important cause of downer cow. Prolonged recumbency results in ischemic necrosis of the muscles. This also leads to increased muscle cell membrane permeability allowing the loss of potassium from the cells resulting in myotonia there by inability of the cow to get up. Damage to the underlying nerves like sciatic plexus due to prolonged recumbency is also attributed as the cause. The other complications due to prolonged recumbency are decubital ulcers over sternum, stifle joint and perennial region and mastitis.

The most important clinical finding is, cow is down and unable to get up. Majority of the downers are bright and alert, ruminating, defecating and urinating normally and make attempts to get up on their forelimbs. Such cows are referred as creeper cows. Usually the temperature is normal and heart rate may be elevated to 80-100/min.

As the condition advances decubital ulcers develop on sternum, hock joint, elbow joint and gets complicated with maggots. The condition of the animal also goes down with marked emaciation. The prognosis is poor in the recumbent animals even after 7 days. The death ensues in such cases due to myocarditis.

The calcium, phosphorus, magnesium and glucose levels are usually within the normal range which poses a challenge for the diagnosis. Recent studies indicate 10-50 times elevation of serum enzymes like creatinine phosphokinase (CPK) and serum glutamic oxaloaecetate transaminase (SGOT) within 24-48 hours after onset of recumbency are characteristic biochemical findings of the downer cows. Marked proteinuria in early recumbency stage is also diagnostic feature. Elevated blood urea nitrogen (BUN) and serum creatinine usually suggest poor prognosis.

Prolonged recumbency resulting due to physical injuries, ruptured gastrocnemius, dislocation of hip and fracture of major long bone of the hind limb usually should be differentiate from true downers. Slippery floor, no relevance to the recent calving and physical examination of such cases will help in the confirmation of the above.

Treatment of downer cow poses a challenge to the physician and requires lot of patience during the course of treatment. Therefore good nursing care and the management of a downer cow is a priority along with medical management. Provision of good bedding preferably sufficient quantity of straw or soft ground surface like sandpit helps in early recovery. Hot water beds have been used in the downer cows with good success rates.

Medical management includes supplementation of composite solution containing calcium, magnesium, phosphorus and glucose about 400-600ml I/V balanced electrolyte solution 1000 ml I/V., for 3-5 days. Oral supplementation of potassium acetate @ 30-60 gms once in 3 days helps in relieving hypokalemia. Use of nervine tonics containing vitamin B_1 , B6 and B12 @ 10-12ml I/M for 5 days gives encouraging results in downer cow syndrome. A recent report recommends use of nervine tonics with double dilution with distilled water by epidural route in line with treatment of paresis cases of human beings. Half of the dose is administered high epidural (lumbo-sacral) and remaining half is given low epidural (sacro-coccygeal). This treatment regimen is repeated once in four days for one or two times.

Early detection and treatment of parturient paresis helps reducing the incidence of downer cow syndrome. Periparturient cows should be provided good bedding and once recumbent treatment should be initiated at the earliest. Frequent rolling from side to side on hourly basis should be done till animal makes attempt to get up.

Bovine ketosis

(*Estate Acetonemia*)

It is a production disease resulting due to negative energy balance and represent incomplete combustion of fatty acids which results in accumulation of ketone bodies. Biochemically it is characterized by Ketonemia, Ketonuria, hypoglycemia and low hepatic

Recent classification of the disease based on etiopathogenesis is as follows:

1. **Alimentary ketosis:** It is due to exclusive feeding of silage which is rich not only in butyrate, a highly ketogenic substance, but also decrease the feed intake due to poor palatability.
2. **Starvation ketosis:** Poor quality of food stuffs which are low in proprionate and proteins are the causative factors. Famines and drought conditions also contribute in certain area due to non availability of fodder.
3. **Production ketosis:** Also referred as primary ketosis. This form of ketosis is most common. High yielding animals have constant drain of lactose in milk which should be appropriately replaced by dietary input of glucose precursors like propionate and amino acids. This deficiency compounds during the peak of lactation which accelerates the rate of gluconeogenesis which contributes for accumulation of ketone bodies.
4. **Secondary ketosis:** The disease conditions like traumatic reticulitis, metritis, mastritis displacement of abomasum cause decrease in apetite thereby resulting in a state of hypoglycemia which triggers gluconeogenesis ultimately leading to accumulation of ketone bodies.

The incidence is high during first two months postpartum. Mature cattle in their 3rd to 4th lactation are highly susceptible as they attain maximum milk yield at this stage. The disease is more commonly seen during the winter as the animal is in need of extra energy. Cross-bred cattle and buffaloes are more susceptible.

Ruminants mainly depend on hepatic gluconeogenesis. The ingested carbohydrates are fermented into volatile fatty acids. In this propionate is the important glucose precursor whereas acetate and butyrate are long chain fatty acids precursor. High yielding dairy animals are usually in negative energy balance in first few weeks of lactation. The dietary inadequate glucose precursors create a state of hypoglycemia which triggers accelerated gluconeogenesis. Mobilization of fat for this purpose results in accumulation of Acetyl coA which is normally oxidized via TCA cycle. But its oxidation depends on adequate oxaloacetate for which propionate is precursor. Therefore, deficiency of propionate and resultant inadequate oxaloacetate limit the aceyl coA oxidation via TCA cycle. The accumulated acetyl coA combines to form acetoacetate which is referred as parent ketone body. Acetoacetate further breaks down to form β-OH butyrate, acetone and isopropyl alcohol. The nervous signs in ketosis are due to the damage caused by isopropyl alcohol to the brain.

The rapid loss of body weight (wasting) in ketosis is due to the mobilization of sub-cutaneous fat for the purpose of gluconeogenesis.

Ketosis occurs in two forms. Majority of the cases show the wasting form that is characterized by sudden, severe drop in milk yield, refusal of concentrate (selective appetite for roughages), licking of inanimate objects, passing of hard pelleted faeces covered with mucus, woody appearance with prominent rib cage. The animals have pronounced sweetish odour due to presence of acetone in the breath. The temperature, pulse and respiratory rates are usually in normal range.

About 5% of the cases of ketosis show nervous form. Nervous signs like walking in circles, salivation, Vigorous licking of skin and inanimate objects, hyperasthesia, aimless wandering, convulsive episodes are commonly encountered.

The tentative diagnosis can be made on history of recent calving and important clinical signs. The laboratory diagnosis can be made by urine analysis. Rothera's Test is employed for the detection of ketone bodies in the urine. The urine is saturated with ammonium chloride in which 2 - 3 drops of 5% sodium nitro-prusside is added. Ammonium hydroxide or strong ammonia solution is added over this urine nitro-prusside conjugate. The development of purple colour at the junction is positive test. The intensity of purple colour indirectly reflects the amount of ketonuria. Alternatively, the test can also be performed on milk.

Hypoglycemia (upto 40 mg %), ketonemia (upto 100 mg %), elevated levels of free fatty acids, triglycerides are important biochemical changes associated with ketosis.

Supplementation of glucose is the preliminary concept in the treatment of ketosis. Hypertonic solutions (10-50%) of glucose at the rate of 0.5 gm/kg body weight by intravenous route are used. As there is a concurrent hypoinsulinemia in ketosis supplementation of insulin has proved more efficacious. Short acting insulin at the rate of 0.5 units/kg body weight is injected sub-cutaneously after 10-15 minutes of initiation of intravenous hypertonic glucose.

Alternatively,oral supplimentation of propionate precursors is also effective. Sodium propionate @ 110 gms along with 8 gms nicotinic acid or 6 gms of DL-methionine by oral route for 7 days. Propylene glycol @ 125 ml can be a replacement for Sodium propionate. Glucocorticoids @ 0.05 Mg/kg body weight alone or in combination with insulin is also effective. Preferential breakdown of proteins for gluconeogenesis by glucocorticoids reduces the acetyl coA accumulation.

Post-parturient haemoglobinuria

(Nutritional haemoglobinuria, Lahu mootna)

Post-parturient haemoglobinuria (PPH) is an acute disease of high yielding buffaloes and sometimes cows which occurs soon after parturition or in advanced pregnancy and is clinically characterised by haemoglobinuria, anorexia, anaemia and death in untreated cases.

Extensive studies have been undertaken and are still in progress to elucidate the exact cause of intravascular haemolysis which is the cause of haemoglobinuria in this disease. Workers so far have not succeeded in pinpointing specific aetiological factors and explaining the exact pathogenesis of the disease. Hypophosphataemia is a consistent finding in diseased animals in India. Experimental production of disease has been reported by workers in cows fed on low phosphorus diets for three pregnancies and therefore prolonged hypophataemia has been considered to be a major predisposing cause of this disease. However in New Zealand, where the disease occurs in cows, both hypophosphataemia and molybdenosis induced hypocupraemia have been reported from different sets of animals respectively. According to one widely accepted hypothesis advanced by workers, inorganic phosphorus deficiency decreases glucose utilization rate in glycolytic cycle and adenosine triphosphate (ATP) production by the erythrocytes leading to decrease in synthesis as well as reduction of glutathione which predisposes the erythrocytes to oxidative damage by oxidant radicals (free radicals) like hydroxyl (OH), superoxide (O_2) and hydrogen peroxide (H_2O_2). These radicals are normally produced during the metabolism of erythrocytes but are taken care of by an antioxidant mechanism consisting of catalase, glutathione peroxidase and superoxide dismutase enzymes as well as biological antioxidants like vitamin C (ascorbic acid) and E (-tocopherol). In the face of an excessive oxidative insult (as probably encountered in PPH) this antioxidant mechanism is overwhelmed and the erythrocytes are oxidatively damaged. This damage is in form of lipid peroxidation of lipid bilayer of erythrocyte membrane. Probably some damage is also caused to erythrocyte membrane skeleton composed of proteins.

A massive increase in malondialdehyde levels in the blood of diseased buffaloes, as reported by many workers clearly points towards the possibility that erythrocytes are oxidatively damaged by free radicals. A further support to this assumption is lent by the fact that in erythrocytes the conditions that favour peroxidation are optimal. Some of these conditions are richness in polyunsaturated fatty acids, continuous exposure to high oxygen tension and presence of haemoglobin (a potential catalyst of such reactions).

Here it is important to point out that hypocupraemia has not been reported from majority of cases of PPH in buffaloes from India. An inter-relationship of hypophosphotaemia and related effects is presented in the schematic flow diagram. The factors enlisted are based on various reports by workers. This is the best possible explanation of intravascular haemolysis which is the cause of haemoglobinuria in this disease. All these reports pertain to buffaloes since from India the reports of occurrence of this disease in cows are rare.

The disease is almost worldwide in distribution. There are reports of the disease in both cows and buffaloes perhaps due to different aetiology, respectively. It occurs only in high producing adult female animals usually following parturition but some cases do occur in advanced pregnancy. Majority of the cases are encountered in animals in their 3rd to 6th lactation corresponding to peak lactation phase. In buffaloes the disease is encountered only in animals reared in village conditions. There are no reports of disease from organised dairy herds. In India the disease is more common in arid and semi-arid tracts of Haryana, Punjab, Rajasthan, Uttar Pradesh and adjoining areas. The first ever reports of this disease in buffaloes were from Egypt in 1963 and from India in 1968. In cows the disease is known to occur since 1853 when first reported from Scotland.

The disease is clinically characterised by passage of red or coffee coloured urine. Initially the animals continue to take feed and water normally but after 1-2 days there is gradual loss of appetite and anorexia sets in. A typical clinical sign in buffaloes is "straining during the act of defaecation" inspite of normal consistency of faeces. Anaemia is evidenced by pale mucous membranes and increase in respiration rate alongwith laboured breathing. There is gradual loss of body condition and recumbency is followed by death in untreated cases. Course of the disease is variable but usually ranges between 5-8 days. Moderate fever (up to 40° C) may be seen in cows but no fever is seen in buffaloes. Occasionally gangrene and sloughing of the tip of the tail or digits is observed in cows. While occasional recovery with long convalescence of several months is seen in cows, buffaloes usually die if treatment is not done.

Diagnosis of the disease is based on history, clinical signs and Laboratory diagnosis such as –

(i) Haematology – Haemoglobin, total erythrocyte count and packed cell volume levels are drastically reduced. In advanced cases of buffaloes, haemoglobin levels as low as 3-4 g % is observed.

(ii) Urinalysis – Microscopic examination shows absence of erythrocytes (rules out haematuria) while benzidine test (for presence of blood) is positive.

(iii) Serum biochemical examination - levels of serum inorganic phosphorus are low (0.5-3.0 mg %) against a normal of 4.0-7.0 mg %.

Therapy of PPH is based on the aetiological factors supposedly involved. Administration of copper salts is beneficial especially in cows if there is hypocupraemia. Copper sulphate @ 5.0g dissolved in drinking water orally or @ 500 mg dissolved in normal saline is given intravenously. Here the molybdenosis also needs to be corrected if evident. Recently copper glycinate has also been tried with success (1.5 mg/kg b.w., max. 500 mg dissolved in normal saline intravenously, single dose).

Following supportive therapy is recommended in clinical cases for speedy recovery and prevention of relapse:

(i) Vitamin B complex - Parentrally

(ii) Blood transfusion - May be necessary in cases in which haemoglobin level is alarmingly low.

(iii) Oxygen releaser - Inosine @ 500 mg alongwith parentral therapy is beneficial in cases in which dyspnoea is pronounced.

(iv) Mineral mixture @ 100 g orally once daily × 5 days followed 50 g orally once daily × 20 days.

Haematinic drugs containing iron are contraindicated in this disease as long as haemoglobinuria is there. This is due to the reason that iron is known to enhance the oxidative stress, which is believed to be the significant factor in the causation of disease. However, these drugs can be given following stoppage of haemoglobinuria.

Hypomagnesaemic tetanies

These diseases are common in ruminants and are characterized universally with depression of blood/serum magnesium levels. Cattle, buffaloes, sheep, goats and calves are commonly affected but the cause of hypomagne-saemia is somewhat different in calves. On this basis the tetanies can be subdivided into two major groups:

(a) **Hypomagnesaemic tetany of calves :** It is caused by primary deficiency of magnesium in diet.

(b) **Hypomagnesaemic tetanies of cows/buffaloes/sheep/goats :** In these tetanies there may or may not be a primary deficiency of magnesium in feed but some nutritional or metabolic factors reduce the availability of magnesium to the body or increase its excretion leading to a secondary deficiency.

In adult ruminants the main cause of magnesium deficiency is an inadequate absorption of magnesium from the digestive tract, Magnesium is a very important intracellular cation and participates in some major enzymatic reactions in the body. This element is absorbed from forestomachs and large intestine and not from the small intestine. No active mechanism controlling the magnesium homeostasis in the body has yet been identified and its level in serum/plasma varies within a wide range depending upon magnesium intake and availability. Magnesium from body reserves in bones (about 70%) is not readily mobilised as evident from the report that normal levels of bone magnesium are maintained even in the face of chronic hypomagnesaemia of several months standing.

Magnesium is normally excreted from the body in milk, faeces and urine. Under normal circumstances dietary magnesium is sufficient enough for body requirements and excess amounts are excreted in urine. Kidney is supposed to act as a modulator of magnesium homeostasis as magnesium reabsorption readily takes place across renal tubules and the degree of reabsorption is dependent on the level of blood magnesium. It has been reported that excretion of magnesium in urine stops when the serum levels fall below 1.8 mg%. Harmonal control of this magnesium conservation is partially performed by parathyroid gland through paratharmone.

During lactation there is an increased requirement of magnesium since its levels in milk (12-16 mg/kg) are significant and any dietary deficiency or absorption restriction makes the animal prone to hypomagnesaemia. Magnesium transport across the epithelium of forestomachs is carried out by sodium linked ATPase dependent transport system. The degree of absorption is dependent on sodium and potassium ratio (Na : K ratio) in the rumen. Absorption is significantly reduced if the ratio is less than 3:1 while it increases to plateau at a ratio of 5:1. Following factors have a significant bearing on the magnesium intake and absorption for the animal:

(a) Lush green rapidly growing grass is low in sodium and high in potassium and if fed to animals, causes impairment of absorption of magnesium.

(b) Normally saliva has a high Na and K ratio but if diet is low in sodium this ratio is disturbed and potassium replaces sodium, decreasing the magnesium absorption.

(c) High ammonia content coupled with increasing amounts of readily degradable carbohydrates in rumen decreases magnesium absorption. Mechanism is unknown.

(d) Hypomagnesaemia may occur in pregnant/lactating animals feeding on pastures having magnesium content less than 2.0 mg/kg dry matter, the safe content being 2.5 mg/kg dry matter.

(e) Pastures having potassium and nitrogen content greater than 30 g/kg and 40 g/kg dry matter respectively is hazardous to animals.

(f) Pastures top dressed with nitrogenous fertilizers are dangerous and their toxicity is enhanced by potash application.

(g) Cereal grazing may lead to hypomagnesaemia and is related to its lower magnesium and higher potassium content.

(h) Highly leached, acidic and sandy soils are magnesium deficient and therefore fodders grown on these soils are also low in this element.

(i) Reduced dry matter and energy intake makes the animals prone to hypomagnesaemia.

(j) Concurrent hypocalcaemia is often seen in cases of hypomagnesaemic tetany and it is believed that this triggers off the clinical disease. However, hypocalcaemia in these cases is of less severe degree than that seen in parturient paresis. This phenomenon has been explained on the basis of the fact that chronic hypomagnesaemia depresses paratharmone action which in turn negatively affects calcium homoeostasis.

(k) In cases of excitement a sudden fall in magnesium levels of blood has been reported. This is related to action of epinephrine released.

(l) A number of outbreaks of this condition are associated with variable periods of starvation (due to any reason).

Therefore, one or more of the above mentioned factors may precipitate the clinical disease and the worst possible combination of factors may be feeding on lush green pastures (low magnesium and energy intake) by recently parturited animals in wet, cold and windy weather.

Morbidity or mortality rates are variable but on average may be around 12% and 30% respectively.

In case of sheep, hypomagnesaemia outbreaks are sometimes preceded by pregnancy toxaemia and often accompanied by ketosis. Usually sheep and goats on poor ration and in inclement weather are prone to this condition.

The clinical syndrome associated with hypomagnesaemia has been explained on the basis of its metabolic requirements. Magnesium has a significant role in the impulse transmission along neuromuscular junctions and also on the release of acetylcholine. Impairment of these functions in cases of deficiency is believed to be the cause tetanic spasms and hyper-excitability seen in this condition. However, there are reports indicating that several animals in a herd do not exhibit clinical signs inspite of hypomagnesaemia. It is also significant to note that cerebrospinal fluid levels of magnesium are more predictive of clinical disease than serum levels indicating that there are alterations in central nervous system functions as well as peripheral nerve functions.

The syndrome can be subdivided into acute, sub-acute and chronic forms on the basis of severity and duration :

Initially some animals (cattle/buffaloes) appear to be hyperalert, hyperaesthetic and uncomfortable. On even slightest stimuli they start frenzied movements and bellowing. Cessation of grazing accompanies these clinical signs. This is followed by staggering gait and sudden fall on the ground. In many cases no perceptible clinical signs are noticed before the animals fall dead. Usually death is preceded by clonic convulsions along with nystagmus, opisthotonus, champing of the jaw and frothing from mouth. There is fever also (104-105oF) accompanied with increased respiration rate and tachycardia. Intensity of heart sounds is so enhanced that it can be heard from a distance. Death usually occurs within an hour and case fatality rate is very high since death usually occurs before any treatment can be attempted. However whenever treatment is given in time, the response is

Diagnosis is based history of feeding on lush green rapidly growing grassy or cereal pastures top dressed with nitrogenous and potash fertilizers. Serum magnesium levels in healthy ruminants range between 1.7-3.0 mg%. In case of ruminants clinical signs appear when the levels fall to 0.5 mg% or less. In case of sheep and goats, clinical symptoms appear when the level are even lesser than 0.5 mg%. Against this, clinical signs are not seen in some animals even with serum magnesium levels as low as 0.4 mg%. This is probably due to the fact that there are individual susceptibility differences. Also, the level of ionisation of magnesium alongwith differences from CSF levels plays their role. Low serum calcium (5-8 mg %) levels also have triggering effect on development of clinical signs and serum potassium levels are dangerously high in hypomagnesaemic animals.

The blood of affected animals should be tested and the therapy should be based on results. For example, in cases of concurrent hypomagnesaemia and hypocalcaemia, combined therapy with calcium and magnesium is recommended while magnesium therapy alone may be done in cases where there is normocalcaemia. However some workers recommend combination therapy for all cases and it appears to be the safest option also. A preparation containing 25% of calcium borogluconate and 20% of magnesium hypophosphite may be given intravenously (500 ml in cattle/buffaloes and 50 ml in sheep/goats). Magnesium sulfate alone @ 300 ml of 20% solution given intravenously in large ruminants leads to a rapid rise in serum magnesium levels which return to pre-injection levels within 3-6 hours. Subcutaneous injection of 50% magnesium sulfate solution (200 ml for large ruminants) give optimum results since the rise in magnesium levels is slow and sustained. Subcutaneous injections are recommended alongwith intravenous therapy. Precautionary measures are needed while magnesium therapy is being done since cardiac and medullary disturbances may precipitate respiratory failure. In some acute cases, bouts of convulsions may occur during therapy which should be prevented by a large intramuscular dose of a suitable tranquilizer before the commencement of therapy.

Azoturia in horses

(Paralytic myoglobinuria, exertional rhabdomyolysis, tying up, chronic intermitten rhabdomyolysis, monday morning sickness)

It is a disease of horses which may occur during rigourous exercise following a period of inactivity or otherwise and it is clinically characterised by stiff gait, hard/painful muscles and sometimes myoglobinuria.

The exact reasons for the causation of this disease are not known. However some factors associated with some cases of the disease are said to be anomalies of muscle metabolism like polysaccharide storage myopathy and defects in sacrolemmal function. Some other possible factors are harmonal imbalance, high carbohydrate diets, exercise at irregular intervals, and hypothyroidism.

The disease is worldwide in distribution. It commonly occurs in thoroughbred racehorses and endurance horses. The disorder is almost always observed during or immediately after exercise but some cases do occur after short period of rest following exercise. However, some degree of unpredictability is always associated with the disease occurrence. Some familial and genetic susceptibility is also seen. Some workers have suggested increased susceptiblity in female race horses of 2-5 years age.

Predisposing cause is usually exercised but exact pathogenesis is unknown. Earlier the disease was supposed to be due to lactic acid accumulation myopathy but recent reports have ruled out this possibility. In any case, muscular degeneration does occur and myoglobin is released leading to myoglobinaemia and in some cases myoglobinuria.

Degree of severity of disease is variable. In milder cases there are recurrent instances of anorexia, stiff and short stepping gait, reluctance to move and frequent shifting of weight on legs following exercise.

In severe cases the animals are not able to continue exercise, sweat profusely and gluteal muscles are hard and painful to touch. Mild to moderate fever may also occur alongwith evidence of abdominal pain and tachycardia. Myoglobinuria may occur but is not always seen. Later on the animals may become recumbent and die. Case fatality rates are usually high.

Diagnosis is based on history and clinical signs and laboratory findings as:

(i) Increased serum myoglobin levels

(ii) Increase levels of enzymes like serum creatinine kinase, as– partate amino transferase and lactic dehydrogenase.

(iii) Decrease in serum sodium and increase in serum potassium values in severely affected horses.

(iv) Urine may contain myoglobin when myoglobinuria is present.

Further exercise should be immediately stopped and the animal rested in standing posture. Analgesia may be provided by phenylbutazone (2.2-4.4 mg/kg b.w.

intravenous) or ketoprofen (2.2 mg/kg b.w. intravenous) every 12-24 hours. Adequate water should be provided to animals. Ringer's lactate solution or other isotonic fluids should be administered parentrally to severely affected animals to correct dehydration. To minimise the chances of myoglobinuric nephrosis, urine alkalizers should be given (sodium bicarbonate 1.3% solution intravenously or 50-100 g orally). Mild sedatives (xylazine 0.1 mg/kg intramuscularly) are helpful in decreasing muscular pain and anxiety. If the animal is recumbent, sufficient soft bedding should be provided and periodical shifting every 3-4 hours is helpful in preventing bedsores.

Frost bite

Frost bite is a condition caused by exposure to extreme cold weather resulting in destruction of superficial tissues. If the animals are deprived of their coats or do not receive nutritious diet, their body temperature may go down and the muscles become stiff rendering the animal to move with difficulty. Immobility further lowers the temperature. Under these conditions, the blood from skin is driven to the interior of the body and so the heart and lungs are engorged and over burdened. Rate of metabolism slows down and the cardiac and respiratory centers are depressed. If the temperature of the body falls below 70°F, the heart stops. Dry cold below 0°C freezes local exposed parts of the body which are poorly insulated such as nose, ears, scrotum, snout, tail, teats, extremities, comb and wattles, especially at high altitude where oxygen demand is high and availability is low. In horse the lower parts of the limbs are mostly involved in frost bite due to prolonged contact with ice cold mud. A breach of surface in the integument favours the injurious effects of cold on the tissues beneath. It causes various degrees of dermatitis. Sometimes gangrene supervenes, constituting "gangrenous dermatitis". The gangrene may extend into a joint or important synovial sheath; the toxaemia may be so severe that the animal succumbs to its effects.

Radiation injury

Radiation injury in animals occurs during atomic explosion due to blast, heat and irradiation. Irradiation is the major cause of mortality in animals. There are two types of radiation injuries, one from acute direct exposure to irradiation and the second from cumulative effect due to accumulation of radioactivity in tissues resulting from continued exposure to small but toxic levels of radioactive material. Radiation fallout may contaminate pastures and leaves of the plants and cause direct radiation. Radioactive iodine causes damage to the thyroid gland and strontium causes destruction of bone tissue, depression of leukocyte and platelet production and terminal depression of erythropoiesis. Both are excreted in the milk of the animals and may cause deleterious effects in humans. The radiation exposures also result from diagnostic as well as therapeutic radiation devices.

Radiation even below the permissible dose once absorbed by the tissue, bring some change within the living tissues. As results of radiation the cells within the tissue come to a state of high chemical reactivity due to the atoms losing their electrical charge. The cellular injury causes pathological and physiological changes leading to radiation sickness and other manifestations of radiation damage to the body.

Burns

Burns is a thermal injury occurs on excessive absorption of heat. The thermal injury caused by hot liquids or steam is known as scald. The extent and depth of injury due to burns depends on the temperature of the object and the period for which it is applied. The nature of heat modifies the character of lesions. A scald is likely to be more severe than a burn because the hot liquid may penetrate deeper into the tissues. Dry heat causes desiccation and charring while moist heat causes boiling of tissues resulting in opaque coagulation. Being the first tissue exposed to the full intensity of heat, the epidermis is the most severely affected. Tissues lying deeper are injured less. The excessive heat and corrosive substances cause tissue protein denaturation (50-60°C) and protein coagulation (<60°C). Usually burns in animals occur when a building in which the animal is housed catches fire. The burns have been classified Wilson into three degree, according to the depth of the tissue involved. In first degree burn only the epidermis is affected. There is erythema of the skin due to hyperemia of the cutaneous vessels. No other damage is noticed and the cells show no changes. Mild acute inflammatory exudates may be present in the dermis. After a few days, desquamation of surface epithelium may occur. In second degree burns, partial thickness of the skin is affected with complete loss of the epidermis. In this type along with erythematous changes coagulative necrosis of the epidermal cells takes place. Bullae or vesicles are formed due to infiltration of blood serum into epidermis or below it separating it from the dermis. The vesicles contain a pink-staining granular debris, fibrin and neutrophils. Healing is prompt and complete by the regeneration of epithelium unless there is secondary infection. In third degree burn full thickness of the skin is destroyed. In addition subcutaneous fascia and deeper tissue are also involved. Repair is by scar formaton preceded by sloughing of the necrotic tissue.

Heat stroke

In this condition the heat regulating centre located in the hpothalamus is not able to cope up with the situation and hence the body temperature rises. The onset of the condition may be sudden but usually preceded ataxia, weakness, hot skin, impaired sweating, rapid pulse, pupil contracted and high temperature. The major cause of hyperthermia is high environmental temperature and prolonged muscular exertion especially when humidity is high. Fatty animals having heavy hair coat when confined in houses with inadequate ventilation are susceptible. Buffaloes have less heat tolerance and hence are more prone. Other causes of hyperthermia include: dehydration, iodism, neurogenic hyperthermia and malignant hyperthermia in the porcine stress syndrome. During hyperthermia, the metabolic rate increases by 40-50%, liver glycogen stores are rapidly depleted and extra energy is derived from the endogenous metabolism of proteins. In anorexia occurs, there will be considerable loss of body weight and lack of muscle strenth accompanied by hypoglycemia and rise in non-protein nitrogen. An increase in heart rate occurs due to rise in blood temperature and also due to fall in blood pressure resulting from peripheral vasodilatation. Respiratory rate and depht increases due to effect on the respiratory centres. Urine secretion is decreased due to reduced renal blood flow. When the body temperature rises above the critical temperature, the blood is directed to the periphery depriving organs of oxygen, causing brain and liver damage, rhabdomyolysis and disseminated intravascular coagulation. An elevation of body temperature in most of the species is the first observable clinical reaction to hyperthermia. The rectal temperature exceeds 103oF (39.5°C). In most instances the temperature exceeds 107oF (42°C) and may reach 110oF (43.5°C). There is increase in heart and respiratory rates with weak pulse, sweating and salivation. The animal becomes restless but soon becomes dull and depressed. There is dyspnoea, congestion of mucous membranes, trembling, convulsions and followed by death. In early stages, there is increased thirst and the animal seeks cool places often lying in water, attempting to splash itself. Death occurs in most of the cases when the temperature is above 107-108°F.

Mastitis

(Mammitis, Mammite)

Mastitis is the term that denotes inflammatory condition of the udder irrespective of causes. It is a global problem. It is characterized by physical, chemical and microbiological changes in the milk and pathological changes in the glandular tissues of the udder. The changes in the milk include change of colour, change of consistency (clot) and presence of abnormally large number of leukocytes.

A large number of species of microorganisms have been implicated as cause/s of mastitis. They are bacteria, fungus, mycoplasma and virus. In most countries, surveys of the incidence of mastitis, irrespective of cause, show comparable figures of about 40% morbidity rate amongst dairy cows and quarter infection rate as measured by an indirect test, of about 25%. Infection of each mammary gland occurs via the teat canal, the infection originates from either an infected udder or the environment, and in dairy cattle the infection originating from infected udders is transmitted to the teat skin of other females by milking machine liners, milkers' hands, washcloths, and any other material that can act as an inert carrier. *Streptococcus agalactiae* and *Staphylococcus aureus* are causes of contagious mastitis. Escherichia coli is a common inhabitant of the cow's environment, and an unusual inhabitant of cow tissues, which it invades by chance. Its source as an infection is bedding, or other inert materials, so that it is the principal cause of environmental mastitis. All breeds of dairy cows, buffaloes, goats, sheep, pig, horse are susceptible. High yielding dairy cows are more commonly affected than low yielder. Infection rate is more in successive lactation than the first lactation. Exotic and crossbred cows are more prone to mastitis than the Indian zebu cows.

Infection of the mammary gland always occurs via the teat canal and on first impression the development of inflammation after infection seems a natural sequence. However, the development of mastitis is more complex than this and can be most satisfactorily explained in terms of the three stages of invasion, infection, inflammation. Invasion is the stage at which organisms pass from the exterior of the teat to the milk inside the teat canal. Infection is the stage in which the organisms multiply rapidly and invade the mammary tissue. After invasion a bacterial population may be established in the teat canal and, with this as a base, a series of multiplications and extensions into mammary tissue may occur, with infection of mammary tissue occurring frequently or occasionally depending on its susceptibility. This in turn causes inflammation, the stage at which clinical mastitis appears or a greatly increased leukocyte count is apparent in the milk.

Mastitis may clinically be classified as (i) Acute mastitis, (ii) Sub acute mastitis, (iii) Chronic mastitis and (iv) Sub-Clinical Mastitis

In acute form systemic reactions are not seen. The udder becomes swollen and there are changes in the milk. The milk may be replaced by curd like material yellow or brown fluid with flakes or clots. Infection may be localized in one quarter or the entire udder may be involved. The infection may render the udder non-functional or recovery may be achieved by judicious treatment. This form is caused by *Str. agalactiae* or *Staph. aureus*.

Sub acute mastitis remains for longer period of time with above-mentioned symptoms. In most of such cases udder reduces and milk yield decreases for 2-3 weeks. Few flaks may be present in the milk. Sometimes little fibrosis may also occur in the udder. Treatment response is less in this type of mastitis.

Chronic mastitis is the terminal stage of disease. Udder becomes hard due to fibrosis. The supra-mammary lymph nodes become palpable. The quarters may become thickened, firm and nodular and at time atrophic. The parenchymatous cells are replaced by connective tissues. The milk change varies considerably. Some time it may look as green or yellow-green and foul smelling.

Sub-clinical mastitis denotes absence of apparent gross abnormalities in the mammary gland but with recovery of bacteriological pathogens from the milk. In herds in which staphylococcal mastitis is a problem, 50% or more of the cows may have chronic sub clinical infection. Infection if persists for more than few months does not respond to treatments because of development of a tissue barrier between the antibiotics and the organisms.

The diagnosis of the disease is depended upon -

(a) Physical examination of the udder

(b) Tests for milk abnormalities such as Strip cup test, Bromothymol blue test (BTB), Bromocresol purple test, Chloride test, Hotis test, California mastitis test (CMT), White side test and Catalase test.

Depending on the type of infection and clinical severity, first and foremost task for a Veterinarian is to decide whether to adopt parenteral or intramammary or both therapeutic approaches.

The parenteral therapy is to be adopted in case having marked systemic reactions and / or severely inflammed mammary gland interfering with proper diffusion of antibiotic given intramammary to all parts of glandular tissues. Any antibiotics such as Penicillin (10,000-16,500 U/kg), enrofloxacin (2.5 mg/kg), oxytetracycline (10 mg/kg), tylosin or erythromycin (12.5 mg/kg) and sulfadimidine (200 mg/kg) may be given for 5 days, depending on the sensitivity to microorganisms. Udder infusions are conventional choice. Disposable tubes containing drug in water-soluble ointment base are available for intra-mammary infusions. Strict hygiene is necessary to avoid introduction of organisms. Blockage of lactiferous ducts and alveoli with inflammatory debris may impair the diffusion of drugs.

11

Livestock for Sustainable Agriculture

Manish Kumar, Sunil Kumar and Teena Rani

Introduction

"Sustainable development as a concept represents the latest step in a long evolution of public concerns with respect to both natural resources and environment". It is defined as "development that meets the needs for the present without compromising the ability of future generations to meet their own needs". This definition indicates that to meet their needs, the present generations have no right to overuse natural resources and the environment, which will ultimately threat the living standards of the future generations. Therefore, stewardship of both natural and human resources is of prime importance (Freenstra 1997). Advocates of sustainable development perceive that the biosphere imposes limits on economic growth, express a lack of faith in either science or technology as leading to human betterment, are extremely averse to environmental risks, support redistributive justice and egalitarian ethics, profess concern over population growth and have faith in the wisdom of human capital development, and have survival of species and protection of the environment and of minority cultures, rather economic growth *per se*, as goals.

Sustainable agriculture is defined as a dynamic set of practices and technologies that minimize damage to the environment while providing income to the farm over a long time span (Flora 1992). It integrates three main goals - environmental health, economic profitability, and social and economic equity. Although sustainable agriculture is generally defined with the negative term as low input agriculture, this may not be fair because sustainable agricultural system use the best technology in balanced, well managed and environmentally responsible system (Hess 1991). The operational goals of sustainable agriculture, as stated by Benbrook (1991), include the following: (a) more through incorporation of natural processes such as nutrient cycling, nitrogen fixation and beneficial pest-predator relationships into the agricultural production process; (b) reduction in the use of off-farm inputs with the greatest potential to harm the environment or the health of farmers and consumers; (c) productive use of the biological and genetic potential of plant and animal species; (d) improvement in the match between cropping patterns and the productive potential and physical limitations of agricultural lands; and (e) profitable and efficient production with emphasis on improved farm management, prevention of animal diseases, optimal integration of livestock and cropping enterprises, and conservation of soil, water, energy and biological resources.

In order to achieve the goals of sustainable agriculture and rural development it is necessary that crop and livestock production should be accompanied with proper farm management and maintenance of publicly owned natural resources. Farm level practices may include, but are not limited to, control of chemicals, pests management, proper irrigation, crop rotation, legume rotation, use of animal manure, proper grazing of range lands, land consolidation, proper use of agricultural machinery, proper care of plant and animal health, and so on. Considering the ecology, farm system and socioeconomic structure of a region, similar farm level practices, which aim to protect and long term utilization of farmland and other natural resources, can also be applied.

Livestock production is considered as a part of agricultural production and makes substantial contributions to economic development. It provides the basic meat and dairy products needed by the urban and rural households; makes significant contributions to textile and leather industries; increases cash availability for the family farms; increases organic matter of soil by providing manure; makes it possible for rural families to utilize land which is not appropriate for cultivation, as well as, to utilize publicly owned resources such as meadows and range lands; requires fodder crop production which increases soil fertility when included in a rotation pattern; increases labour productivity within a family farm; provides animal power when tractor purchase is not economical; and in terms of climatic and environmental conditions it is a less risky facility as compared to crop production alone.

Livestock sector plays an important role in socio-economic development of rural households. It contributes about 6 percent to the Gross Domestic Product and 25 percent to the Agricultural Gross Domestic Product. Over the last two decades, livestock sector has grown at an annual rate of 5.6 percent, which is higher than the growth of agricultural sector (3.3 percent). This suggests that livestock is likely to emerge as an engine of agricultural growth in the coming decades and has become important for sustainable agricultural development. It is also considered as one of the potential sector for export earnings. The importance of livestock goes beyond its food production function (Birthal *et al.* 2002). It provides draught power and organic manure to crop sector and hides, skin, bones, blood and fibers to the industrial sector. Livestock sector also makes significant contributions towards conservation of environment. Livestock sector supplements income from crop production and other sources and absorbs income shocks due to crop failure. It generates a continuous stream of income and employment and reduces seasonality in livelihood patterns particularly of the rural poor (Birthal and Ali 2005).

Rural Poverty is largely concentrated among the landless and the marginal households comprising about 70 percent of rural population (Taneja and Birthal 2004). Several empirical studies indicate that livestock rearing has significant positive impact on equity in terms of income, employment and poverty reduction in rural areas (Birthal and Ali 2005) as distribution of livestock is more egalitarian compared to land (Taneja and Birthal 2004). In India, over 70 percent of the rural households own livestock and a majority of livestock owning households are small, marginal and landless households. Small animals like sheep, goats, pigs and poultry are largely kept by the land scarce poor households for commercial purposes because of their low initial investment and operational costs (Birthal 2002).

Livestock serve a valuable role in sustainable agricultural systems. The rumen serves as a reserve containing the microbial enzyme cellulase, the only enzyme to digest the most abundant plant product, cellulose (CAST 1975). Cattle, sheep and goats are particularly useful in converting vast renewable resources from rangeland, pasture and crop residues or other by-products into food. With ruminants, land that is too poor or too erodable to cultivate becomes productive. Also, nutrients in by-products are utilized and do not become a waste-disposal problem. Some question the use of cereals for animal production. About one-sixth of the energy required by livestock is derived from grains. Poultry and swine consume about 59% of these cereals to produce about 39% of the human food energy from livestock and ruminants (dairy and beef cattle, sheep and goats) consume about 37% of grains fed to livestock yet produce 61% of human food energy from animal agriculture (Wheeler *et al.* 1981). Optimal integration of crop and animal agriculture must be based on rigorously derived and quantitatively defensible data and analysis (Baldwin *et al.* 1992). Since livestock provides many benefits both for family farmers and for the economy as a whole, it makes significant contributions to sustainable use of agricultural potential in rural areas.

Livestock population dynamics

India has huge livestock population with 185 million cattle, 98 million buffaloes, 124 million goats, 61 million sheep, 14 million pigs and 489 million poultry birds (Table 1).

Table 1. Livestock population in India (millions)

Year	Cattle	Buffalo	Sheep Million numbers	Goat	Pig	Poultry
1982	192.5	69.8	48.8	95.3	10.0	207.7
1992	204.6	84.2	50.8	115.3	13.0	307.1
1997	198.9	89.9	57.5	122.7	13.3	347.6
2003	185.2	97.9	61.5	124.4	13.6	489.0
2007	199.1	105.3	71.5	140.5	11.1	648.8
Annual growth, %						
1982-1992	0.6	1.9	0.4	1.9	2.4	4.0
1992-1997	-0.6	1.3	2.5	1.3	0.8	2.5
1997-2003	-1.0	1.2	1.2	0.2	1.0	7.0

Source: Livestock Census (various issues), Ministry of Agriculture, Government of India

Out of the total livestock in the country, around 38.2 percent are cattle, 20.2 percent are buffaloes, 12.7 percent are sheep, 25.6 percent are goats and only 2.8 percent are pigs. All other animals are less than 0.50 percent of the total livestock population. The composition of livestock population in broad groups like bovine (cattle and buffaloes), ovine (sheep and goats), pigs and poultry, however, has changed over the last two decades. Cattle population that had been increasing until 1992 has started declining and between 1992 and 2003. The decline in the cattle population is confined to indigenous stock that comprised 87 percent of the total cattle population in 2003, while that of the crossbred increased by 62 percent. Within the indigenous stock, decline was drastic for males (22%). The main reasons for decline in indigenous cattle population are: increasing substitution of draught animals with mechanical power and low milk yield (Birthal and Taneja 2006). The buffalo population has increased from 70 million in 1982 to 98 million in 2003 and to 105.3 million in 2007. There has been a small decrease in total bovines in the country by 1.9% between 1997 and 2003.

Total ovine population has increased from 144 million in 1982 to 186 million in 2003. The number of goats increased from 95 million in 1982 to 124 million in 2003, but at a retarding rate throughout. Sheep population though has been increasing but with considerable variations in the trend. Poultry is gaining importance in India due to growth and availability of poultry feed at reasonable prices. Between 1982 and 2003, poultry population increased more than double, from 207 million to 489 million. Except during 1992-97, poultry population has maintained a steady growth of above 4 percent a year. Between 1997 and 2003, poultry witnessed an all time high growth of 7 percent a year. Pig population has increased from 10 million in 1982 to 14 million in 2003. Growth in pig population, however, has retarded sharply since 1992, due to lack of widespread demand for pork (Birthal and Taneja 2006). As per livestock census 2007, the population of all the livestock has increased except pig.

Diverse and changing roles of livestock

Livestock are capital assets, produced in the past and contributing to future product output. Investment in, or the acquisition of, livestock involves saving or borrowing, justified by the expected future return on capital (Figure 1). Apart from durable capital embodied in the animals, circulating capital is needed to meet current costs of production.

Investment in livestock raises farm production through (a) extension of land area that can be utilized, (b) diversification of the productive activity on a crop farm and (c) intensification, i.e. by raising livestock value of output and hence

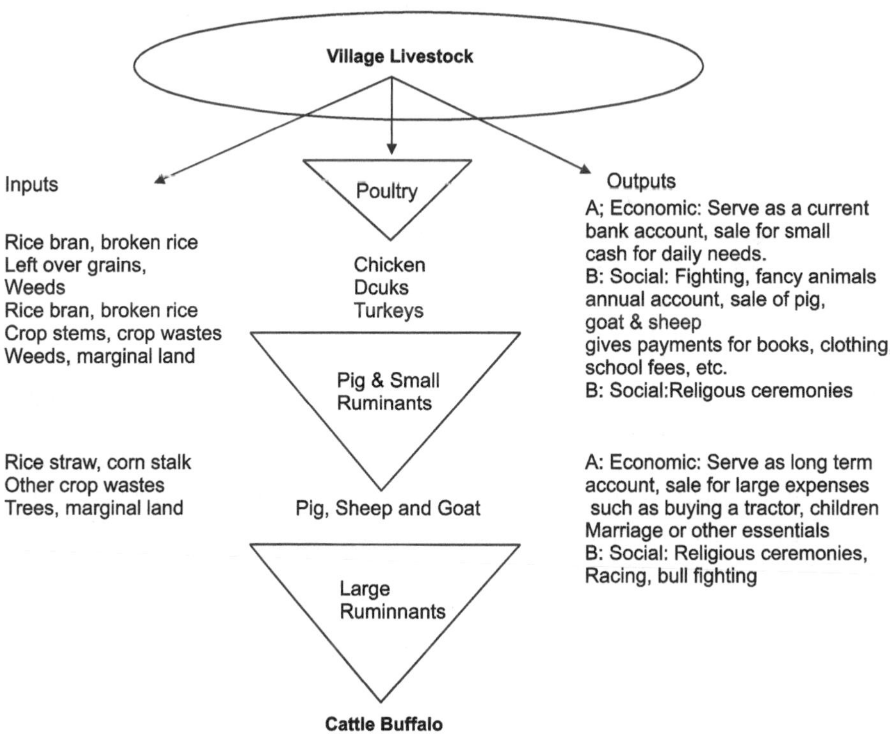

Fig. 1. Role of different livestock species of animals in rural livestock sector

total production per hectare of agricultural land increases. The latter involves increased inputs of labour and/or capital and may be achieved by increasing the stocking rate, increasing yield per head of livestock or changing to a more intensive production system. Scope for extending production onto unused, virgin land is limited since there is little remaining that is usable. There is also limited scope for increasing stocking rates under pastoral systems, since pastoralists already make effective use of available grazing.

Livestock production systems are broadly categorized into

i) 'grassland-based' pastoralism and ranching,

ii) 'mixed-farming', either rain-fed or irrigated, and

iii) 'landless', mainly pig and poultry production systems.

These are listed in order of increasing intensity. The 'landless' production systems are largely responsible for the rapid growth in average meat supply per person in the developing countries, poultry production have doubled over the last 10 years. Reproduction and growth rates are faster in pigs and poultry than in the ruminant species of livestock. However, housing and hand feeding increase capital requirements and labour costs. Yields per head of these species and of cattle, in developing countries, are well below those in developed countries and increase depends upon improvement in animal health, nutrition, breeding and management.

The 'landless' livestock production systems represent labour-using technology, in that labour requirements per hectare (devoted to feed production) are higher than for other systems. Conversely, the use of animal draught power is labour-saving, in reducing hand-labour requirements particularly at peak work periods. Use of the plough may allow a larger proportion of the farm area to be cultivated each year, to increase cropping intensity.

Landless livestock systems provide most of the world's production of pig and poultry meat. The majority is produced in developed countries and from large-scale commercial enterprises, now spreading in the developing countries. These products make up two thirds of all meat production world-wide, while, in the developing countries, poultry meat now accounts for more than half of all meat produced. Ruminant fattening is less important. Concerns arise regarding limited benefits to the rural poor, risks of environmental pollution and use of cereals to feed animals. Poor producers may participate, possibly by co-operation or vertical integration. Use of feed-grains does not compete directly with human consumers, while pig and poultry production converts feed efficiently and provides a cheap source of animal protein.

Mixed crop-livestock production systems are important as the source of the bulk of ruminant livestock production and the home of the majority of the world's poor. Complementary relationships exist with livestock, fed on crop by-products and other plant material, contributing draught power, manure, additional sources of food and income, savings and buffer against risk. As intensity and livestock number rise, crop-livestock interactions become increasingly competitive, for the use of land and other resources. There is little, or no, interaction between crops and landless livestock systems.

Inter-regional differences in livestock production systems depend upon agro-ecological features, human population density and cultural norms. A comparison of the main developing country continents, shows that Sub-Saharan Africa, Latin America and the Near East, with reasonably large areas of land per person engaged in agriculture, have a greater proportion dependent on grassland-based ruminant livestock systems than do the more densely populated, land-scarce, regions of South and East Asia. Nonetheless, in all the continents listed, most of the agricultural populations are engaged in mixed farming systems. These are mainly rain-fed in Africa and Latin America but, in South and Eastern Asia and the Near East, about half are irrigated. Poultry production is a key enterprise in Latin America, particularly Brazil, and in East and South East Asia, mainly China which is also a major pig producing country. South Asia is the largest milk producing region.

Analysts have suggested that production systems intensify and evolve, from pastoral livestock keeping to specialist crop production, then to mixed farming and eventually to independent crop and 'landless' livestock systems. The change from pastoralism to cropping is seen as a land-saving technology, or the substitution of labour for land. The move to mixed farming is capital using and land-saving; capital is substituted for land. Animal draught power, as an innovation, involves the substitution of capital for labour. Industrial development also affects technological change in agriculture, in different ways. In America and Oceania industrialization occurred while population was sparse; labour-saving mechanization technology change was induced. In Japan industrial innovations occurred with a dense rural population, and therefore a land-saving bias, with improved seeds and agri-chemicals was induced.

Sustainability of land and the production enterprise

Non-cereal resources for ruminant feed are enormous. Constituting 55% of the world's land, pastures, rangelands, meadows, forests and woodlands have the potential of producing 5.8 trillion Mcal of metabolizable energy annually. Crop residues from other lands contribute up to 2.9 trillion Mcal and processing by-products about 0.6 trillion Mcal (Wheeler *et al.* 1981). In the United States alone, 250 billion Mcal of crop residues and by-products result from crop agriculture. This, if fully converted in today's production systems, could yield

4.5 billion kg of beef or 750 billion kg milk (Smith 1980). Ruminant livestock production takes advantage of forage production on approximately 25% of potentially arable land to minimize water and soil erosion. Vast hilly or mountainous areas discourage cultivation and crops (CAST 1982). Highly erodable land may have no other sustainable economic use than for grazing ruminant livestock. If soil is the focal point for sustainability (Hauptli *et al.* 1990), then minimization of erosion, compaction and oxidation also helps to maintain its organic state. This is better achieved on any soil by keeping some type of continuous cover as much of the year as possible (Ely 1994). In arid areas, pastoral use of ruminants in agriculture is utilized, as opposed to the integrated crop and livestock system in more humid areas. Livestock are essential for family farm agriculture. Crop and livestock together make sustainable practices such as crop rotations and production of legume or grass forages. Ruminants convert low-quality feeds from land not suited for cultivation into a salable product for farmers and high-quality food for people. Including livestock on the farm also improves living standards on family farms, providing more employment opportunities in rural areas. It is not always true, as implied by Waggoner (CAST 1994), that cropland used for animal feed production could produce as many or more calories and protein as when that cropland is used to produce human food crops. For example, in California, alfalfa yields average 15 t/ha annually whereas wheat averages 5.4 t/ha (CDFA 1993). Alfalfa is about 20% protein; wheat is about 12% protein. Hence, a hectare yields 3,010 kg of protein from alfalfa, or 647 kg of protein from wheat (21.5% that of alfalfa). Dairy cows convert the protein in alfalfa to milk protein at about 25% efficiency, giving 753 kg of food protein per hectare, compared to 647 kg for wheat. In addition, milk and milk protein is of much higher quality and biological value. The wheat crop will take less water but will require considerable nitrogen fertilizer for the yield assumed, whereas alfalfa requires no nitrogen fertilizer. Alfalfa adds nitrogen to the soil, which is used by wheat in normal crop rotation. Wheat produces vast amounts of straw, a crop residue, which can also be converted to human food by ruminants as part of the production system.

Competition with human for food

Cereal grains are not necessary for ruminant production and ruminants need not compete for human foods. However, feed grains and other concentrate feeds increase productivity and efficiency, resulting in economical production and maximal use of the forage resource. There are a wide range of strategies in which various proportions of non-competitive feed resources, grains and other concentrates can be used for ruminants. Each of these strategies has a different efficiency. A common error is to assign a single efficiency to all types of ruminant meat production. Questions have been asked, however, about the use of humanly edible foodstuffs (grains, protein sources, etc.) in ruminant diets. Does this use create a net loss of nutrients for human consumption? What level of use is necessary or desirable? Does the use of some of these improve the nutrient (e.g. protein) quality or product value? Too often the opponents of animal agriculture evaluate the desirability of animal production on gross caloric efficiency or protein intake/output values. However, in many cases the feeds used in animal production are not humanly consumable and in order to determine the true efficiency of animal production, humanly consumable energy and protein intake should be used for efficiency comparisons.

Animal/plant interactions: nutrient acquisition and use by livestock

Nutrient intake constitutes the interface between animals and plants within the soil–water– atmosphere–plant–animal–human (SWAPAH) continuum. This interface also represents a major impact point of human management decisions affecting the animal's well being, ecosystem productivity for human endeavours, biodiversity of plants and animals, nutrient cycling, water quality/yield and soil stability. The hierarchical nature of herbivore diet selection processes and associated physiological needs, relative to the spatial configuration of natural and man-made landscape features, set limits regarding where animals harvest nutrients and how effectively nutrients are ingested, absorbed and recycled in the ecosystem. The selected mix of animal species and their dietary requirements, relative to the available diversity of plant species, determines how efficiently nutrients are acquired at the landscape level.

Value of livestock output

Agriculture is the major source of livelihood in rural India accounting for about one-fourth of Gross Domestic Product (GDP). In fact, the growth in livestock sector has always remained higher than the growth in crop sector. Though in absolute terms, value of output of various livestock products increased during 1980-81 to 1990-91 and 1990-91 to 2003-04, the annual growth has retarded during 1990-91 to 2003-04 for almost all the products (Table 2). The annual growth in pork, poultry meat and eggs is higher than the growth in milk and meat from other animals.

The major share of livestock output comes from food based products, which constitutes 86.7 percent of total value of livestock sector (Table 3). Milk and milk products have contributed 62.2 percent to the total value of livestock sector in 1980-82 which has increased to the extent of 66.2 percent in 2001-03 due to various policy interventions for development of dairy sector in the country. The contribution of meat and meat products has increased from 14.9 percent in 1980-1982 to 16.7 percent in 2001-03. However, the composition of meat output has been changing significantly towards poultry sector.

Table 2: Value of output from livestock sector in India

Commodity	Value, Rs. in Bllions at 1993-94 prices			Annual growth rate, %	
	1980-82	1990-92	2001-03	1980-90	1990-2003
Food based products	305.6	519.7	806.2	5.46	3.94
Milk Group	241.2	399.7	615.7	5.22	3.94
Meat	51.7	96.2	150.6	6.26	3.77
Beef and Buffalo meat	7.7	12.5	14.6	4.82	1.45
Goat meat/ Mutton	21.1	38.2	46.8	5.11	1.42
Pork	2.7	5.5	10.4	7.83	5.97
Poultry Meat	20.2	40.0	78.8	7.61	5.81
Meat Product	2.4	3.4	4.5	1.96	2.27
Egg	10.3	20.4	35.3	7.35	4.80
Non-food based product	82.3	92.5	123.4	1.52	2.41
Livestock, Total	387.9	612.3	929.6	4.72	3.72

Source: National Account Statistics (various issues), CSO, Ministry of Statistics and Program Implementation, GOI.

Table 3: Composition of value of output from livestock sector in India (%)

Commodity	1980-82	1990-92	2001-03
Food based products	78.8	84.9	86.7
Milk Group	62.2	65.3	66.2
Meat	13.3	15.7	16.2
Beef and Buffalo meat	2.0	2.0	1.6
Goat meat/ Mutton	5.4	6.2	5.0
Pork	0.7	0.9	1.1
Poultry Meat	5.2	6.5	8.5
Meat Products	0.6	0.6	0.5
Eggs	2.7	3.3	3.8
Non-food based products	21.2	15.1	13.3
Livestock, Total	100.0	100.0	100.0

Source: National Account Statistics (various issues), CSO, Ministry of Statistics and Program Implementation, GOI.

Waste products of animal origin

Organic farming is a form of agriculture that relies on crop rotation, green manure, compost, biological pest control and mechanical cultivation. Organic farming maintains organic matter at an optimal level in soil and it sustains reasonable productivity. Livestock plays an important role in relation to the general principles of organic agriculture and supports biological cycles like carbon cycle, water cycle, oxygen cycle and nitrogen cycle within the farming system and diversifying production. Sustainable agriculture is considered as a salubrious relation among soil, plant, livestock, man and environment. In organic farming farmers use natural processes to enhance productivity, maintain the nutritive status of the soil and to be less dependent on external resources to keep cost down. Organic farming uses natural materials which are by products of the farm and these include the production of compost which is key elements of organic farming. Organic matters improves water holding capacity of light textured soils and enhances supply of nutrients particularly nitrogen, phosphorus, sulphur and at the same times makes them less susceptible to leaching. On the other hand, organic matter makes heavy soils more porous, permits better aeration and also reduces crusting. The iron-organic complexes such as Fe-humate make iron more available to plants. Organic matter protects the boron against leaching which is required for maintaining the structure of plants. It is also essential for embryonic development and organogenesis in plants. Polysaccharide, lignin and protein are major constitutes of organic matters. The benefits to soil from organic matter is ceaseless decomposition of plant and animal residue into simple inorganic compounds like carbon dioxide, water, nitrate, phosphate and sulphate which can be easily recycled by plants. The organic methods can help small farm family to survive, increase in farm productivity, repair of environmental damage, knit communities in to smaller more sustainable distribution networks, all leading to improve food security around the world. Several waste products of animal origin are used in organic farming can be listed below.

Compost

Composting is an aerobic decomposition technique of biodegradable organic matter. The decomposition is performed primarily by facultative and obligate aerobic bacteria; yeast in the cooler initial phase and in the end phase decomposition is performed by a number of larger organisms like oligochaete worms, springtails, ants and nematodes. It is environmentally safe and it enhances the nutritive value of the soil. It nurtures the organism in the soil which is generally destroyed by the use of the chemical fertilizer and pesticides. Compost significantly reduces cost of production. Use of compost improves the soil fertility. Application of compost may have reducing effect on attack of soil pathogens.

Vermicomposting

Vermicomposting is process of breakdown of organic matter by the joint action of earthworms and microorganisms which does not involve the generation of high heat as in case of composting. It is the process of transforming farmyard manure into the compost with the help of earthworm. Cattle shed manure supplies nearly 10 times as much nitrogen and phosphoric acid to the soil as other manures and fertilizers. Vermicomposting results in bioconversion of the waste streams into two useful products, the earthworm biomass and the vermicompost. During vermicomposting important plant nutrients like nitrogen, phosphorus, potassium and calcium present in feed material are converted into the form that are much more soluble and easily available to plants.

Slurry

Slurry is the mixture of faeces and urine it contains approximately 60% of faeces and 40% of urine. The main component in urine is urea. Some salts are also present with vitamins, hormones and enzymes. Phosphorus is absent in urine and potassium is found in mineral form which is easily available to the plants. Slurry should be stored at least for two month especially under cold conditions to reduce the parasitic load.

Cow urine

It is being considered as a natural disinfectant and pest repellent. Since very long time, it is being used in combination with various plants or dung to manufacture plant growth promoting products. Cow dung and urine are two bio-matters which can be used methodically to get better result in controlling several plant pathogenic fungi. It can be also substituted for phenyl. If 10 liters of cow urine mixed with one kg of neem and 100g of garlic and boiled till its volume comes to half filter this mixture can be used for controlling insects in vegetables, legumes and in cotton at the rate of 50ml liquid in per liter of water.

Biogas

Biogas refers to a gas produced by the biological breakdown of organic matter in the absence of oxygen. Biogas originates from biogenic materials and is a type of biofuel. The biogas plants have become a boon to the rural farmers. The two main products of it are enriched compost manure and methane. Biogas is used for cooking, lighting purpose and in case of large plants as motive power for driving small engines. The wide scale utilization of biogas and compost making technology offers the most appropriate answer to solve three national interconnected major problems which the countries are facing today called three F's food, fertilizer and fuel. The biogas is an ideal inexhaustible source of energy that could serve the rural area very well. Utilization of biogas reduces the consumption of commercial energy sources like coal, kerosene which results in reduction of family fuel budget. Forest wealth of nation is also saved. Valuable organic fertilizer can be obtained thus reducing the dependences on chemical fertilizers.

Thus, livestock by utilizing the natural resources produces high value nutritious food for human kind which helps to earn the profit for farming communities and also help in food security. As the livestock byproducts, if utilized properly, not only provide organic manure for agricultural production but also contribute the fuel energy which maintains the soil fertility and increase the productivity. Also, by utilizing livestock byproducts the environmental protection, safeguarding the public health and reduction in environmental pollution can be achieved. For this we only need awareness activities to be carried out in rural farming communities. If it is being so, we need not to worry about uncertain shortage of commercial chemical fertilizers, now and then.

Trade in livestock products

Livestock products account for about one-fifth of the global trade of agricultural products. However, India does not have a significant presence in global livestock trade. India shares only 0.3 percent of world exports and 0.4 percent imports. The global market for animal based foods has been expanding rapidly (Birthal and Taneja 2006). Table 4 represents net trade in livestock and livestock products in various parts of the world.

In value terms, export of livestock products have remarkably increased from US$90.8 million in 1980-82 to US$ 469.6 million in 2002-04 whereas, import declined from US$ 261.8 million to US$ 257.5 million during the same period. India was a net importer of livestock products during 1980s. This sector now holds good export potential as its export exceeded total imports significantly. The situation turned around during early 1990s due to increasing export of meat and meat products and significant decline in import of milk and milk products. The process of trade liberalization and the efforts made by the government in recent times have certainly boosted the country's export of livestock products to newer heights.

Table 4: Net trade in livestock and livestock products by continent

Net trade in livestock and livestock products, overall	Sub-Saharan Africa Net importer	Near East and North Africa Net importer	South Asia Net exporter	East & South East Asia Net importer	Latin America & Caribbean (Inc. China) Net exporter
Main net import by value	Cattle Beef Poultry meat MilkEggs	Cattle Sheep & goats Poultry -meat Hide & skin	Wool	Cattle Beef Lamb wool Hides & skins	Pig meatBeef
Main net export by value (Principle\ source country in parenthesis)	Sheep & goat (Sudan) Camel (Somalia)	None	Buffaloes (Pakistan) Buffalo meat (India)	Pigs (China) Poultry meat (Thailand) Rabbit meat (China) Honey (Vietnam) Eggs (China)	Cattle (Mexico) Poultry meat (Brazil) Honey (Brazil)

Source: FAOSTAT, 2003

In 1980-82, livestock products accounted for 3.7 percent of export and 18.2 percent of import of agricultural products. Share of livestock import during pre-liberalization period was quite fluctuating with increasing trends but after liberalization it has drastically come down. The share of livestock export during pre-liberalization period was steady which has significantly increased during post liberalization period. Interestingly, the share of livestock products in agricultural export almost doubled since 1989-91, while that in import fell drastically from 16.0 percent to 5.5 percent.

Employment generation

The share of agricultural sector in gross domestic product is significant. Out of total 4.9% persons are engaged in livestock production (Table 5). The contribution is more of female counterpart. It is evident from Table 6 that as compared to males, females allocate more time to livestock production activities. Females supply bulk of the labour needed for livestock production activities. During 1980s, 65 percent of the labour requirement in livestock sector was contributed by females and this increased to 71 percent during 1990s.

Table 5: Sectoral distribution of rural workers in India during 1999-2000

Sector description	Total persons (%)	Male (%)	Female (%)
Crop production	68.1	66	72
Livestock production	4.9	2.2	9.8
Plantation	1.4	1.3	1.6
Forestry and logging	0.3	0.2	0.4
Fishing	0.3	0.4	0.1
Agricultural services	1.2	1.1	1.3
Non-farm sector	23.8	28.9	14.7
Total	100	100	100

Source: Results on Employment and Unemployment Situation in India (various issues), NSSO, Ministry of Statistics and Program Implementation, Govt. of India

Table 6: Female labour employment as percent of total employment by sector

Sector description	1983	1993-94	1999-2000
Crop production	37.6	37.5	37.0
Livestock production	64.9	70.8	70.6
Plantation	33.8	32.0	39.9
Forestry and logging	29.0	35.9	51.9
Fishing	17.8	15.7	11.9
Agricultural services	24.9	43.9	38.9
Non-farm	24.8	23.1	21.6
Total	37.3	36.0	35.0

Source: Results on Employment and Unemployment Situation in India (various issues), NSSO, Ministry of Statistics and Program Implementation, Government of India.

On the other hand, their contribution to crop production labour requirement has almost remained unchanged at 37 percent and to the non-farm sector it has declined from 25 percent during 1980s to 22 percent during 1990s. This suggests that livestock enterprise is women oriented and growth in livestock production would help to improve gender equity. For upliftment of the rural workers, the development of livestock is also important along with agriculture and other allied sectors.

Rural poverty and livestock sector

The challenges for redressing poverty in rural areas is mainly related with expansion of economic opportunities, empowerment of poor to take advantage of new opportunities and an effective safety net to reduce vulnerability and protect poorer of the poor (Kozel and Parker 2003). Poverty is primarily caused by low level of assets holding, coupled with low and uncertain returns. Land and livestock are considered to be the major assets of rural households for livelihood support in general, and small and marginal households in particular. As distribution of land holdings in rural India is highly skewed towards medium and large framers, rearing of livestock particularly small ruminants, pigs and poultry are considered to be the potential options for poor households to earn their livelihood on sustainable basis. Table 7 shows the trends in rural poverty in major states of India over the period of 1983 to 1999-2000. Rural poverty in India has declined from 45.6 percent in 1983 to 27.1 percent in 1999-2000. The rate of decline varied across states.

Table 7: Rural Poverty across major States (%)

States	1983	1987-88	1993-94	1999-2000
Andhra Pradesh	26.5	20.9	15.9	11.1
Assam	42.6	39.4	45.0	40.0
Bihar	64.4	52.6	58.2	44.3
Gujarat	29.8	28.7	22.2	13.2
Haryana	20.6	16.2	28.0	8.3
Karnataka	36.3	32.8	29.9	17.4
Kerala	39.0	27.1	25.8	9.4
Madhya Pradesh	48.9	41.9	40.6	37.1
Maharashtra	45.2	40.8	37.9	23.7
Orissa	67.5	57.6	49.7	48.0
Punjab	13.2	12.6	12.0	6.4
Rajasthan	33.5	33.2	26.5	13.7
Tamil Nadu	54.0	45.8	32.5	20.6
Uttar Pradesh	46.5	41.1	42.3	31.2
West Bengal	63.1	48.3	40.8	31.9
All India	45.6	39.1	37.3	27.1

Source: Planning Commission, Government of India

A faster decline is observed in the states of Punjab, Haryana, Andhra Pradesh, Gujarat, Karnataka, Kerala, Maharashtra, Rajasthan and Tamil Nadu. These states have experienced faster growth in agriculture and/or livestock sectors. On the other hand, in states like Bihar, Orissa, Uttar Pradesh and West Bengal the intensity of poverty is still high. There is significant inverse relationship between poverty and value of livestock output. The states with higher livestock share have low level of poverty and vice versa.

Tables 8 and 9 showing distribution of livestock production across different size group of rural households during the year 1991-92 and 2002-03. About 22 percent of the households did not have access to land for crop production in 1991-92

Table 8: Distribution of livestock holdings in India 1991-92

Category	Landless, <0.002ha	Marginal, 0.002-1.0 ha	Small,1 .0-2.0 ha	Medium, 2.0-4.0 ha	Large,> 4.0 ha	All
% households	21.8	48.3	14.2	9.7	6.0	100.0
Distribution of livestock, %						
Bovine	2.5	43.8	23.3	17.7	12.7	100.0
Ovine	5.1	46.2	19.3	15.0	14.4	100.0
Poultry	6.4	54.9	19.0	14.4	5.3	100.0
Pigs	7.7	49.9	20.4	13.9	8.1	100.0
Size of livestock holdings, no/100 households						
Bovine	23	180	324	361	418	198
Ovine	20	81	115	131	203	85
Poultry	49	190	223	247	147	166
Pigs	2	4	6	6	5	4

Source: NSS Report No. 408, Livestock and Agricultural Implements in Household Operational Holdings 1991-92, Ministry of Statistics and Program Implementation, GOI.

Table 9: Distribution of livestock holdings in India 2002-03

Category	Landless, <0.002ha	Marginal, 0.002-1.0 ha	Small, 1.0-2.0 ha	Medium, 2.0-4.0 ha	Large, >4.0 ha	All
% households	31.9	47.1	11.2	6.2	3.4	100.0
Distribution of livestock, %						
Bovine	0.6	51.3	21.2	15.0	11.9	100.0
Ovine	2.1	61.5	15.7	9.6	11.0	100.0
Poultry	4.4	62.7	17.4	6.8	8.6	100.0
Pigs	3.2	76.2	12.0	5.5	3.0	100.0
Size of livestock holdings, no/100 households						
Bovine	3	169	293	374	535	156
Ovine	4	84	90	99	203	64
Poultry	17	164	191	136	306	123
Pigs	0.3	5.3	3.5	2.9	2.9	3.3

Source: NSS Report No. 493, Livestock Ownership Across Operational Land Holding Classes in India 2002-03, Ministry of Statistics and Program Implementation, GOI.

which has increased to 32 percent in 2002-03. About 48 percent of the households have land holdings less than one hectare. The size of land holding particularly of the marginal landholders is too small to provide an adequate livelihood. These households seek supplementary livelihood opportunities from livestock and allied activities. Livestock being less capital-intensive is an important option for them, because of less land requirement, low initial investment and operational cost (Birthal and Ali 2005).

A comparison of livestock holding among landless households between 1991-92 and 2002-03 shows that landless households significantly increased during the period whereas, share of livestock holding among landless households declined for all type of livestock. Livestock rearing amongst landless households largely depends on common property resources such as common grazing lands, common water resources, forests, wastelands and roadsides for feed and fodder. The decline in availability of these resources has affected the landless households to quit livestock rearing. In case of marginal households, the share of livestock holding has substantially increased in 2002-03 for all types of livestock. Therefore, minimum availability of land for feed and fodder is an important determinant of size of livestock holding. But, given the resources to the land scarce households, the utility of livestock as provider of livelihood opportunities is far greater for them (Birthal and Ali 2005). Thus, the flow of income from livestock is also expected to be more favourable to low income groups comprising the landless, marginal and small landholders.

To assess the impact of livestock on rural poverty, a log-liner regression equation was estimated by pooling time series and cross-section data with state as a unit of observation (Table 10) (Ali 2007). Though, there are a number of factors that influence incidence of poverty, the per capita state income, share of livestock in agricultural output, share of agriculture in gross state domestic products, landless households and marginal households are included as explanatory variables in the regression equation.

Table 10: Log-linear estimates for poverty and livestock sector

Explanatory variables	Dependent variable: Log of poverty, %	
	Regression coefficient	t-value
Constant	25.945	4.41*
Per capita GSDP, Rs/annum	-1.253	-2.93*
Share of livestock output in agricultural output, %	-1.351	-3.46*
Share of AgGSDP in GSDP, %	-0.788	-1.69
Landless Households, %	-0.455	-2.04**
Marginal Households, %	-0.768	-1.46
Log-likelihood function	-14.669	

* Significant at 1 percent level
** Significant at 5 percent level

The coefficient of all the explanatory variables is negative, which implies that improvements in these variables will lead to reduction in poverty level (Table 10). The coefficient of per capita income and share of livestock in agriculture output is significant at one percent level. However, the magnitude of regression coefficient on share of livestock is larger, indicating that growth of livestock sector contributes more to poverty reduction.

Socio-economic constraints to crop–livestock integration

Population pressure on a limited land resource base, well-developed markets and adequate purchasing power appear to be prerequisites for the integration of crop and livestock production. Cash income permits investments in soil fertility and feed production. Yet, even in areas where further intensification of crop and livestock production is feasible and desirable, a variety of social and economic constraints may inhibit greater integration.

A recurring situation is periodic and prolonged drought and flood. This can lead to large livestock loss due to limited availability of forage and the weakened state of livestock arising from low body reserves. The loss of animals through death or distress reduces the quantity of manure available for recycling. Reconstructing herds, particularly cattle, may take several years resulting in multiple year reductions in manure availability and cropland productivity. Inadequate livestock holdings due to lack of capital to purchase animals may also hinder crop–livestock integration. Indeed, under open access and common property grazing, the principal constraint to livestock holdings and further integration at the individual household level may not be the shortage of feed, but the liquidity position of the farmer that prevents animal purchase or necessitates the frequent sale of livestock. There is evidence to suggest that beyond certain critical levels of herd size and cultivation densities, competition increases between crops and livestock for scarce resources, particularly labour and land. Land competition varies across the agro-ecological zones. It is lowest in the humid zone where livestock are restricted due to pests and diseases, high in parts of the semi-arid zone and particularly prominent in the highlands where population density and stocking rates are already high. The competition for land and labour between crops and animals for the production of food and feed continues, even in fully integrated crop–livestock systems such as those found in the highlands and densely settled areas. It may be expected that when such numbers are reached, the more extensive systems of crop–livestock production will give way to the more intensive, highly integrated mixed farming systems. More intensive, integrated systems are labour-demanding in terms of fodder production, crop-residue harvesting and storage, feeding work oxen and manure spreading. Major bottlenecks may develop, for instance, where labour needs for manuring conflict with labour demands for crop production. Although, the point at which labour conflicts appear will differ in land-scarce and land-abundant zones, the potential for conflict between cropping and livestock for labour is nonetheless real in both systems. It is also clear that the roles of men, women and children in crop and animal production vary and are highly influenced by ethnicity, religion and stage of agricultural development. Whereas, in crop production men may be responsible for the more laborious tasks of land preparation, women are involved in planting, weeding, harvesting, manual threshing, grain processing and marketing. Most of the herding and corralling of animals on cropland for manuring is done by men, while women may be major providers of fodder and water, and haul manure from stalls in intensively managed mixed farming systems. While, the low literacy rates may be considered a constraint in implementing technological improvements, it is clear that farmers and herders have vast experience and knowledge of crop and livestock production. Such indigenous knowledge can be a valuable resource in designing appropriate technologies and extension programmes for improving agricultural productivity. Studies on traditional practices have shown that indigenous knowledge pertaining to crop and livestock production is generally scientifically sound. Methods of learning from and incorporating indigenous knowledge into technology development are needed. For instance, although farmers and pastoralists have no past experience in feed production *per se*, their indigenous knowledge of natural pastures and feeding systems needs to be understood and incorporated into research aimed at improving animal nutrition.

Promoting livestock development

What needs to be taken into account?

Livestock production makes an important contribution to economic development, rural livelihoods, poverty alleviation and meeting the fast growing demand for animal protein in developing countries. The case for promoting increased livestock production is pressing given the rapidly growing demand for animal products and the global aim to halve, by 2015, the proportion of the world population living in abject poverty, most of whom are dependent, at least in part, on food and income derived from livestock. The landless and rural women are important sub-groups of the rural poor likely to get benefit from livestock development. Meeting these needs for promoting livestock production "while, at the same time sustaining the natural resource base (soil, water, air and bio-diversity), is one of the major challenges facing world agriculture today" (Bruinsma 2003).

Livestock are self sustaining capital investments, since replacements may be bred and reared within the flock or herd. However, establishment of foundation stock, breed improvement or purchase of specialized equipments involve new investment from outside the system. Traditional foundation stock may be inherited or transferred at marriage, but where re-stocking is needed, following drought, disease outbreak or other disaster, or for innovations, such as new livestock enterprises, upgrading of stock, or change of system, funding is needed. For the poor lacking the necessary finance and paucity of credit facilities is a serious constraint. Credit-in-kind, such as that provided by 'heifer in trust' schemes, may alleviate the problem. However, additional circulating capital is likely to be needed with the introduction of intensive livestock systems, requiring purchased inputs. Vertical integration with large-scale processor and marketing agencies may allow inter-linkage of credit provision with other services, while independent producers must still rely upon informal local sources.

Successful implementation and spread of innovations in livestock production, requires an increase in human capital in the form of new knowledge of appropriate husbandry and management methods. Farmer participation in testing and development, together with effective extension support, is essential for the successful introduction of new livestock enterprises or methods of production.

There is a need to ensure that proposed new technologies are appropriate, accord with the producers' objectives and constraints, and match with consumer demands in accessing market outlets. Sustainability of a livestock development project must depend upon availability of inputs, particularly fodder resources, but also delivery systems for concentrate feeds, genetic material (e.g. male breeding-animals, semen or day-old chicks) and disease control measures. The physical infrastructure of roads and other communications has an important influence, but an effective and appropriate institutional framework is a prerequisite.

While these broad generalizations apply to all types of livestock production system, there are such large differences between them in terms of physical, social and market environments, that different policies are needed for promoting development. Extensive grassland-based systems remain in the arid and semi-arid regions of Sub-Saharan Africa, Latin America, Near East and Central Asia, as the only feasible form of agricultural land use and source of livelihood for the local population. The climate is harsh and rainfall is very unreliable. Ruminant livestock are kept to produce milk and meat for consumption and sale. There are risks of land degradation and loss of pasture species due to open access to rangelands. However, with effective communal ownership and control, apart from dangers of drought, sustainable production is possible. Delivery of animal health and other services is difficult in remote pastoral areas but local markets for milk and meat operate effectively other than in periods of severe drought, when meat prices collapse while grain is scarce and expensive. Possible options for improvement include provision of water supplies and measures for drought relief.

Ranching, practiced in parts of Latin America as in land-rich developed countries of North America and Oceania, is capital-using and labour-saving when compared with pastoralism. Few of the poor are supported by this system but a substantial marketed surplus of meat is produced, some for export. Attempts to replace pastoral systems by group ranching, in East and Southern Africa have had little success.

Mixed crop-livestock farming systems in the humid, sub-humid and temperate regions of all the main continental blocks, makes the main contribution to ruminant livestock production and livelihoods for the rural poor. These more favourable climates permit food-crop production and further intensification by the introduction of livestock. Irrigated systems, involving buffaloes in place of cattle, are prevalent in Asia. Complementary relationships and nutrient recycling, based on crop products fed to animals and manure returned to the soil are beneficial, while diversification provides a buffer against risk and livestock serve as a form of savings.

Small stock, such as poultry, sheep and goats, are more likely to be owned by the poor, particularly women, since they cost less to purchase, are of a more convenient size for home consumption or for sale in times of distress and reproduce and grow faster. They may survive by foraging on harsher terrain and vegetation than cattle. Investment in fencing or housing for livestock is likely to be needed to protect crops from damage. Changes in land tenure institutions may be required to permit enclosure of what was previously common land. With increasing stocking intensity, competition grows for land, between fodder and other crops or for cereals and oil seeds between animals and human. In either case the value added by raising livestock is generally sufficient to justify the incremental costs.

Ruminant livestock may be kept mainly for meat, for milk or for both or, in the case of sheep, for wool. Traditional breeds are dual-purpose but overall increase in productivity is generally sought by breeding for specialized production of meat or milk. Specialized milk production generally yields more than enough for consumption within the family and markets must be found. Co-operative group action may be needed for processing and marketing, or delivery to a large commercial dairy. The availability of animal draught power is often listed as an additional advantage of mixed farming. However, only the large ruminants are suitable for use as draught animals, while production of draught power competes with milk and meat production. The main benefit in Asia appears to be the saving of labour in major field operations and transport. In parts of Africa and Latin America, where land is less scarce, animal draught may allow an increase in the frequency of cultivation. Provision of animal health services, with cost recovery, may be feasible and sustainable in areas with a high livestock population density particularly where productivity is high. Services may be extended to areas of less intensive livestock production with the employment of para-veterinarians. There is much scope for productivity improvement through research and development in animal health, nutrition, adaptive breeding and switches in production system.

The landless production systems particularly pig and poultry enterprises are potentially the most intensive since their expansion is independent of the area of land available to the farm household, if household wastes or purchased feeds are used. These systems are found in all parts of the world, although production is concentrated in the industrialized countries. Within the developing country group, they are most important in the land-scarce countries of East and South East Asia with China and the Near East, although poultry production is fast expanding in Latin America. Systems vary from traditional 'backyard' production to large-scale industrial type systems most common in the developed countries. These large systems are increasingly being established in peri-urban areas of developing countries, with advantages of greater productive efficiency from use of hybrid stock, pre-mixed concentrate feeds, low transaction costs through vertical integration, together with economies of scale in processing and marketing. Dangers of disease outbreaks are increased by keeping large numbers of birds or other animals in close proximity but delivery of animal health care and general hygiene are facilitated in these circumstances.

The spread of intensive, commercial poultry and pig production systems is viewed with alarm, in some quarters, as only benefiting large scale producers and middle-class urban consumers, creating water and air pollution, reducing bio-diversity by the dependence on exotic hybrid stock and relying on feed grains and oilseeds which might have contributed directly to human diets (Cox and Varparma 2000, Fresco and Steinfeld 1998, Durning and Brough 1991). Against these arguments must be set the fact that poultry and pig production are the most economically efficient enterprises, or cheapest sources for meeting human needs for animal protein. Increased production in the developing countries promises to reduce the growth of imports and save scarce foreign exchange. Profitability depends on the price of feed grains, which make up a major part of the costs of production. Debate surrounds the relative costs of importing poultry and pig-meat versus the importation of the necessary feed grains. Increased productivity of both intensive livestock and cereal systems will benefit the national economy. Controls may be needed to limit pollution and loss of bio-diversity. It is important for small-holder producers to participate in the spread of poultry and possibly pig production by the promotion of semi-intensive methods of production and vertical integration with processors and marketing agencies, which might be achieved through co-operative group action.

Similar issues arise in relation to intensive small-holder milk production. Animals are housed and may be zero-grazed and fed some concentrates, cross-bred with exotic breeds, is commonly practiced and there are economies of scale in processing and marketing dairy products. Producer co-operatives have been successful for processing and marketing in some countries.

Trade policies are coming to protect domestic producers in developed countries, lower the prices on world markets, particularly for beef and dairy products. Pressure is being applied, through the WTO, for reduction of these trade barriers, since despite benefiting developing country consumers; they create unfair competition for their producers. However, it is predicted that the effect of the complete removal of such barriers would be small. Non-tariff barriers, such as sanitary regulations imposed by developed countries, are likely to become increasingly important. Compliance raises costs for developing country exporters, for which foreign aid may be justified. There are national and global benefits to be gained from improved sanitary regulations and other quality standards for both exporters and importers.

How can it be done?

In the face of the fast growing demand for livestock products in developing countries, there are opportunities for major gains to be derived from improvements in the productivity of livestock systems. Increase in domestic livestock production contribute to growth in national income, reduced dependency on imports with savings in scarce foreign exchange and in rural areas, additional employment, improved livelihoods and poverty alleviation. There is considerable scope for improvement in livestock production and productivity but increased funding will be needed. The current decline in development funding for agriculture must be reversed. It will require more resources to be provided by national governments, but also increased bilateral and multilateral aid and technical assistance.

Development policy instruments fall into three main groups:

1. Price policies, including trade and exchange rate policies
2. Institutional policies
3. Promotion of technological change

Price policies are the responsibility of national governments, although they may be influenced by international agencies, such as customs unions, World Bank or WTO. However, institutional and technological changes are introduced not only by national and local governments but also by private individuals or associations and Non-Government Organizations.

Price policies fall into four main categories of

(i) trade policy,

(ii) exchange rate policy,

(iii) tax and subsidy policy, and

(iv) direct interventions such as floor prices and fixed prices.

Trade policy, for the developing countries, should include continued pressure, through international form such as the WTO, on developed countries to reduce tariffs and other barriers aimed at supporting their own producers. Possibly greater benefits might be achieved by reducing levels of protection for industrial sectors within the developing countries; as such protection raises input costs and effectively taxes agricultural producers. Taxes, subsidies and direct market interventions by government are also less common today, having failed to bring lasting benefits and possibly having contributed to the international debt crisis. There remains a case for limited use of subsidies for disaster relief and to promote the use of beneficial new inputs, such as vaccines or drugs. Alternatively, moderate taxes on livestock producers might be used to recover costs of providing public goods such as disease control or eradication programmes.

Policies for the promotion of appropriate institutions can have a major impact on livestock projects. The authors of a review of about 800 livestock development projects found that most had failed to bring about significant sustainable improvements in livelihoods of the poor. They conclude that 'The key lesson to emerge from our review is the importance of institutions in defining the success of pro-poor measures' (LID 1999). Benefits would accrue to livestock producers, as to all members of society, if along with improvements to the physical infrastructure of communications and transport routes, electricity and other services, the institutional infrastructure of law and order, respect for property rights and contractual agreements with legal support. In relation to grassland-based and mixed-farming systems, strengthening of property rights in land and water supplies may bring major benefits. Legal methods of excluding non-members of the community from enjoying common property rights, prevent inter-community strife and regression of rangeland use to one of open access. Secure property rights provide incentives for land conservation and improvement.

Institutional development is needed for the provision of credit, animal health services and genetic material, for example, by artificial insemination. The introduction of new technology must be accompanied by the development or strengthening of the institutional framework needed for its implementation. The other key area, where institutional change is essential for the success of livestock development projects and programmes is that of marketing, including transport, processing and selling. As there are economies of scale in these marketing activities, large commercial operations are most likely to be cost-effective. However, because of the high transaction costs of individual spot market sales by small-scale producers to processing and marketing companies, there is a need for formal contracting or vertical integration. In negotiating contracts, small-scale producers are in a weak position, lacking market power and information on patterns of supply, demand and prices. Thus, in promoting institutional development, there is a need for dissemination of market information and encouragement of co-operative group action and participation by small-scale producers to strengthen their bargaining position. This might result, as in the case of the Indian dairy industry, in producer co-operative unions managing the processing and marketing operations. Additional benefits may be achieved by the development of linkages between different input markets, and between product marketing agencies and the delivery of inputs.

Technological change may be promoted by supporting research and development and the dissemination of information, or extension to farmers. Public funding for agricultural research and particularly that for livestock research has declined over recent decades. The decline in public sector and NGO funding must be reversed to allow faster growth in livestock production. National agricultural research systems are expected to concentrate effort on the adaptive and applied research, development and testing end of the chain. An appropriate institutional framework must be developed to integrate a farmer participatory systems approach with science based adaptive and applied research, necessitating collaboration between producers, and natural and social scientists. The national research organizations must take responsibility for research prioritization, ensuring that it is appropriate for relative resource availability, taking account of the needs of the poor and coordinating donor assistance.

Areas of research deserving attention include veterinary public health measures and disease control, improvements in forage crops and utilization of crop by-products, and improvements in husbandry and management of production systems. Local breed improvement is a long-slow process and increase in production are generally more readily achieved by cross-breeding with, or adopting exotic breeds. Equally important is the need for socio-economic research into the institutional framework for the allocation of natural resources, credit and labour hire, the delivery of inputs and the processing and marketing of livestock products. Research is needed to describe and analyze the strengths and weaknesses of existing institutions and to propose and test alternatives for improvement where necessary. In addition socio-economists are needed to contribute to the research prioritization process, by assessing likely costs and benefits of proposed research projects.

What could the benefits be?

The primary benefit to be derived from increase in livestock productivity is a sustainable improvement in the livelihoods of livestock producers, many of whom are resource poor, many of these being women and some of whom are landless. Some of the benefits will be reflected in improved levels of nutrition, while increase in market sales will provide income for other uses.

Increase in domestic production and supply of livestock products may result in reduction of prices. This will benefit consumers and accelerate the growth in demand. However, the fall in price is unlikely to be large enough to cancel out the benefits to producers of the increased productivity. The main effect for most developing countries will be the substitution of domestic products for imports. This effect will bring additional benefits by saving scarce foreign exchange.

Improvement in animal and veterinary public health not only save farm costs, increase productivity and incomes, but also reduce risks of losses and for those countries producing more than enough for home consumption, improve access to world markets.

Summary

Livestock have served and will continue to serve a valuable role in sustainable agricultural systems. They are particularly useful in converting vast renewable resources from rangeland, pasture and crop residues or other by-products into food edible for humans. With livestock, land that is too poor or too erodible to cultivate becomes productive. Also, nutrients in by-products are utilized and do not become a waste-disposal problem. The need to maintain ruminants to utilize these humanly inedible foodstuffs and convert them into high-quality foods for human consumption has been a characteristic of advanced societies for several thousand years. Further, ruminant livestock production is entirely consistent with good agronomic practices in which forages are grown on 25% of arable land to minimize water and soil erosion. In addition, the protein resulting from livestock production is of better quality with a greater biological value than protein in the substrate feeds. The evidence that livestock belongs to sustainable livestock production systems, is convincing.

Conclusion

Livestock are an integral component of agriculture in India and make multifaceted contributions to the growth and development of the agricultural sector. Livestock help in improving food and nutritional security by providing nutrient-rich food products, generate income and employment and act as a cushion against crop failure, provide draught power and manure inputs to the crop subsector and contribute to foreign exchange through exports. Also, by using crop residues as feed, livestock save land for food production that would otherwise be used for fodder production. Additionally, livestock make substantial contributions to environmental conservation, supplying draught power and manure for fertilizer and domestic fuel that helps in saving the petro-products for future generations. save on the use of petro-products. This paper assesses the contribution of livestock subsector to the growth of agriculture and socioeconomic development.

Livestock sector in India has experienced remarkable growth during the last two decades in terms of production, value of output from livestock and trade. Livestock sector contributes nearly 25 percent to the gross value of agricultural output at the national level. At the household level, its contribution is much more in the case of small landholders who comprise a sizeable proportion of rural households and control bulk of the livestock resources (Birthal and Taneja 2006). Livestock production is more women oriented as women contribute more than 70 percent to the labour requirement. Livestock asset is more equitably distributed than land. The regression analysis of relationship between poverty and livestock income has shown that growth in livestock sector contributes more towards poverty reduction. Animal agriculture is an integral part of food production systems, using lands and products not usable by humans for the production of human food, making useful contributions to crop rotations and soil conservation through the growing of forages, serving as a buffer for feed grain supplies and prices, and improving nutritional quality of the diet, as well as adding variety and palatability. In addition, the protein resulting from livestock production is of higher quality with a greater biological value than protein in the substrate feeds. Most careful analysis conclude that if food requirements of the expanding world population are to be met, both in terms of quantity and quality, all available food production resources must be used effectively and efficiently. Livestock is integral in those systems and as sources of human food. Thus, livestock can be used as an effective tool for reducing rural poverty and is essential for sustainable agricultural development.

References

Ali J 2007. Livestock sector development and implications for rural poverty alleviation in India. Livesto. Res. Rural Dev. 19(2). http://www.lrrd.org/lrrd19/2/ali19027.htm

Baldwin RL, Donovan KC, Beckett JL 1992. An update on returns on human edible input in animal agriculture. In: Proceedings California Animal Nutrition Conference, Pp 97-111.

Benbrook CM 1991. Introduction. In: BANR's Sustainable Agriculture Research and Education in the Field: A Procedings. National Academic Press, Washington D.C.

Birthal PS 2002. Technological Change in India's Livestock Sub-sector: Evidence and Issues, In: Technology Options for Sustainable Livestock Production in India. National Centre for Agricultural Economics and Policy Research, New Delhi, International Crops Research Institute for the Semi-Arid Tropics, Patancheru, Andhra Pradesh, and International Livestock Research Institute, Addis Ababa.

Birthal PS, Taneja VK 2006. Livestock sector in India: Opportunities and Challenges, presented at the ICAR-ILRI workshop on 'Smallholder livestock production in India' held during January 24-25, 2006 at NCAP, New Delhi.

Birthal PS, Ali J 2005. Potential of livestock sector in rural transformation, In: Rural Transformation in India: The Role of Non-farm Sector Institute for Human Development. Manohar Publishers and Distributors, New Delhi.

Bruinsma J 2003. World agriculture: towards 2015/2030, an FAO Perspective. London: Earthscan.

CAST 1975. Ruminants as food producers: Now and for the future. Council for Agricultural Science and Technology. Westview Press, Boulder, CO. Spec. Publ. No. 4. pp 1-13.

CAST 1982. The U.S. sheep and goat industry: Products, opportunities and limitations. Council for Agricultural Science and Technology. Rep. No. 94.

CAST 1994. How much land can ten billion people spare for nature? Council for Agricultural Science and Technology. Spec. 121: 1-64.

CDFA 1993. California Field Crop Review. California Department of Food and Agriculture.

Cox J, Varpama S 2000. 'The 'Livestock Revolution' - Development or Destruction? ' Compassion in World Farming's Report into Factory Farming in 'Developing' Countries. June 2000. www.ciwf-livestock-revolution.co.uk/report%20cover.htm

Durning AB, Brough HB 1991. Taking stock: animal farming and the environment. Worldwatch Paper No. 103. Washington D.C. Worldwatch Institute.

Ely DG 1994. The role of grazing sheep in sustainable agriculture. Sheep Res. J. Spec. Issue. pp 37?51.

Faostat 2003. FAO statistics database on the world wide web . http://apps.fao.org/(accessed December, 2003).

Flora CB 1992. Building sustainable agriculture: A new application of farming system research and extension. J Sustain Agr. Volume 2 Article # 3

Freenstra G 1997. What is sustainable agriculture? Sustainable Agriculture Research and Education Program. University of California. Retrevied December 14, 2004 from http://www.sarep.ucdavis.edu/staff/staff7.html

Fresco LO, Steinfeld H 1998. 'A food security perspective to livestock and the environment', in A.J. Nell (ed.) Livestock and the Environment. Proceedings of the International Conference on Livestock and the Environment held in Ede/Wageningen, Netherlands, 16-20 June, 1997. Wageningen: International Agricultural Centre.

Hauptli HD, Katy BR, Thomas RM, Goodman 1990. Biotechnology and crop breeding for sustainable agriculture. In: C. A. Edwards, R. Lal, P. Madden, R. H. Miller, and G. House (Ed.) Sustainable Agricultural Systems. Westview Press, Boulder, CO. pp 141-156.

Hess CE 1991. The US Department of Agriculture Commitment to Sustainable Agriculture. In: BANR's Sustainable Agriculture Research and Education in the Field: A Procedings. National Academic Press, Washington D.C.

Kozel V, Parker B 2003. A Profile and Diagnostic of Poverty in Uttar Pradesh, Economic and Political Weekly, January 25, 2003. pp 385:403.

LID 1999. Livestock in Poverty-Focussed Development. Crewkerne, Somerset: Livestock in Development.

Smith NE 1980. Opportunities for forage, waste and by-product conversion to human food by ruminants. In: R. L. Baldwin (Ed.) Animals, Feed, Food and People. pp 52?62. Westview Press, Boulder, CO.

Taneja VK Birthal PS 2004. Role of Buffalo in Food Security in Asia, Asian Buffalo Magazine. 1(1): 4-13.

Wheeler RD, Kramer, GL, Young, KB, Ospina E 1981 The World Livestock Product, Feedstuff, and Food Grain System. Winrock International, Morrilton, AK.

12

Nutraceuticals in Livestock Practices

Ravindra Kumar and Prabhat Tripathi

Several centuries ago, Father of medicine Hippocrates, has said "Let food be your medicine". Our traditional therapies or alternative medicine effectively integrated this concept and include herbal remedies, nutritional interventions like vitamins/minerals supplementation specialized diets etc, and other techniques such as acupuncture, relaxation regimens, and homeopathic medicines. While, in modern medical science medicines are used to cure diseases which may have some inherited side effects. From the final decade of the last century, a renewed interest was seen in traditional therapies and people are thinking of integrated medicine with a focus on prevention of disease rather than its cure. In this context many terms like "dietary supplements", "nutraceuticals" and "functional foods" were frequently used to address this issue. The term "nutraceutical" was coined from "nutrition" and "pharmaceutical" in 1989 by Stephen DeFelice, MD, founder and chairman of the Foundation for Innovation in Medicine (FIM). Nutraceutical can be defined as, "a food or part of a food, that provides health benefits, including the prevention and treatment of a disease." However, the term nutraceutical as commonly used in marketing has no regulatory definition. There is confusion between the closely related terms nutraceuticals, functional food and dietary supplement. Kalra (2003) differentiated between these terms as: Functional food provides the body with the required amount of vitamins, fats, proteins, carbohydrates, etc, needed for its healthy survival. When functional food aids in the prevention and treatment of disease(s) and disorder(s) other than anemia, it is called a nutraceutical. Since most of the functional foods act in one or other way as antianemic, the anemia was considered as a better index to have a clear distinction between the two terms i.e. functional food and nutraceutical. Thus, a functional food for one animal may act as a nutraceutical for another animal. Dietary supplement is the product intended to supplement the diet that contains one or more of these dietary ingredients i.e. vitamin, mineral, and amino acid, while nutraceuticals must not only supplement the diet but should also aid in the prevention and treatment of disease and disorder. Kramer *et al.* (2001) defined nutraceutical as a dietary supplement administered in large dose to obtain desired pharmacological effects, however functional food and supplements are used to deliver approximate the same dose that is delivered by diets. *Veterinary Nutraceutical:* As defined by the newly created North American Veterinarian Nutraceutical Council, Inc. (NAVNC), "a substance which is produced in a purified or extracted form and administered orally to patients to provide agents required for normal body structure and function and administered with the intent of improving the health and well-being of animals." Nutraceuticals has created a new era in animal health and nutrition. In recent years, research and practice has emphasized the pharmacological effects of numerous nutritionally based compounds to enhance health in farm animals as well as companion animal.

Classification of Nutraceuticals

Hathcock (2001) has grouped nutraceuticals used in humans/animals into following three categories

1. **Nutrients:** Substances which are having established nutritional functions such as vitamins, minerals, amino acids and fatty acids
2. **Herbals:** Herbs or botanical products as concentrates and extracts
3. **Dietary supplements:** Substance which are derived from other sources *e.g.* pyruvate, chondroitin sulphate, steroid hormone precursors and are serving specific functions, such as used in sports nutrition, for weight-loss and as meal replacements.

1. Nutrients

Antioxidant minerals, water and fat-soluble vitamins are most commonly used nutrients that are used in humans and animals as nutraceuticals. Supplementation of these nutrients in livestock enhances the immunity against diseases, such as mastitis, by increasing resistance to infections and by decreasing severity of infections. This can be related to the stimulation of the immune system. Supplemental Zn, Cu, Se, and other trace elements can alter immune function of newly born calves. Some field trials have also shown decrease in Bovine Respiratory Disease (BRD) morbidity rate with supplementation of zinc, copper and selenium. Among minerals zinc, copper, selenium play a vital roles in maintaining health of animals. Zinc is necessary in production and maintenance of the "killer cells" of immune system, used to fight off infection. Zinc deficiencies in the animal have resulted in a decreased ability to produce these various immune system components. Similarly, a large number of disorders are attributed to deficiency of Cu in cattle. Symptoms including anemia, severe diarrhea, depressed growth, change in hair color and weak, fragile bones are some important characteristic signs of clinically depressed Cu status. One of the more sensitive indicators of low level of blood Cu is achromatrichia or loss of hair pigment. Black hair pigments of cattle tend to develop a red tinge to the hair coat whereas the red hair of cattle becomes lighter considerably. Loss of hair around the eye is also seen in Cu deficiency. Likewise, deficiencies in copper have been shown to reduce the effectiveness of the immune system. Selenium and Vitamin E are integral and essential part of body for prevention of oxidative damage of cellular and subcellular membrane. Researchers had shown that supplementation with vitamin E decreases the incidence of mastitis. Combining selenium and Vitamin E supplements appear to result in the greatest increase in defense against mastitis. Deficiencies of vitamin E and selenium also have been found to increase the incidence of retained placenta. Selenium deficiency alone can increase the incidence of embryonic death and uterine infections and can decrease fertility. In case of companion animals nutrient such as essential fatty acids are often used to cure and prevent skin problems. The list of common nutrients with their most important health benefits are summarized in Table 1:

Table 1: Health benefits of different common nutrients.

A. Vitamins	Functions
1. ***Fat Soluble Vitamins***	
Vitamin A	Acts as antioxidant, essential for growth and development, maintains healthy vision, skin and mucous membranes, may aid in the prevention and treatment of certain cancers and in the treatment of certain skin disorders
Vitamin D	Essential for formation of bones and teeth, helps the body to absorb and use calcium
Vitamin E	Acts as antioxidant, helps to form blood cells, muscles, lung and nerve tissue, boosts the immune system
Vitamin K	Essential for blood clotting
2. *Water Soluble Vitamins*	
Vitamin C	Acts as antioxidant, necessary for healthy bones, gums, teeth and skin. It helps in wound healing, may prevent common cold and attenuate its symptoms
Vitamin B_1	Helps to convert food into energy, essential for neurological functions
Vitamin B_2	Helps in energy production and other chemical processes in the body. It also helps to maintain healthy eyes, skin and nerve function

Contd.

Vitamin B_3	Helps to convert food in to energy and maintain proper brain function
Vitamin B_6	Helps to produce essential proteins and convert protein in to energy
Vitamin B_{12}	Helps to produce the genetic material of cells, helps in formation of red blood cells, maintenance of central nervous system and synthesize amino acids. It is involved in metabolism of fats, protein and carbohydrates
Folic acid	Necessary to produce the genetic materials of cells, essential in first three months of pregnancy for preventing birth defects, helps in red blood cell formation, protects against heart disease
Pantothenic acid	Aids in synthesis of cholesterol, steroids and fatty acids, crucial for intraneuronal synthesis of acetylcholine
B. Minerals	
Macrominerals	
Calcium	Essential for building of bones and teeth. It also help in maintaining bone strength and is very important in nerve, muscle and glandular functions
Iron	Helps in energy production, also essential to carry and transfer oxygen to tissues
Magnesium	Essential for healthy nerve and muscle function and bone formation
Phosphorous	Essential for building of strong bones and teeth. It also helps in formation of genetic material, energy production and storage
Trace elements	
Chromium	With insulin it helps in conversion of carbohydrates and fats into energy
Cobalt	Essential component of vitamin B_{12}, but ingested cobalt is metabolized *in vivo* to form the B_{12} coenzymes
Copper	Essential for hemoglobin and collagen production, healthy function of the heart, energy production and absorption of iron from digestive tract
Iodine	Essential for proper function of thyroid gland
Selenium	Antioxidant, essential for healthy functioning of heart muscle
Zinc	Essential for cell reproduction, normal growth and development in neonates, wound healing, production of sperm and testosterone hormone
C. Vitamin like compounds	
L-Carnitine	It helps in oxidation of fatty acids, promotion of certain organic acid excretion and enhancement of the rate of oxidative phosphorylation
Choline	It acts as lipotropic agent and is used to treat fatty liver and disturbed fat metabolism
Vitamin F	Involved in proper development of various membranes and synthesis of prostaglandins, leukotrienes and various hydroxy fatty acids
Inositol	It acts as lipotropic agent. It is also necessary for amino acid transport and movement of potassium and sodium
Taurine	It helps in retinal photoreceptor activity, bile acid conjugation, white blood cell antioxidant activity, CNS neuromodulation, platelet aggregation, cardiac contractility, sperm motility, growth and insulin activity

2. Herbals

Among all categories of nutraceuticals herbal nutraceuticals have a quite important place. In Ayurveda they are the source of the treatment and prevention of many diseases. Theses herbs are store house of some phytochemicals which act as nutraceuticals in preventing and treatment of diseases. In the past some of these phytochemicals found in fruits and vegetables were classified as vitamins.

Flavonoids were known as vitamin P, glucosinolates and indoles present in cabbage were termed as vitamin U and ubiquinone as vitamin Q. But later on their vitamin status was changed because their specific deficiency symptoms could not be established. Different plants containing these phytochemicals are in practice in veterinary use. Their popularity is gaining day by day because they concentrate a therapeutic principle (different herbs with potentially different mechanisms of action for the same problem), have unique actions and have nutritional properties. A list of commonly known herbal products with their therapeutic activity is presented in Table 2.

Table 2. Therapeutic activity of common herbs

Herb	Therapeutic activities
Aloe vera	Dilates capillaries, anti-inflammatory, emollient, wound healing properties
Chamomile	Antiinflammatory, spasmolytic, antimicrobial, wound healing
Echinacea	Immunostimulant, treatment of cold and flu symptoms
Ephedra	Bronchodilator, vasoconstrictor, reduces bronchial edema, appetite suppressant
Evening primrose Oil	Dietary supplement of linoleic acid, treatment of atopic eczema
Feverfew	Treatment of headache, fever and menstrual problems, prophylactic to reduce frequency, severity and duration of migraine headaches
Garlic	Antibacterial, antifungal, antithrombotic, hypotensive, fibrinolytic, antihyperlipidemic, anti-inflammatory
Ginger	Carminative, antiemetic, cholagogue, positive inotropic, treatment of dizziness
Ginseng	Adaptogen
Ginkgo	Vasodilation, increased peripheral blood flow, treatment of post- thrombotic syndrome, chronic cerebral vascular insufficiency, short term memory loss, cognitive disorders secondary to depression, dementia, tinnitus, vertigo
Goldenseal	Antimicrobial, astringent, antihemorrhagic, treatment of mucosal inflammation, dyspepsia, gastritis
Horehound	Expectorant, antitussive, choleretic
Licorice	Expectorant, secretolytic, treatment of peptic ulcer
Melissa	Topical antibacterial and antiviral
Plantago seed	Cathartic
Slippery	Mucilaginous demulcent, emollient and nutrient, used to sooth elm irritated mucous embranes, ulcerations of the digestive tract
St. John's	Anxiolytic, antiinflammatory, antidepressant, monoamine oxidase wort inhibitor
Valerian	Spasmolytic, mild sedative, sleep aid
Willow bark	Anti-inflammatory, analgesic, antipyretic, astringent, treatment of rheumatic and arthritic conditions, headache and gout

Out of above mentioned herbs following are most commonly used herbs in pet animal practice.

Echinacea: Echinacea is a well- known immune stimulant. It improves the immune system and helps pets fight off infections, diseases and even cancer.

Aloe: This thick-leaved plant is used for animals much like it is for humans as a soothing, itch relieving ointment that is applied directly to the skin. It is also said to help heal cuts and protect them from infection.

Ginger: Roots of ginger are used for centuries as an anti-inflammatory and an aid for stomach problems. Ginger is said to help animals with car sickness and digestive problems like gas and diarrhea.

Milk thistle: Pets having liver problems like hepatitis may be given this supplement. It is supposed to protect the cells of the liver from toxins.

Saint John's wort: Mainly used in human medicine as a treatment for depression. This herb is used as a treatment for viral infections and neural disorders in veterinary medicine.

Ginkgo: This herb work by expanding the blood vessels in the brain, thereby increasing blood flow. In human medicine it is used to increase the memory and improve brain function. Veterinarians are using it to treat animals that exhibit cognitive dysfunction of animal equivalent to senile dementia.

Slippery Elm: The bark of this tree is used as an aid to the digestive system for pets that suffer from constipation and upset stomach. It has also been used as a cough suppressant and a poultice.

3. Dietary supplements

Dietary supplements have been developed to manage a variety of disease conditions and syndromes. Ketogenic diets, comprised of high fat and low protein and carbohydrate content, have been reported to improve seizure control in man and dogs. The two nutraceuticals namely glucosamine sulfate and chondroitin sulfate are effective and safer to alleviate symptoms of osteoarthritis (Reyes 2000). Once it was thought that cartilage could not be rebuilt once it is degenerated. But now, it is believed that some supplements can help the body in repairing the damaged cartilage. Two such chondroprotective agents which are included in the majority of nutritional supplements are glucosamine and chondroitin sulphate. A number of controlled, short-term studies have found glucosamine and chondroitin sulphate, either alone or in combination have significant degree of chondroprotective activity. Numerous casein and whey protein have been identified to form angiotensin-I-converting enzyme inhibitory peptides and hydrolysates. These peptides/hydrolysates may be classified as nutraceuticals due to their ability to provide health benefits (Korhonen *et al.* 2000). Conjugated linoleic acid present in ruminant's milk and meat products have been found to exhibit a number of beneficial health effects in cancer, cardiovascular diseases and diabetes in animal model (Bauman *et al.* 2001). Direct-Fed Microbials (DFM) has been of great interest in recent years. These are live microorganisms and their metabolites (compounds that they produce as waste) to improve general health and production of animals. There have been several hypotheses put forth to explain the usefulness of DFM. One of the most believed explanations for improved animal's health and production suggested that the addition of beneficial bacteria prevent the colonization of pathogens in the lower gut by competing for space and nutrients. Production of antimicrobial end products such as acids and antibiotics has also been found to add beneficial effects. Some of the important mechanisms of action that how DFMs acts include:

1. Production of antibacterial compounds (acids, antibiotics).
2. Competition with undesirable organisms for space and nutrients in the digestive tract.
3. Production of nutrients (e.g. amino acids, vitamins) or other growth factors which stimulate growth and reproduction of other microorganisms in the digestive tract.
4. Production and stimulation of enzymes.
5. Breakdown and detoxification of undesirable compounds
6. Stimulation of the immune system in the host animal.

Because of these reports and findings, nutraceuticals are becoming more widely accepted as an adjunct to conventional therapies for enhancing general health and reproduction of humans and animals. As a result of extensive research on these nutraceuticals, various branded products of nutraceuticals are available in National and International market. Nutraceuticals and functional foods represent one of the fastest growing markets in the developed world.

Conclusion

Nutraceuticals have a wide beneficial effects in human and animals to protect their health, reproduction and production. Utilization of these products are mainly based on studies in animal model, traditional history and epidemiological studies. A lot of research is still needed as to how these compounds actually function in the animal and what is their mode of action. The intervening results, however, have been good enough that producers and researchers are giving desired attention towards this direction.

References

Bauman DE, Corl BA, Baumgard LH, Griinari JM 2001. Conjugated linoleic acid (CLA) and the dairy cow. pp 221-250. In: Recent advances in animal nutrition-2001. PC Garnsworthy, J Wiseman (Eds.) Nottingham University Press, Nottingham UK.

DeFelice SL. FIM Rationale and Proposed Guidelines for the Nutraceutical Research & Education Act - NREA, November 10, 2002. Foundation for Innovation in Medicine.

Hathcock J 2001. Dietary supplements: How they are used and regulated. J Nutrition 131: 1114-7.

Kalra EK 2003. Nutraceutical - Definition and Introduction. *AAPS* PharmSci, 5 (2): article 25.

Korhonen H, Marnila P, Gill HS 2000. Bovine milk antibodies for health. Br J Nutr 84:135-46.

Kramer, Hoppe P-P, Packer L 2001. Nutraceuticals in Health and Disease Prevention, Marcel Dekker, Inc, New York 318 pages,

Reyes GC, Koda RT, Lein EJ 2000. Glucosamine and chondroitin sulfates in the treatment of osteoarthritis: a survey. Prog Drug Res. 55:81-103.

13

Goat Improvement for Milk Production

Rajendra Kumar Pandey

Historical background of dairy goats

Small ruminants sector plays an important role in the national economy. This enterprise is associated with social and cultural fabric of millions of resource poor farmers. Small ruminant especially goats gain importance mainly on account of their short generation intervals, higher rates of prolificacy and the ease with which they can be marketed. They arc vcry uscful in scmi-arid and arid zonc where they can sustain themselves on sparse vegetation and esteemed climatic conditions.

There are four main breeds of dairy goats commonly found in Western Country, namely, Alpine, Nubians, Saanens and Toggenburgs. They are considered to be the best milking breeds of the world. All these four breeds have recently been introduced in India for improvement of milk quality of local goats. These breeds have been improved by selection, upgrading, and the wise use of pedigree records and registration over a period of more than half a century in Europe and North America.

Some of the original research work to improve the dairy goats of the United States, were done at the Agriculture Research Center at Beltsville, Maryland. Work began with the breeding of common American dose to bucks of the same type in order to improve the exiting strain of goats the use of selection. Progress was found to be difficult and slow. Perhaps the same would be true in India. To shorten the process, India has adopted the same practice used in United State of upgrading the local dairy type goats. USA stated with the Saanens and Toggenburgs, whereas in India, Alpines have been imported in large number in Ludhiana District of the Punjab; some Nubians and Toggenburgs are also being used.

Dairy goat production in Indian scenario

There are 20 well defined breeds of goats in India apart from the non-descript goats. "Barbari" a dual purpose breed from south western semi-arid zone of Uttar Pradesh is considered to have good potential for others semi-arid parts of India. Similarly "Jamunapari" from Etawah district of Uttar Praesh, "Beetal" from Punjab and "Suarti" from Gujarat, for milk and "Sirohi" from Rajasthan for milk and meat.

The goat population in India during last two decade has increased at the fastest rate of 2.71 per cent among all major livestock species in spite of annual slaughter rate of nearly 41 per cent and natural death rate of 15.5 per cent per annum. The annual output of milk from goats is 2.55 MT (Hindu Survey 2004) contributing about 3 pr cent to the total milk production but almost strategizing at this level.

Better species of goats are efficient converters of feed to milk. They can produce milk at a cheaper cost than either the cow or the buffalo. Goats in India contribute about 4.3 million tonnes milk which is 5% of the total milk production (84.6 million tonnes), (Economic Survey 2004).

At present the goats annually produce 948 million kg or milk (FAO 1981, Anonymous 1988), which though resemble in composition to cow milk but us considered superior in certain respect viz., easy digestibility due to smaller fat globules and soft coagulation of proteins and of therapeutic value for those who are allergic or intolerant to cow, buffalo milk products The goat meat (Chevon) which is most expensive of all meat in Indian people, provide 840 g. protein 100 g. fat and 3078 gm on the basis of per kg dry matter and chevon is completely fit religiously among the Hindu religion Studies conducted on comparative economics of sheep, cattle and goats rearing under arid zone range conditions revealed that goats are 123 per cent more economical than sheep and 130 per cent more economical than cattle and ensure a profit upto Rs. 250 per head per year. Goats are known to have a cost benefit ratio of 1: 1.17 and an I.R.R. (Investment Return Rate) value of over 30 per cent, which is by far more than any bank interest. Further the export of meat to other countries has increased from Rs. 273 crores in 1983-84 to Rs. 821 crores in 1984-85 registering three times increase within a year. Besides it is reported that British market is likely to be developed for export of canned meat from India where in goats have to play the main role.

Indian goats are prolific breeder and a single goat is capable of producing over 100 milking descendents in 10 year time as against only about 6 by a cow or buffalo a feature which is of crucial significance from the meat production point of view.

Importance of goats

The Indian goats possess high quality carcass, heat resistance and water stress also, then other kind of livestock. Goat is most converter of rough feed in to goat cash product and fertilized land. The goat plays a major role in animal production by way of producing meat, milk, fiber, skin and manure, hence it is popularly called "poor man cow" and also know for quick return.

Goat is a versatile animal and perhaps amongst the most adoptable. It is distributed right across the globe, in almost all ecological region of Asia to Europion Union, U.K, Russia and Africa main land etc. and is main stay for existence of several settlement and societies and in India, particularly the tribes and nomads.

With the high yielding milch breeds of goats, which continue milking for over 10 months, it is advantageous to dry them up two months before kidding. In other words, a dairy goatman tries to gets 10 months' milk production from each doe and then gives her two months' rest period before she kids again. This means that the doe will need to be bred during the seventh month of her lactation period and dried off at the end of the tenth month; in another two months she should be ready to kid.

Indian improver breeds and goat development experience

A development scheme on Beetal was taken up in the country in 1939 at the Government Livestock Farm, Hisar, to improve the milk production through selection. The effort however, dwindled as there was a decline in the lactation yield, daily yield and yield per day of kidding-interval by 6.8, 19.5 and 21.8%, respectively, and an increase in the age at first kidding, lactation length and kidding interval by 12.6, 22.5 and 12.9% respectively, over the foundation stock after 5 generations of selection. The decline of performance was ascribed to the poor breeding value of the sires used in the successive generations. The few state governments, during the post Independence years prior to the organization of the State Agricultural Universities, undertook research and development work at the state farms on Beetal, Jamunapari, Sirohi and Black Bengal flocks. Occasional efforts were also made to produce the exotic crosses with Saanen-Alpine, Anglo-Nubian and Toggenburg. A few missionary, voluntary and non-governmental organizations undertook the sporadic developmental work in some pockets of the country in seclusion, introducing Swiss-Alpine. Israel-Saanen and Boer goats. Such goat development efforts in the different states have been rather meager except in Rajasthan, Gujarat, Maharashtra, Uttar Pradesh, Himachal Pradesh and Jammu & Kashmk The early experience of crossbreeding under field conditions, except in the Western Himalayan and southern plateau and hill regions, is discouraging, for the crossbreds generally failed to sustain the nutritional and environmental (including disease) stress and endured high mortality.

In the absence of reasonable data, on comparison among the breeds, having identical production trait(s) and on compatibility with improve breeds in diverse geo-climate, a few Indian breeds have since been identified as improver breeds, fascinated often by the adult body size and milk yield. Jamunapari, Beetal and Barbari have been employed more often as improver breeds in India and the other neighboring countries to-upgrade the regional nondescript, small and dwarf breeds. The valuable traits, like fecundity, disease resistance, grazing habit, meat quality and skin characteristics became secondary though crucial in the development programme of goats. The short-term crossbreeding experiments in the institutional farms really demonstrated the transmissibility of some of the production traits of the Indian improver breeds to the breeds of lower productivity. Jamunapari x Bengal and Beetal x Bengal crosses showed an improvement of 76.3 and 90.8%, respectively, in the milk production over purebred Bengal. The improvement was obviously spectacular due to the large genetic diversity between the parental breeds. The effect of the heterosis was rather conspicuous when the high-yielding dam of Jamunapari was crossed with low-yielding sire breeds, e.g. Bengal. The reciprocal crosses did not show the heterosis. The improvement was imperceptible when the Sirohi breed was crossed with Beetal where the genetic divergence between the two breeds was limited. The situation was similar when Black Bengal was crossed with Barbari though the milk yield was considerably increased in the crosses over the Black Bengal.

Management practices for improvement of dairy goat for better milk production

The science and art of goat management, like any livestock management incorporates the management of breeding behaviour and reproductive performances, feed habits and feeding, disease surveillance and prevention, prenatal prudence of does and neo natal kid care, besides management of personnel shelter and other infrastructures at the farm complex. Such exuberant management theme has been circumscribed here to deal in the aspects of goat improvement for milk production.

Identification

Each goat should be identified in some manner. The use of tattoo marks in the ears makes a permanent record. Metal labels or notching the ears are two other systems used identifying goats.

Housing

Roofing material

Roofing material in the goat shed is an important factor, which determines the cost of construction and microenvironment within the shed. A wide range of roofing materials is available in different climates, which differ to a great extent in their thermal characteristics. The worth of over 50 materials were tested for use as roofing material in California, ranging from hay or straw, galvanized steel and plywood to several types of plastics. On a summer day, the difference in the radiant heat load, under sheds covered with straw versus galvanized iron or plastics, was of the order of 163 k call hr/m2 of the animals' surface. In a study at the CIRG, Makhdoom, it has been observed that the improved thatch provided the coolest micro-environment during hot part of the year. Different roofing materials commonly used in India were tested and they can be graded as, the most to least effective in that order, improved thatch, thatch, asbestos, lite roof with top surface painted white and non–painted literoof. The life of the improved fire-proof thatch is reported to be 8 years as against about I to 2 years for untreated thatch.

Height and shape of roof

The height at centre in 'A' shaped roof is suggested to vary between 3 to 3.5 m. A height of less than 3 m interferes with the proper ventilation resulting in reduced convective heat loss from animals. In temperate and hot-humid climate, where more height does not provide any additional benefit, a height of 3 m is expected to suffice. The heat loss through radiation from goats to cool sky is curtailed in low roof sheds. The shadow size is not affected by height, since higher the shed the faster the shadow moves. In hot-dry environment, radiant heat load plays a vital role. Within the boundaries of the shed, the radiant load on goats comes from the atmosphere, the roof, the shaded ground and sunny ground. The heat load on goats will be minimized within the shed when the distance from the roof is maximum.

'A' shaped roof is indeed better for hot climatic regions. In the hot weather one side of' A' shaped roof saves the other half from direct solar radiation by casting its shadow. This helps in cutting down heat gain from the roof of the shelter.

Double roof with both roofs of same or different materials are effective in reducing the heating of shed in hot weather conditions. The additional cost of construction in double roof, however, restricts its use to very high producing animals.

Floor type and space

The surface upon which an animal lies may potentially be a source of thermal and physical discomfort, injury and infectious diseases. The ideal bed, therefore, needs to be hygienic, dry, resilient and reasonably temperature resistant. The relative importance of these four criteria differs markedly for different species and classes of farm animals. Deep, clean, dry straw can provide an ideal bed for weaners and growers during the cool period, but a thin layer .of straw is likely to be more suitable during warm or hot weather conditions. An inadequate bed of wet, filthy straw fails on all accounts. Perforated or slatted floors are mostly drier and more hygienic than Solid floors with limited bedding and thus more suitable for humid and high rainfall areas. The type .of floor which provides both comfort and cleanliness with minimal risk .of injury should be given preference. The floor space for different age groups should be slightly more in hot weather conditions in comparison to cool weather. This helps in radiant heat losses and convective removal .of heat and water from the animal body. In tropical climates, huddling is disadvantageous for health of the animals and their productivity. The optimum floor space requirements to provide healthy and clean micro-environment have been recommended later .on the basis .of scattered information.

Shelter surroundings

Surfaces around the shelter are very important in view of the radiation exchange between surrounding surface and the shelter. The temperature of different surfaces varies significantly at the same air temperature. It is well known that the green surface do not heat up as much as other surface like gravel or lose loam. It should be possible to maintain the green vegetation adjacent to the goat shelters except the paddock.

Optimum floor space requirement per goat

The knowledge on air space requirement per animal is thought to be important, as the cleanliness of ambient air, with regard to air-borne wastes from animal body, must be directly proportional to the actual volume of air into which those wastes are dispelled. Conversely, concentration of pollutants is directly proportional to stocking density. Systematic studies to delineate the optimum floor space requirements of different goat breeds in different environmental conditions have been carried out at the CIRG.

Essential appliances

Feeders: The wastage and contamination .of feed with faces and urine has been the major problems with existing, feeders in goat farms. To rectify some of these shortcomings, new feeders were developed at the CIRG. The rectangular and hexagonal feeders have been developed with provision of feeding green roughage, straws as well as concentrate. It has been possible to drastically reduce the wastage of all 3 types of feed with use of these feeders. The animals are not able to void faces or urine in the feeding troughs nor can they keep their legs in the feeders thus avoiding the chance to contaminate the feed. The newly developed rectangular as well as hexagonal feeders are effective in saving time, labour and space, and seem to reduce the competitive struggle among animals to reach the feed.

Water troughs: Contaminated water is an important source of infection. The water tanks or troughs should be covered and require regular cleaning. Water troughs 3 to 4 cm in length per goat, when raised in groups are sufficient. Water tanks with flat valves may be suitable for large-scale intensive goat production. Availability of water in goat house round the clock will be beneficial for attaining optimum productivity under any system of production.

Kidding pens: Individual spacious pens are essential to house does in late pregnancy. Movable hurdles can also be used for preparing kidding pens. Individual kidding pens are contaminated very quickly, and need frequent cleaning and disinfections. Otherwise, they may constitute an important source of naval infection-to kids.

Rearing of young kids needs special attention while planning goat housing and a fly proof shed for kids up to 3 months of age will prove worthy. The pens for kidding should be at the warmest part of the goat- house complex if the kidding is expected in cool weather. The protection of newborn kids from low ambient temperatures is essential to reduce the kid mortality.

Shed hygiene

The shed management includes constant vigil to general cleanliness, timely repairs and renovations. The sheds would require seasonal spraying to protect the flock from ectoparasites and disinfections particularly prior to kidding operation. The earthen floors, where applicable, are needed to be replaced every 3 months besides disinfection with carbolic acid which takes care of snakes and other reptiles. Annual white washing with lime prior to winter are suggested: Painting of sheds are recommended prior to monsoon. Application of thatch panels to control the excessive airflow during summer and winter through the sheds in the hot-arid environment is expected to improve the shed micro-climate.

Breeding operation

In order to achieve the high efficiency of reproduction in the flocks, the management has to ensure heat detection, preferably twice daily, morning and evening at fixed hours, during the entire breeding months. The AI or natural services are recommended at 12 hours interval twice/oestrus and records are to be maintained for future use. The display boards indicating the does (with identity numbers) daily coming to heat and their coverage schedule are useful for efficient monitoring of the operation. Prior preparation in terms of man and material is essential for the kidding operation which will obviously demand round the clock vigilance. Pre- and post-parturition care and attention during actual process of parturition are obligatory for each doe and the attending staff has to ensure feeding of colostrums within 10-30 minutes of the birth of kid(s). At times help to the new- born is necessary to promote the first act of breathing. Placement of kids in boxes inside the kidding cage - on day I have been found to be helpful. The management of breeding bucks should invariably be done in separate sheds and during breeding season, they should be prevented from infighting and exhaustion. Exposure to high ambient temperature and conductive heat gain in the scrotal surface should be avoided.

Kid care

The first few days of life of kids are rather vital in the management of kids. With multiple births weaker kids often require special attention particularly at the time of suckling. It is a good practice to clean the udder with KMn04 solution each time, prior to suckling. Overfeeding and underfeeding are both deleterious to the health of the kid(s). It becomes essential to detect and isolate the kid suffering from diarrhoea or other ailment and to place the cases under the attending veterinarian. Healthy kids become playful within 72 hr and they require soft bedding and clean gassy paddocks. The kids are very susceptible to cold and special attention is required to monitor the shed temperature with simple maximum-minimum thermometer and to impose correction of ambient temperature to about 18°C in winter.

Grazing schedule and supplementary feeding

Goats naturally enjoy grazing and foraging to their choice. Except in very intensive system of management, tight scheduling of time and grazing area is obligatory. Depending on the nutrient potential of the grazing area at a particular season, supplementary feeding with concentrate mixture, particularly to lactating/pregnant does and growing kids, are necessary. With extensive grazing in the good pasture and in migratory flocks, such supplementation may not be necessary though grazers prefer to carry ground seeds/cereals and salt for feeding in adversities. A good feeding management is rested on storage of feed and hay and the efficiency of management is soon reflected in the general health of the flock.

Health management

The susceptibility of goats to viral, bacterial and parasitic diseases has been dealt already. Good management will, however, invariably ensure timely vaccination according to schedule to all members of the flock. Routine dipping will indeed render the flock free of ectoparasites. The administration of anthelmintic drugs before and after monsoon is always advisable. Trimming of hoofs is absolutely essential for those goats, which do not undertake long walking/ grazing on the hard and rocky terrains. Even with the best health care and management, a few animals in the flock may develop illness at any time and a good manager will not fail to detect the unhealthy behaviour of these animals. Isolation of these animals in the sick shed and prompt Treatment by a competent veterinarian is obligatory. Routine testing of the Dock for brucellosis and Johne's disease should under no circumstances be deferred to as precipitation of these diseases in the flock will be highly derrogatory in future. Vaccination of pregnant does against FM and colibacillosis and antibiotic administration to control insidious mycoplasmal infection should receive imminent priority. The kids are highly susceptible to FM, colibacillosis and coccidiosis for which preventive vaccination and administration of drugs where indicated, are necessary.

Problems and constrains for preventive health

Several new vaccines and other preventive measures have lately become available. The implementation of adequate preventing measures in a country like India requires n augmented infrastructural facility at the village level.

There is a paucity of skilled veterinary assistants (labour) in India and most of the problems of goat farmers remain unattended and unsolved, perhaps due to an overemphasis on the health of cattle and indifference to goat diseases.

Diagnostic facilities and Immunobiologicals

The disease prevention and efficient health control require adequate laboratory facilities and diagnostic antisera and antigens. There is also a need to develop an early diagnostic test to detect sub clinical cases of *M.paratuberculosis* in goats. Disease-monitoring and surveillance system have not been adequately geared to cover the goat diseases.

Vaccination schedule

In intensive and semi-extensive systems of goat rearing, it becomes essential to rigidly follow the preventive vaccination schedule for mock-health security. While raising goats under extensive systems particularly associated with migration, the vaccination schedule would require period adjustment to avoid vaccination *en route.* The migratory flocks are often exposed to endemic and epidemic diseases which might be prevailing on the migratory routes.

Land degradation and goats

Out of 350 Million ha (Mha) of land in India, about 174 million could be graded as 'good' but the remaining land is either partially or fully degraded. The classification and distribution of the ecologically unstable lands (often refer to as wasteland) have been shown by different authors. Five major factors are generally responsible for the degradation of land mass:

1. Erosion of the top soil by the action of water or wind. The loss of top soil by water action alone is estimated to be around 12,000 Million tonnes per annum,
2. Water logging and consequent salinization and alkalinization mainly due to lack of adequate drainage in canal irrigated areas,
3. Floods, which affect an average area of around 9 million every year,
4. Diversion of good agricultural lands to urban, industrial and other non-agricultural uses. It is estimated that land under such uses will increase from the present level of 18 million to around 24 million ha around the year 2000 AD, and
5. Deterioration of our most productive soil under conditions of multiple cropping with the aid of perennial irrigation and the application of large quantities of inorganic fertilizers.

The grim effects of the land degradation include massive soil erosion, floods, siltation of reservoirs, droughts, salinity, changes in the climate and consequent infertility of even cultivable lands. The process of denudation of soil cover is being further aggravated by growing livestock population coupled with unchecked grazing, de-vegetation and illicit felling of trees. The reclamation of these wastelands thus demands an integrated 3 dimensional farming system for the concurrent sustainable production of agriculture, forestry and animal husbandry. It is not a mere coincidence that goats are reared in India primarily on wastelands and a significant positive correlation (r=0.78) exists between goat population and available wastelands in different states. All scientific observations revealed that the contribution of goats to the fragility of the unmanaged wasteland is only marginal.

Breeding Management Practices

The normal breeding season of the imported goats extends from September through February and even early in March. The Barbari and Bengal goats can be bred at different times of the year. They generally have a short lactation period of four or five months, especially of they are bred twice in 12 to 15 months. The Beetal and Jamanapari goats tend to bred in July to September, but individual does vary considerably.

During the usual breeding season, the heat period in the does occurs every 18 to

21 days. During the earlier months of September through November, the does tend to be in heat for two to three days, and then as the season progresses the length of the heat periods diminishes. The general consensus is that the conception rate is highest when the does are bred between the tenth and fifteenth hours after the first appearance of the signs of heat.

Care and management kid and doe

An experienced goatman will be familiar with the signs of the doe that is about to kid. Common symptoms are: the doe becomes restless and nervous, makes a low bleating sound has a hollow appearance in the flank region and on either side of the tail. A doe about to kid often paws at her bedding and gets up and lies down at frequent intervals. Another signs before kidding is the appearance of an opaque or slightly yellowish vaginal fluid. This is a sign of that parturition has begun. If the doe has difficult in delivering one or more kids, the veterinarian should be called after the doe has been in labor for not more than two hours.

As soon as the kids are delivered, the navel should be treated with tincture of iodine.

Generally goats' dairymen, who are interested in producing milk for sale, remove the kids from their mother at birth. In this way, the doe comes into milk production more satisfactory than when the kid is permitted to nurse a few days and is then removed. Generally, after 2 or 4 hours the kids will gain strength and begin to show signs of hunger. They should be fed colostrums from the mother.

Feeding of kid and doe

Feed the fresh doe lightly the first two days after kidding and gradually increase the amount of feed as her milk yield increases. Milk the doe completely dry twice a day; in the case of a high yielding doe is may be necessary to milk her three times a day at 8-hours intervals. It takes about a month for doe to reach her peak milk production.

When hand-feeding of milk to the kids, you my use either a bottle or nipple or practice pan feeding. Regardless of the method adopted, attention must be given to cleanliness and sanitation.

Start feeding the kids on a milk replacer (kid-starter) about 10 days after the kid are born. They will gradually start eating the 'kid starter' themselves. Feed good quality legumes such as Lucerne or Berseem in a manger for it is helpful in getting the kids started properly.

Unwanted male kids should castrate when they are ten days to three weeks of age. It is also recommended that the horns be removed from stall-fed goats to prevent them from injuring each other or the attendant. It is best to disbud the kids with caustic potash when they are 5 days old.

Exotic breeds and crossbreeding experience of dairy goats

The elaborate list of exotic breeds of goats, grouped under meat. Dairy, meat and dairy fine fibre and hairy fibre is indicative of the dimension of genetic resource and distribution all over the world. The description of these breeds, based on physical conformation and body size with certain indication of the major production traits is available in the Mason's World Dictionary of Livestock Breeds, Types and Varieties. There is very little information with respect to the exact areas of habitation. Population size and trends over time, physical environment, feed resources, flock size and performances e.g. reproduction growth, survival, milk production, fibre production, meat production etc. The knowledge on the adaptation, disease resistance and grazing behaviour is dangerously scanty.

Specialized exotic dairy breeds of the temperate zone, viz. Alpine, Saanen and Mohair breeds, Angora had been used in controlled experiments in India during 70's and 80's to assess the improvement in the milk yield and Mohair production. The crossbreds were expected to inherit the eco--adaptability and disease-resistance traits from the native dam breeds and production characteristics from the exotic sire breeds. The level of production of crosses of high-yielding indigenous dam 'breeds, e.g. Beetal, were superior to those of the crosses of relatively low-yielding Malabari in spite of similar body size. Further, Saanen had apparently a better combining ability with the indigenous breeds than Alpine, even under unfavourable agro climate conditions. The milk yield of the crosses increased significantly with increase in exotic inheritance up to 75%. The higher crosses of Malabari did not excel half-breeds in milk production at Kerala, probably due to environmental limitations of the western costal region. The progenies of F, hybrids born out of *inter-se* mating had significantly lower yield in subsequent generations (F2 and F). The advantage of improvement in milk yield of exotic crosses at all the experimental stations has been further discounted by abysmally low fertility and high mortality of kids in institutional flocks.

References

Acharya RM 1988. Goat breeding and meat production, Goat meat production in Asia. Proceedings of a workshop held in Tando Jam, Pakistan, pp.14-29.

Acharya RM, Patnayak BC 1974. Rali of sheep in the desert ecosystem and draught prooving through improved sheep production with special reference to Rajanthan, Migraph report, Central sheep and goat research institute, Avikanager, Rajasthan.

Acharya RN Mishra RK, Patil VK 1982. Breeding strategy for goat in India, ICAR Pub, New Delhi.

Al'kov GV 1983. A new breed of down-producing goat. Ovtsevodstvo, No 8, pp 29-32

Amir P 1988. Economics of goat meat production. International Development Research Centre, No IDRC-268e, pp 202-211.

Anonymous 1996. Livestock census data. Livestock Policy and Information Unit (L.P.I.U.), Accra.

Anonymous Economic Survey 2003-04, published by Govt of India, Ministry of Finance and Company Affairs, Economic Diversion pp. 156.

Anonymous Hindi Survey 2004. The Hindu Survey of India Agriculture pp. 116.

Aydin U, Arik IZ 1999. Comparison of slaughter and carcass characteristics of hair goat kids in intensive and pasture conditions. Ziraat Fakultesi Dergisi, Akdeniz Universitesi, Vol 12, No 1, pp 75-86.

Azamel AA, El-Sherbiny AA, Bata SS, Mokhtar MM, Shehata MF 1992. Effect of early castration on performance, carcass characteristics and meat acceptability in Baladi goat kids. : Indian Journal of Animal Sciences, Vol 62, No 8, pp 723-727.

Barbiker SA, Mglad M, Koudoda ME 1985. Effect of castration on performance and carcass characteristics of male Sudan desert goat. World Review of Animal Production, 21(1): 11-13.

Baxendell SA 1984. Diseases of goats. In : Copland IW (Ed.), Goat production and research in the tropics processing series no 7, Australian Centre for International Agricultural Research, Canberra, Australia pp 94-102.

Bhargava VN 1982. "Goat farming. A key to rural prosperity" Farmer and Parliament XVII (10) pp 15-19

Bhattacharyya NK, Khan BU 1988. Goat meat production in India. International Development Research Centre, No IDRC-268e, pp125-134, 21 ref.

Birnkammer H 1986. The Boer goat. A highly productive meat-type breed. Tierzuchter, 39(11): 465

Dawa O, Tawah CL, Mawaina T 1996. Carcass characteristics of adult sheep and goat breeds in the subhumid and semi-arid zones of Cameroon. Bulletin of Animal Health and Production in Africa, Vol 44, No 3, pp 145-149.

Devendra C, Owen JE 1983. Quantitative and qualitative aspects of meat production of goats. World Animal Review 47:19-29.

Dhanda JS, Taylor DG, McCosker JE, Murray PJ 1999. The influence of goat genotype on the production of Capretto and Chevon carcasses. 1. Growth and carcass characteristics. SO: Meat Science, Vol 52, No 4, pp 355-361.

Djajanegara A, Chaniago TD 1988. Goat meat production in ndonesia. International Development Research Centre, No IDRC-268, pp 135-139.

El-Muola IHA, Babiker SA, El-Khidir OA, Ibrahim SE 1999. Meat production from female goat kids compared with males.Journal of Agricultural Science, 133(2): 223- 226.

FAO1981. Food and Agriculture organization of the United National Production Year Book, Rome, Italy.

Goonewardene LA, Day PA, Patrick N, Scheer HD, Patrick D, Suleiman A 1998. A preliminary evaluation of growth and carcass traits in Alpine and Boer goat crosses. Canadian Journal of Animal Science, 78(2): 229-232.

Huang JC, Shieh RC, Chang HJ, Su AK, Wen SH, Li PH 1994. Special Publication - Taichung District Agricultural Improvement Station, No.35, pp.981-986.

Huq MA 1988. Goat meat production in Bangladesh. International Development Research Centre, No IDRC-268e, pp 112-118.

Ibarra PI 1988. Goat meat production in the Philippines. International Development Research Centre, No IDRC-268e, pp 170-178.

Isani GB, Soomro GH 1988. By-products from goat meat production and their marketing in Pakistan. International Development Research Centre, No IDRC-268e, pp 238-246.

Juarez Lozano A, Peraza Castro C 1991. Profitable goat meat production Memorias del VIII Congreso Nacional, AZTECA

Kharel M, Pradhan SL 1988. Goat meat production in Nepal. International Development Research Centre, No IDRC-268e, pp 152-160.

Kirton AH 1982. Carcass and meat qualities, In World Animal Science CI, Sheep and goat production pp. 259-74 (Ed) Coop IE. Elsevier Scientific Publishing Co., Amsterdam.

Kirton AH 1988. Characteristics of goat meat including carcass quality and methods of slaughter. Goat meat production in Asia. Proceedings of a workshop held in Tandojam, Pakistan, 13-18 March, pp 87-99.

Kirton AH 1989. Principles of classification and grading. Meat production and processing occasional publication no.-11, pp-143-57(Eds) Purchas RW, Butter-Hogg BW and Davies A.S., New Zealand Society of Animal Production, New Zealand.

Koyuncu M, Tuncel E Akman 1996. N Fattening performance and carcass characteristics of male Angora goat kids under intensive and pasture conditions. Turk Veterinerlik ve Hayvancilik Dergisi, 20(3):57-161.

Kuchtik J, Chladek G, Hosek M 1999. Growth ability and carcass value of dairy kids of White Short-wooled goat and cashmere goat. Acta Universitatis Agriculturae et Silviculturae Mendelianae Brunensis, 47(5):79-86.

Kumar R, Kumar A, Singh H 1980. Nate on the effect of castration on meat production in goats. Indian Journal of Animal Science. 50: 1160-62.

Lana K, Suarna M, Putra S, Sukanten W, Suata K 1994. Carcass composition of local goat fed natural grass, Gliricidia and Ficus leaves. Sustainable animal production and the environment. Proceedings of the 7th AAAP Animal Science Congress, Bali, Indonesia, Volume 2: contributed paspers, 1994, pp 131-132.

Louca A, Economides S, Hancock I 1977. Effect of castration and growth rate, feed conversion efficiency and carcass quality in Damascus goats. Animal Production 24 (3):387-91.

Mexicana de Zootecnistasy Tecnicos en Caprinocultura, AC 1991. 1-4 Octubre 1991, Universidad Autonoma Metropolitana Iztapalapa., pp.85-104.

Mioc B 1998. Relationships of breed with growth intensity and chemical composition of goat meat: birth weight, daily gain and some carcass traits. Poljoprivredna Znanstvena Smotra, 63(4):179-186.

Moniruzzaman M, Hashem MA, Akhter S, Hossain MM 2002. Effect of different feeding systems on carcass and non-carcass parameters of Black Bengal goat. Asian-Australasian Journal of Animal Sciences, 15(1):61-65.

Ndamukong KJN, Sewell MMH, Asanji MF 1987. Production of Sheep and Goat under three management system at Bamenda, Cameroon. Trop. Anim-Health Prod. 19:237-244.

Niekerk WA, van, Casey NH 1988. The Boer goat. 2. Growth, nutrient requirements, carcass and meat quality. Small Ruminant Research, 1(4)355-368.

Raghavan GV 1988. The influence of sex on goat meat production. Goat meat production in Asia. Proceedings of a workshop held in TandoJam, Pakistan, pp 63-71.

Rajaguru ASB 1988. Goat meat production in Sri Lanka. International Development Research Centre, No IDRC-268e, pp 179-187.

Rana CL 1966. Dairy goat husbandry in India, Proc Seminar held in dec 1966 at InT House, New Delhi.

Rao KS 1988. By-products from goat meat production and their marketing in India. In goat meat production in India. Proceeding of a workshop, 13-18 March, 1988 Tandojam, Pakistan, pp 230-37 International Development and Research Centre, Canada.

Rekib A 1985. Newsletter, Dec. 1985, Published by Indian Society of Sheep and Goat Production and Utilization, Avikanagar, Rajasthan.

Saithanoo S, Milton JTB 1988. Goat meat production in Thailand. International Development Research Centre, No IDRC-268e, pp 188-196.

Sengar OPS 1976. Study on the anatomy and histology of digestive system in ruminant (*Bas bubalis* L), PhD thesis Agra, University of Agra.

Sharma K, Ogra JL 1985. Antibacterial property of goat milk. Journal of Animal Research 2, (9):99

Shatunov VA 1973. Increasing the production of goat down. Ovtsevodstvo, No.9, pp.19-22

Tahir MAH, Abdulla AHH, Al-Jassim AF 1994. Effect of castration and weight at slaughter on carcass traits and meat quality of goat. Indian Journal of Animal Sciences, 64(7): 778 - 782.

Tuah AK, Buadu MK, Obese FY, Brew K 1992. The performances, potentialities and limitations of west African dwarf goats for meat production in the forest belt of Ghana, In : Ray B, Labbil SHB, Reynolds L (Eds.) Small ruminant production and development in Africa, Adadis Ababa, Ethiopia, pp 435-443.

Vasil'ev NA 1982. The Altai Mountain breed of down goat. hivotnovodstvo, (6): 42-43

Wenxiu H 1988. Goat meat production in China. International Development Research Centre, No IDRC-268e, pp 119-124.

Willson RT 1982. The economic and social importance of goat and their products is semiarid area of northan tropical Africa Proc of 3rd InT conference on goat production and diseases, Jan-10-15. Arijon USA pp 186-191.

Yadav RL, Dwivedi BS, Singh VK, Avind K Shukla 2001. Nutrient meaning and apparent balances in different zones of UP, fertilizers news, April-2001, pp 13-31

Younis AA, Seudy AM, Galal ESE, Ghanen YS, Khishin SS 1975. Effect of place of nutrition on feed lot performance and carcass traits of desert sheep. Tropical Agriculture 52:233-42.

14

Guinea Fowl as an Alternative Source of Chicken

Brijesh Singh

Introduction

The present world wide situation of food items calls for a broad range of poultry alternatives to meet increasing demand of quality protein. Besides, complication involved in managing chicken, emergence of new diseases, increase in feed cost and increasing demand of poultry meat bound us to search for alternative species of poultry. In this circumstance rearing of Guinea fowl framed for centuries to meet poultry production requirements of diversified agricultural milieu. The traditionally reared Guinea fowl could become a significant source of food and also a source for substantial supplement income.

History

Guinea fowl a bird of West African origin came to India during mediaeval age. Now guinea fowl are being raised in the districts of eastern Uttar Pradesh and in some parts of Tamilnadu. The birds are very popular with marginal farmers and poor classes and are being largely kept as foragers and insectivorous under semi-range and range conditions.

Guinea fowl specialities

Guinea fowl are the natural germplasm for the sub-continent and have better resistance against general diseases as compared to chicken. These are hardy birds with a reputation for resistance for most of the poultry diseases. These birds are having gregarious habits which in turn reduce the feed cost. The long neck and legs make them sustainable for grazing. They can eat insect, earth worms, variety of green leaves, vegetables and fruits and variety of feeds in harvested land also. In this way it is helpful in development of low input grain saving aviculture system in Uttarakhand. It has got better tolerance of afla-toxin in comparison to chicken and hence eat somewhat bad ration to produce good quality of meat. Guinea Fowl eggs have less weight but are hardy and hence have better storage capability. The egg and meat have less cholesterol and hence heart patient can also utilize it. The bird provides variety to existing food basket. Guinea Fowl is the only domesticated bird, which can provide sport quality meat in this era of ban on hunting. It can be raised under a wide range of climatic conditions without any special type of housing etc. Guinea fowl thrive well under adverse climatic conditions, such as poor feeding, poor housing and poor management with variable humidity and temperature. Guinea fowl can eat everything which chickens does not like to consume like grasses, insects, fruits, green leaves and vegetables etc. They don't cause any loss to the field but increase the productivity. Its gregarious habits, flocking instinct, long neck and strong and sufficiently long legs are some of the special advantageous points in favor of the species.

Adaptability and disease resistance

General disease resistance measures were comparable or higher than that obtained for indigenous fowl (Kadaknath) and commercial broilers (Saxena 1993).

Srivastava and Singh (1988) showed higher resistance of guinea fowl to *Argus persicus* ticks infestation. Similarly, guinea fowl showed lower susceptibility to poultry lice (*Menonpon gallinae* and *Lipeurus tropicals*). Guinea fowl also seemed to possess high innate tolerance to aflatoxin as compared to chicken and other poultry species. Very few adverse effects were observed among guinea fowl fed 1.0 ppm of aflatoxin (Johri *et al.* 1988 and Bedi *et al.* 1996).

Behavior of birds

Survival and performance potential of a poultry species in different conditions is influenced by its inherent behavioral characteristics such as escaped tendencies, broodiness, aggressiveness, social, foraging and feeding behaviors etc. Guinea fowl have gregarious habits and lives in small groups.

Housing and feeding management

Guinea fowl does not require any special type of housing. The traditional extensive system is most popular in rural areas through out country. It can also be raised in modern intensive system on complete feed, good hygiene and high input. However, its unique status of low input grain saving aviculture can be explored better on extensive system. It has been experimentally proved that there are no significant differences in growth rate between the birds reared on complete feeding and feeding with foraging. This is because of its better digestibility of carbohydrates including pentosan and lignin component of carbohydrate on foraging. Moreover, higher afla-toxin tolerance indicates its potential to utilize lower quality feed ingredient.

In intensive rearing system Guinea fowl keets require 2700 k cal energy/ kg feed and 24 – 26% crude protein to attain optimal body weight up to 12 week of age. However, these birds are not efficient converter of feed as compared to chicken (ratio 3:1) and hence extensive system of rearing Guinea fowl is more popular.

Reproductive performance of guinea fowl

Guinea Fowl is a seasonal breeder. Its reproductive performance is affected by environmental causes like temperature, humidity, light, management, feed and diseases etc. Female starts laying from March to April and ended in August to September. The male mature 15 days later than female. The ratio of male to Female should be 1:5 for optimum fertility.

Age at first egg was varied from 250 to 280 days. Egg production was found to be 45-50 eggs in Ist 100 days laying cycle under backyard system and 80 to 110 eggs under intensive system. Guinea fowl eggs are relatively smaller in size than those of the chicken and have an average egg weight about 40g (Singh *et al.* 1993, Porval *et al.* 2002). The shell color is light pink and is due to the porphyrin compounds/ pigments and it is not affected the nutritive value of the egg. The guinea fowl eggs have shape index ranging from 76 to 79 and specific gravity of about 1.20, which was comparatively higher than these parameters in chicken eggs (Singh *et al.* 1993 and Sajjnar 1994). The values of fertility and hatchability were observed 91.52± 0.528 and 61.82±0.921%. Keet weight was recorded 26.87±0.12 g at hatching. The shell of guinea fowl eggs is much thicker 0.42 to 0.45 mm than chicken eggs 0.32 to 0.34 mm (Singh *et al.* 1993, Song *et al.* 2000). Thicker shell with lower porosity makes guinea fowl eggs relatively stronger against shocks, better resistance against contaminations and thus more wholesome (Sajjnar 1994, Sachdeva *et al.* 2006).

Growth and conformation traits of guinea fowl

Guinea fowls are reared generally for meat. Growth rate is very important trait for meat production. Body weight was observed 26.87 ± 0.12, 208.02 ± 1.32, 619.78 ± 4.06, 951.00 ± 4.87 and 1190.14 ± 5.30 g. at day old, 4, 8, 12 and 16 weeks of age, respectively. Shank length was observed 3.22 ± 0.01, 4.27 ± 0.01, 5.37 ± 0.02 and 6.98 ± 0.01 cm at 4, 8, 12 and 16 weeks of age, respectively. The keel bone length was observed 4.16 ± 0.01, 7.41 ± 0.03, 8.81 ± 0.02 and 9.96 ± 0.02 cm in corresponding age group. The breast angle was recorded 20.74 ± 0.10, 25.17 ± 0.07 and 36.19 ± 0.06 ^{0} at 8, 12 and 16 weeks of age, respectively. Growth in guinea fowl is low in comparison to domestic fowl (Sharma and Singh 2005). The body weight is largely affected by various genetic and non-genetic factors. Among genetic factors, sex and varieties are two important factors (Sharma and Singh 2005).

Carcass characteristics and sensory evaluation

For meat production, carcass characteristics are also important. Guinea fowl seem to have about 77% dressing percentage (Mahapatra *et al.* 1986b, Singh *et al.* 1999, Sharma *et al.* 2000) which is higher than those for commercial broilers. Effect of sex is significant on carcass traits and males have better carcass characteristics (Singh *et al.* 1999, Baeza *et al.* 2000).

Socio-economic importance

The land holding is very small in the area and most of them are uneconomic. It has found a special favor with the land less labourers and marginal farmers because of its ability to provide additional income and nutricious food item with out much expenditure on it. There is not exists any type of taboo among poultry farmers regarding the adoption of this newly introduced alternate species of poultry. It reduces the problem of malnutrition to some extent and improved the economic status. It is also a potential source of raising the socio- economic status of poor farmers.

Table 1: Reproductive performance of guinea fowl

S. No.	Character	Mean ±S.E.	Heritability ±S.E.
1.	Age at sexual maturity (day)	250-280	-
2.	Egg production in 100 days (Number)	45-50	-
3.	Egg weight (g)	40.80±0.056	0.344±0.105
4.	Shape index	77.67±0.057	0.167±0.059
5.	Fertility	91.52± 0.528	0.135±0.048
6.	Dead germ	12.46±0.626	0.069±0.03
7.	Dead-in-shell	16.67±0.706	0.009±0.012
8.	Hatchability	61.82±0.921	0.144±0.051
9.	Keet weight (g)	26.87±0.12	0.25±0.14

Table 2: Growth and conformation traits in guinea fowl

S. No.	Character	Mean ±S.E.	Body weight (g) Coefficient of variation	Heritability ±S.E.
1.	Day old wt.	26.87 ± 0.12	8.88	0.25 ± 0.14
2.	4 week wt.	208.02 ± 1.32	12.08	0.21 ± 0.08
3.	8 week wt.	619.78 ± 4.06	15.48	0.17 ± 0.12
4.	12 week wt.	951.00 ± 4.87	9.70	1.08 ± 0.10
5.	16 week wt.	1190.14 ± 5.30	8.45	0.13 ± 0.11
Shank length (cm)				
1.	4 week	3.22 ± 0.01	6.73	0.23 ± 0.13
2.	8 week	4.27 ± 0.01	8.10	0.42 ± 0.18
3.	12 week	5.37 ± 0.02	8.14	0.07 ± 0.09
4.	16 week	6.98 ± 0.01	4.28	0.13 ± 0.11
Keel length (cm)				
1.	4 week	4.16 ± 0.01	5.87	0.10 ± 0.10
2.	8 week	7.41 ± 0.03	8.34	0.08 ± 0.09
3.	12 week	8.81 ± 0.02	6.14	0.20 ± 0.13
4.	16 week	9.96 ± 0.02	5.61	0.23 ± 0.14
Breast angle (^{0})				
1.	8 week	20.74 ± 0.10	9.44	0.09 ± 0.10
2.	12 week	25.17 ± 0.07	5.84	0.10 ± 0.10
3.	16 week	36.19 ± 0.06	3.20	0.15±0.12

Marketing and scope of Guinea fowl as an organic food

Guinea fowl rearing is a viable occupation because of the fact that small holder poultry products have already local market of it over in spite of existing religious taboos. Moreover, increasing demand of more and more diversity in poultry products and consumer preference for native free range fed organic poultry products farmer are in favour of Guinea fowl produce. Eggs are used traditionally popular preparations like omelettes, poached egg, scambled egg, hard boiled egg, egg curry etc. A comparable sensory evaluation study for different egg preparations of Guinea fowl and chicken was conducted and Mohapatra *et al.* (1986a) reported that significant difference could not be observed for color, texture and chew ability of the preparations from guinea fowl eggs and chicken eggs. However, overall acceptability for most of the preparations was comparatively higher for guinea fowl egg preparations. Low total lipids and cholesterol in guinea fowl eggs was reported against the chicken eggs (Mohapatra *et al.* 1986a).

Guinea Fowl production under backyard system finds an important tool to play in the wake of the need of stress free and harmful residue free birds. Uttarakhand has tremendous potential in organic food product because of the existence of traditional backyard system in the area. These birds are mainly kept as foragers and insectivorous under semi-range and range conditions and produce organic products that are fetch high price. Selvam (2004) reported continuous increase in demand of organic meat and eggs due to the introduction of recent concept of organic food.

Relevance of Guinea fowl as an alternative source of chicken in Uttarakhand and future prospect

Guinea Fowl would be a powerful tool to fight against poverty by creating job opportunities and also increase the production of high biological valued nutritious protein. Guinea Fowl have good adaptation in mountain area and show high resistance against many diseases and stress conditions. Guinea Fowl production seems to be one of the most simple and effective ways to support families to raise their income. Guinea Fowl can be used as organic food along with their eggs and meat under backyard system. If provide a suitable technical inputs, credit facilities, motivation to guinea fowl production and extension facilities, the backyard rearing system of Guinea fowl can be converted to successful organic ventures and as an alternative source of chicken not only in the Uttarakhand but also in the other parts of the country.

Summary

Guinea fowl is a bird of African origin still semi feral in nature having certain unique advantages over chicken and other poultry species which makes it ideal bird .There are great possibilities for integration with agriculture, horticulture, fisheries and other livestock rearing systems. The present paper is an overview of work done on Guinea fowl to popularize it as an alternative source of chicken. The experiment conducted so far on farm and at on farmer's field on growth, production and reproduction parameter clearly revealed that the guinea fowl can be successfully and beneficially reared by the small and marginal farmers as subsidiary occupation to supplement family income especially when the profit margins from agriculture are shrinking and hence it can serve as alternative source of poultry for Uttrakhand.

Conclusion

Guinea fowl can be successfully and beneficially reared by the small and marginal farmers of Bhabhar and Tarai region as subsidiary occupation to supplement family income especially when the profit margins from agriculture are shrinking and hence it can serve as alternative source of poultry for Uttarakhand.

References

Baeza E, Juin H, Rebours G, Constantin P, Morche G, Leterrier C 2000. Effect of genotype, sex, and rearing temperature on carcass and meat quality of guinea fowl. British Poultry Science 42(4):470-476.

Bedi PS, Harpreet Singh, Johari TS, Agarwal RK 1996. Effect on aflatoxin on growth and feed consumption in guinea fowl. XXX World's Poultry Congress WPSA held on Sept. 2-8, 1996 at New Delhi.

Johri TS, Agarwal SK, Sadagopan VR, Singh Harpreet 1988. Effect of dietary aflatoxin on the performance of guinea fowl. Indian Journal of Animal Sciences 58: 673-875

Mahapatra CM, Pandey NK, Verma SS, Singh Harpreet 1986b. Yield, quality and composition of guinea fowl meat at different stages of growth. Indian Journal of Poultry Science 21(3): 204-207

Mahapatra CM, Pandey NK, Verma SS, Singh Harpreet 1986a. Physical quality, composition, cholesterol, vitamin A and fatty acid contents of guinea fowl vis-à-vis chicken egg. Journal of Food Science & Technology 24: 168-171.

Porval V, Singh B, Kumar D, Sharma RK, Pandey H 2002. Genetic studies on growth and conformation traits of guinea fowl. Indian Journal of Poultry Science 37(2): 179-180

Sachdeva AK, Sharma D, Gopal R, Singh Harpreet 2006. Quality composition and genetics of= egg. Indian Journal of Animal Genetics and Breeding 27(1&2): 36-41

Sajjnar CM 1994. Genetic studies on non specific antimicrobial defense of guinea fowl eggs. M VSc Thesis, Deemed University, IVRI, Izatnagar

Saxena VK 1993. Genetic studies on immunocompetence defenses of guinea fowl (Numida meleagris) PhD Thesis, Deemed University, IVRI, Izatnagar

Selvam S 2004. An economic analysis of free range poultry rearing by rural women. Indian Journal of Poultry Science 39(1): 75-77

Sharma Deepak, Singh Harpreet 2005. Breeding guinea fowl for improved performance. Indian Journal of Animal Genetics and Breeding 26(1&2): 60-67

Sharma Deepak, Singh UB Nayal LMS, Singh Sunil, Singh RV 2000. Carcass characterstics of improved guinea fowl at different weeks of age. Indian Journal of Poultry Science 35(2): 224-225

Singh B, Husain KQ, Verma SK1999. Effect of strain and sex on the carcass yield of guinea fowl. Indian Journal of Poultry Science 34(2): 277-279.

Singh Harpreet, Pal SK, Raheja KL 1993. Genetic and phenotypic parameters for egg production and egg quality traits in guinea fowl. Indian Journal of Poultry Science 28(1): 12-19.

Song KT, Choi SH, Oh HR 2000. A comparision of egg quality of pheasant, chukar, quil and guinea fowl. Asian-Australian Journal of Animal Sciences. 13(7): 986-990.

Srivastava SC, Singh Harpreet 1988. Inherent tick resistance in guinea fowl. Preliminary observations. In: Proceedings of 5th Poultry Science Conference and Symposium IPSA held on December 8-10, 1988 at CARI, Izatnagar, AGB-67.62

15

Dairy Farm Management for Profit: An Overview

B K Mishra

India has the highest livestock population in the world with 50% of the buffaloes and 20% of the world's cattle population, most of which are milch cows and milch buffaloes. The dairy industry in our country is considered as one of the most successful development programmes in the post-Independence period. In the year 2006-07the total milk production in the country was over 94.6 million tonnes with a per capita availability of 229 gm. per day. The industry had been recording an annual growth of 4% during the period 1993-2005, which is almost 3 times the average growth rate of the dairy industry in the world. Milk processing in India is around 35%, of which the organized dairy industry account for 13% of the milk produced, while the rest of the milk is either consumed at farm level, or sold as fresh, non-pasteurized milk through unorganized channels. Dairy Cooperatives account for the major share of processed liquid milk marketed in the India. Milk is processed and marketed by 170 Milk Producers' Cooperative Unions, which federate into 15 State Cooperative Milk Marketing Federations. Over the years, several brands have been created by cooperatives like Amul (GCMMF), Vijaya (AP), Verka (Punjab), Saras (Rajasthan). Nandini (Karnataka), Milma (Kerala) and Gokul (Kolhapur). Uttar Pradesh, Punjab, Haryana, Rajasthan, Gujarat, Maharashtra, Andhra Pradesh, Karnataka and Tamil Nadu are the milk surplus states in India. The manufacturing of milk products is obviously high in these milk surplus States. Exports of dairy products have been growing at the rate of 25% per annum in the terms of quantity terms and 28% in terms of value since 2001. Significant investment opportunities exist for the manufacturing of value-added milk products like milk powder, packaged milk, butter, ghee, cheese and ready-to- consume milk products. India has emerged as the largest milk producing country in the world with present level of annual milk production estimated as 94.5 million tonnes. We expect a production level of 135 million tonnes by the year 2015. India has a large livestock population base constituting 278 million livestock including 180.5 million cattle, 82.8 million buffaloes, 4 million sheep and 9.2 million goats. The livestock population is projected to increase to 322 million by the year 2015. The large livestock population is raised primarily on crop residues and grazing in the common property including basement. The forest area, which was a major source of grazing, is no longer available to livestock breeder's especially landless people. As a consequence, the available feed resources fall short of the nutritional requirement. The shortfall is estimated as 59.9 million tonnes for the green fodder and 19.9 million tonnes for dry fodder. This shortfall is likely to increase by 2015 to 63.5 million tonnes of green fodder and 23.56 million tonnes of dry fodder. The landless people are, therefore, likely to face severe shortage of resources to raise cattle and other species of livestock. There is a real danger that in the absence of resources to maintain their stock, these under-privilege rural people may give up livestock farming. This could be a serious setback to lakhs of rural families who derive income as well as employment opportunities from livestock sector. It is well recognized fact that India is basically an agriculture country. Majority of its population exist in village whose primary occupation is agriculture. For a farmer dairying is a supplementary source of income. It plays an important role in the agriculture economy.

- Dairying fits well in diversified farming programs. It is highly recommended to have diversification many farms. This diversification will help to use the farm labour efficiently and dairying will be one such diversified activity in a farm. In addition to this it will be possible to have economic use of buildings and equipments. Since it is always risk to have only one source of income, dairying helps in having an alternative source of income to the farmer.
- Milk animals are efficient Consumers of roughages. The different types of roughages such as paddy straw, wheat straw etc. some times go waste on some farms. If the farmer has dairying these roughages can efficiently be converted into most nutrition's milk. These roughages are bulky and it is not economical to transport them over long distances.
- Dairying provides a stable income: The prices of most of the agriculture produce show great fluctuation. The prices of dairy products do not show such wide range of fluctuation in their prices.

- Income is distributed through out the year: If the farmer depends only on agriculture he would be receiving income from his farm only on harvesting the crop. This would mean that the income of the farmer would only be seasonal. Since the produce from a dairy animal is distributed almost through out the year.
- Dairy production improves the family diet and reduces the food cost. Milk is a basic food and an important item in the family food budget. A small dairy enterprise can be justified on many farms if only for the production of milk products for family consumption. This is especially true when the family is big one.
- Dairying helps in the maintenance of soil fertility legumes and grasses are grown on for from providing fodder to the animals. These crops are soil conserving and soil building crops. The manure provided is distributed on the land and return plant food nutrients to the soil.
- Supplies animals required for draft purpose. In our country most of the agriculture operations are still called out by using bullock power. Unless a farmer has dairy animals with him; he cannot get the bullocks for replacing his old animals. Replacement of bullock by purchasing would be uneconomical for a small farmer. Hence dairying is useful as a source for such supplies.

Dairy development indicators

- World cattle population 1270 Millions
- World Buffalo Population 160.0 Millions.
- Cattle population in India 185.2 Millions.
- Buffalo population in India 97.9 Millions.
- It is 51 % of the Asia and 19 % of the total bovine population of the world.
- India is having roughly 1/7th of World Cattle population India.
- Total Milk Production in the world: 580 Millions.
- Total Milk production in India: 115 Million Tones. (2010)
- India ranks 1st in the world in Milk Production.
- Out of the above milk production approximately 52% is contributed by buffaloes, 45 % by cows and 3% by sheep, goat and miscellaneous.
- The per capital availability of milk in India is more than 250 grams per day / per person. But the I.C.M.R. has recommended a minimum of 280 grams (2010)

Breeds of dairy cattle and buffalos

A breed is a group of animals related similar characters like general appearance. i.e. features and configuration etc. A group of individual which have certain common characteristics that distinguish them from other groups of individuals is known as "Species".

Classification of Cattle Breeds

Indian cattle breeds of cattle classified in to three types a)

1. Milch breeds/Milk breeds
2. Dual Purpose breeds
3. Draught breeds

Milch (milk) breeds

The cows of these breeds are high milk yields and the male animals are slow or poor work animals. The examples of Indian milch breeds are Sahiwal, Red Sindhi, Gir and Deoni The milk production of milk breeds is on the average more than 1600 kg. Per lactation.

Dual purpose breeds

The cows in these breeds are average milk yielder and male animals are very useful for work. Their milk production per lactation is 500 kg to 150 kg. The examples of this group are Ongole, Hariana, Kankrej, Tharparker, Krishna valley, Rathi and Goalo Mewathi.

Draught breeds

The male animals are good for work and Cows are poor milk yielders are their milk yield as an average is less than 500 kg per lactation. They are usually white in colour. A pair of bullocks can haul 1000 kg. Net with an iron tyred cart on a good road at walking speed of 5 to 7 km per/ hour and cover a distance of 30 to 40 k.m per day. Twice as much weight can be pulled on pneumatic rubber tubed carts.

Indian buffalo breeds

India is considered as home tract for some of the best buffalo breeds. I t has also been the center of dispersion of good specimen of buffaloes for improvement of species in the world. The most important breeds of buffaloes which are commonly reared are as under-

(a) Murrah

(b) Bhadawari

(c) Jafferabadi

(d) Surti

(e) Mehsana

(f) Nili-Ravi

(g) Nagpuri

(f) Toda

Economic traits in dairy cattle

The various economic traits in Dairy Cattle management are

1. Lactation yield
2. Lactation period
3. Persistency of yield
4. Age at first calving
5. Service period
6. Dry period
7. Inter calving period
8. Reproductive efficiency
9. Efficiency of feed utilization
10. Disease resistance.

Lactation yield

The lactation yield in a lactation period is known as lactation yield. 'The lactation yield in Indian breeds is very low compared to exotic breeds. This is dependent on no. of calving, frequency of milking, persistency of yield (Normally in dairy cattle 30 - 40 % increase in milk production from first lactation to maturity is observed. After 3 or 4 lactation the production starts declining. For comparison of milk yield of different breeds and animals the milky should be converted into fat corrected milk (FCM). 4% FCM = 0.4 total milk + 15 total fat. After parturition the milk yield per day will be increased and reaches peak within 2-4 weeks after calving. This yield is known as peak yield. The maintenance of peak yield for more time is importance for better milk production. The lactation period in Indian breeds is low and so the production is also less and conversion.

Lactation period

The length of milk producing period after calving is known as lactation period. The optimum lactation period is 305 days. The milk production will be less, if this period is shortened". Indian breeds will have less lactation period, but in some breeds this period is more with very little milk production

Persistency of milk yield

During lactation period the animal reaches maximum milk yield per day with in 2-4 weeks which is called peak yield. For high level of lactation yield, this peak yield should be maintained for longer period as far as possible, the maintenance of peak yield for long period is known as persistency, slow decrease in dairy milk yield after reaching peak yield in necessary. High persistency is necessary to maintain high level of milk production

Age at first calving

The age o the animal at first calving is very important for high life time production. The desirable age at first calving in Indian breeds is 3 years; 2 years in cross-breed cattle and 3. ½ years in Buffaloes. Prolonged age at first calving will have high production in the first lactation) but the life time production will be decreased due to less no of calving. If the age at first calving is below optimum, the calves born are weak, difficulty in calving and less milk production in first lactation.

Service period

It is the period between -date of calving and date of successful conception. The optimum service period helps the animal to recover from the stress of calving and also to get back the reproductive organs back to normal For cattle the optimum service period is 60-90 days. If the service period is too prolonged the calving interval prolonged, less no. of calving will be obtained in her life time and ultimately less fife time production.' If the service period is too short, the animal will become weak and persistency of milk production is poor due to immediate pregnancy.

Dry period

It is the period from the date of drying (stop of milk production) to next calving when the animal is in pregnancy, before next calving. The animal should be given rest _period to compensate for growth of foetus. (A minimum of 2 – 2 .5 months of dry period should be allowed) If the dry period is not given or too small dry period, the animals suffer from stress and in next lactation, the milk production drops substantially and also it gives weak calves. On the other hand if the dry period given is too high, it may not have that much effect on increasing milk yield in the next lactation, but it decrease the production in the present lactation.

Inter-calving period

This is the -period between two successive calving. It is more, profitable to have one calf yearly in cattle and at least one calf for every 15 months in buffaloes. If the calving interval is more, the total no. of calving in her life time will be decreased and also total life production of milk decrease.

Reproductive efficiency

The reproductive efficiency means the more number of calves during life time, so that total life time production is increased, The reproduction or breeding efficiency is determined by the combined effect of hereditary and environment Several measures of breeding efficiency like number of services per conception, calving interval, days from first breeding to conception are useful.) Reproductive efficiency has generally a low heritability value indicating that most of the variations in this trait is due to non genetic factors. In adverse environmental conditions, the poor milk producing animals may not be much affected compared to high effect in high milk yield.

Efficiency of feed utilization and conversion into milk

The animal should utilize the feed efficiently to convert into the milk.

Disease resistance

Indian breeds are more resistant to majority of disease compared to exotic cattle. Cross - breeding helps to get this character.

Dairy Cattle Housing Management

Site selection

The basic points which should be careful before the construction of dairy buildings are as under:

Topography

A dairy building should be at a higher elevation than the surrounding ground to offer a good slope for rainfall and drainage for the wastes of the dairy to avoid stagnation within. A leveled area requires less site preparation and thus lesser cost of building. Low lands and depression and proximity to places of bad odour should be avoided.

Soil type

Fertile soil should be spared for cultivation. Foundation soil as far as possible should not- too dehydrated or desiccated. Such a soil is susceptible to considerable puffiness during rainy season and exhibit plentiful cracks and fissures.

Exposure to the sun and protection from wind

A dairy building should be located to a maximum exposure to the sun in the north and minimum exposure to the sun in the south and protection from prevailing strong wind currents whether hot or cold. Buildings should be placed so that direct sunlight can reach the platforms, gutter and mangers; in the cattle shed. As far as possible, the long axis of the dairy barns should -be set in north- south direction to have the maximum benefit of the sun.

Accessibility

Easy accessibility to the buildings is always desirable. Situation of a cattle shed by the side of the main road preferably a distance of about 100 meters should be aimed at.

Durability and attractiveness

It is always attractive when the buildings open up to a scenic view and add to the stateliness of the scenery. Along with this, durability of the structure is obviously an important criterion in building a dairy.

Water supply

Abundant supply of fresh, clean and soft water should be available at a cheap rate.

Surrounding

Areas infested with wild animals and 'daoits; should be avoided. Narrow gates, high manger curbs, loose hinges, protruding nails, smooth finished floor in the areas where the cows move and other such hazards -should be eliminated.

Labour

Honest, economic and regular supply of labour be available.

Marketing

Dairy buildings should only be in those areas from where the owner can sell his products profitability and regularly. He should be in a position to satisfy the needs of the farm within no time and at a reasonable price.

Electricity

Electricity is the most important sanitary method of lighting a dairy. Since a modern dairy always handles electric equipments which are also economical, it is desirable to have an adequate supply of electricity.

Facilities, labour and food

Cattle yards should be so constructed and situated in relation to feed storages, hay stacks, silo and manure pits as to effect the most efficient utilization of labour. Sufficient space per cow well arranged feeding mangers and resting areas contribute not only to greater milk yield of cows and make the work of the operator easier but also minimize feed expenses. The relative position of the feed stores should be quite adjacent to the cattle barn. Noteworthy features of feed stores are given below.

a) Feed storages should be located at hand near the centre of the cow barn.

b) Milk house should be located almost at the centre of the barn.

c) Centre cross-alley should be well designed with references to feed storage, the 'stall areas and the milk house.

System of housing

Loose housing/conventional system

The most commonly prevalent practice in this country is to tie the cows with rope on a Katcha floor except some organised dairy farms belonging to government, co-operatives or military where proper housing facilities; exist. It is quite easy to understand that unless cattle are provided wit good housing facilities, the animals will move too far in or out of the standing space, defeating all rounds and even causing trampling a wasting of feed by stepping into the managers. The animals will be exposed to extreme weather conditions all leading to bad health and lower production.

Dairy cattle may be successfully housed a wide variety of condition, ranging from close confinement to little restrictions except at milking time.

1. The loose housing barn in combination with some type of milking barn

or parlour.

2. The conventional diary barn.

Loose housing may be defined as a system where animals are kept loose except milking and at the time of treatment. The system is most economical. Some features of loose housing system are as follows.

1. Cost of construction is significantly lower than convent
2. It is possible to make further expansion without change
3 Facilitate easy detection of animal in heat.
4. Animals feel free and therefore, prove more profitable with even minimum grazing.
5. Animals get optimum excise which is extremely important for better health production.
6. Over all better management can be rendered.

Sanitation in dairy farm

Sanitation is necessary in the dairy farm houses for eliminations of all micro organisms that are capable of causing disease in the animals. The presence of organisms in the animal shed contaminates the milk produced thus reducing its self life, milk produced in an unclean environment is likely to transmit diseases which affect human health: Dry floorings keeps the houses dry and protects from foot injury. Similarly the presence of flies and other insects in the dairy farm area are not only, disturb the animals but also spread deadly diseases to the animal's egg. Babesiosis, Theileriosis.

Cleaning of animal sheds

The easy and quick method of cleaning animal house is with liberal use of tap water, proper lifting and disposes all of dung and used straw bedding, providing drainage, to the animal house for complete removal of liquid waste and urine. The daily removal of feed and fodder left over in the manger, reduces the fly nuisance. Periodical cleaning of water through eliminates the growth of algae, bacterial and viral contamination and thus keeps the animal healthy.

Sanitizers

Sunlight is the most potent and powerful sanitizer which destroy most of the disease producing organism. Disinfection of animal sheds means making these free from disease producing bacteria and is mainly-carried out by sprinkling chemical agents such as bleaching powder, Iodine and lodophor, sodium carbonate, Washing soda, Slaked Lime (Calcium hydroxide), Quick Lime (Calcium oxide) and phenol.

Bleaching powder

This is also called calcium hypo chloride. It contains upto 39 % available chlorine which has high disinfecting activity.

Iodine and lodophor

This is commercially available as lodophores and contains between 1 and 2 % available Iodine which is an effective germicide.

Sodium carbonate

A hot 4% solution of washing soda is a powerful disinfectant against many viruses and certain bacteria

Slaked lime and quick lime

White washing with these agents makes the walls of the sheds and the water troughs free from bacteria.

Phenol

Phenol or carbolic acid is a very disinfectant which destroys bacteria as well as fungus.

Insecticides

Insecticides are the substances or preparations used for killing insects. In order to control flies and disease transmitting ticks, insecticides are used in dairy farms. Ticks usually hide in cracks and crevices of the walls and mangers. Smaller quantities of insecticide solutions are required for spraying. Liquid insecticides can be applied with a powerful sprayer_ hand sprayer, a sponge or brush, commonly used insecticides are BHC, DDT, Gramaxane wettable powders, malathion, surriithion, Sevin 50 % emusifying concentration solutions. These are highly poisonous and need to be handled carefully and should not come in contact with food material, drinking, water,milk etc.

Precautions while using disinfection/Insecticide

- Remove dung and used bedding completely.
- Avoid spilling of dung and used bedding while carrying it out.
- Avoid the use of dirty water in cleaning the sheds.

- Never put the fresh fodder over: the previous day's left over fodder in the manger.
- Prevent algae to grow in the water troughs
- Use proper concentration of disinfectant/insecticide solutions to avoid any toxic effects poisoning.
- Avoid of the mat the milking time as milk absorbs these quickly.

Procedure

- Remove the dung from the floor and urine channel with the help of a shovel and basket (iron) and transfer it to the wheel - barrow. Remove the used bedding and leftovers from the mangers in a similar way.
- Empty the water trough and scrape its sides and bottom with the help of a floor brush.
- Wash the water trough with clean water and white wash it with the help of lime mixture once a week.
- Scrape the floor with a brush and broom and wash with water.
- Clean and disinfect the splashes of dung on the side walls, railing and stanchions.
- Remove the cobwebs periodically with the help of a wall brush.
- Sprinkle one of the available disinfecting agent in the following concentration. Bleaching powder should have more than 30% available chlorine. Phenol 1-2% solution, Washing Soda (4% solution).
- Allow adequate sunlight to enter in to the shed.
- Spray insecticides at regular intervals especially during the rainy season.

Care and management of dairy animal

Calf

We must give well! Feeding and management for the calves so that they develop well and,; useful for replacement stock. The feeding and care of the calf begins~ before its birth the dam should be dried 6-8 weeks before expected calving and should be fed well. Underfed animals will give weak and small calves.

(A)Early management

- Immediately after birth remove any mucous or phlegm from those nose and mouth.
- Normally the cow licks the calf immediately the birth. This helps' dry

off the calf and helps in stimulating breathing and circulation. When the cow does not lick or in cold climate, rub and dry the calf with a dry cloth or gunny bag. Provide artificial respiration by compression and relaxing the chest with hands

- The Naval should be tied about 2-5 cms away from the body and cut 1cm below the ligature and apply Tr. Iodine or boric acid or any antibiotic.
- Remove the wet bedding from the pen and keep the stall very clean and dry in condition.
- The weight of the calf should be recorded.
- Wash the cow's udder and teats preferably with chlorine solution and dry.
- Allow the calf to suckle the first milk of the mohter i.e. Colostrum.
- The calf will be standing and attempts to nurse within one hour. Otherwise help too weak calves.

(B) Feeding of calves

- Feed colostrum i.e. the first milk of the cow for the first 3 days. The colostrum is thick and viscous. It contains higher proportions of Vit A and proteins. The proteins are immunoglobulin which gives protection against many diseases. Colostrums contain antitrypsin which avoid digestion of immunoglobulin in the stomach and is absorbed as it is.
- Whole milk should be given after 3 days it is better to teach to, drink the milk from the pail or bucket. Feed twice a day which should be warmed to body temperature. For weak calves feed thrice a day.
- The limit of liquid milk feeding is 10 % of it's body weight with a maximum of 5-6 liters per day and continue liquid milk feeding for 6.10 weeks. Over feeding causes 'Calf Scours'.
- The milk replaces can be given to replace whole milk.
- Give calf starter after one month of age.
- Provide good quality green fodder and hay from 4'h month afterwards.
- Feeding of antibiotics to calves improves appetite, increases growth rate and prevents calf scours. E.g. aureomycin, Terramycin etc.

Other management practices

- Identity the calf by tattooing in the ear at birth, and branding after one year.

- Dehorn the calf within 7-10 days after birth with red hot Iron or caustic potash stick or electrical method.
- Deworm the calf regularly to remove worms using deworming drugs, at regular intervals of thirty days
- Fresh water should be given from second weeks onwards.
- House the calves in individual calf pens for 3 months afterwards in groups.
- After six months males and females calves should be housed separately.
- Weigh the calves at weekly interval upto 6 months arid at monthly interval afterwards to know the growth rate.
- Mortality in calves is more in first month due to pneumonia. Diarrhea (calf scours) and worms. House them under warm condition, clean condition to avoid above condition.
- Extra teats beyond 4 should be removed at 1-2 months of age.
- 8-9 weeks of age, males should be castrated.
- Keep the body clean and dry to avoid fungal infection.
- Mineral-blocks should be provided, so that the calves lick and no changes for mineral deficiency.
- Wean the calf from the mother and feed through pail feeding system.

Care and management of heifer

Better Care and Management of heifer will give high quality replacement stock to the dairy farm. The following care and Management practices are recommended for a heifer.

- Feed the heifer sufficiently to produce normal growth. During the early stage relatively more protein than energy is needed. Most heifers grow well if excellent hay is given as much they can eat. The amount of growth depends upon the quality of forage fed.
- The heifers should be provided with a dry shelter free from drafts. A loose housing system with a shelter open to one side is sufficient.
- The size rather than the age of a dairy heifer at breeding time is important.

Breeding under sized animals is never profitable. They may be stunted or slow to reach maximum size. Small heifers are more likely to have difficulty in calving.

Though the heifer that is bred to calve at an older age yields higher milk yield in the first lactation, the total milk produced by such a cow will be less when compared to the heifers that freshens at an early age. Usually the heifer is bred to freshen at 24-30 months of age.

- The heifer should be growing and in good flesh at calving time. This is necessary so that she can produce milk at the most profitable level.
- Place the heifer in a separate shed about 6-8 weeks before she is due to calve.
- Feed 2-3 kgs of concentrate daily and all the forage she eats.
- Before calving let the heifer becomes accustomed to handling and to the procedures used in the milking herd. Always handle her gently and with kindness.
- Maintenance of health among heifers is very important for proper growth. The health among the heifers is maintained by hygiene housing, water balanced feeding and taking necessary preventive steps against common diseases.
- Periodically the heifers in the herd should be checked for their proper growth and other progress. Animals lagging behind below the required standards should be removed from the herd.
- For the heifer the calving is first time and it may have difficulty in calving. So take extra care during calving.

Care and management of milch animal

To get high milk during any lactation, the milch animal should be properly fed and necessary care and management practices should be followed.

- Provide green succulent forage together with leguminous hay or straw to the extent of animal can consume, so that all its maintenance requirements are met with through forage only. Extra concentrate at the rate of 1 kg for every 2 to 2.5 liters of milk should be provided. Salt and mineral supplements should be given to maintain the lactation.
- Never frighten or excite the animals. Always treat them gently and with kindness.
- With proper feeding and care, a cow will come to heat with in 16® days of calving. Do not with hold service unnecessarily after the signs of heat are noticed in a cow. The shorter the interval between calving, he more efficient the animal is as a milk producer. By maintaining proper records of breeding and calving of the animals will ensure a

study flow of milk through out the year.

- Individual attention to feed each animal according to its production is a must. For this purpose maintain individual production records.
- Keep up regularity of feeding. Concentrate mix is fed before or during milking, when as roughages after milking. This practice will avoid dust in the shed.
- Water should be provided to drink at will or at frequent intervals. It is more beneficial, if the animal is maintained on paddy straw as sole rough age.
- Regularity in milking is essential. Increase of milk in the udder will reduce further secretion of milk. Milking thrice is better than twice since 10 -15 % more milk can be produced.
- Rapid, continuous, dry hand milking should be practiced without undue jerking of teats. Milking should be done with whole hand, but not with thumb and index finger.
- Cows should be trained to let down milk without calf suckling. This will held to wean the calves early.
- Loose housing with shelter during hot part of the day should be provided. The animals will get maximum exercise in loose housing system.
- Grooming of the cows and washing of the buffaloes before milking help in clean milk production. Daily brushing will remove loose hair an dirt from the coat. Grooming will also keep the animal hide pliable.
- Wallowing of buffaloes or water spraying on their bodies will keel6 the buffaloes comfortable especially in summer.
- Common ailments should be properly detected and treated.
- Common vices should be properly detected and care should be taken such as kicking, licking, suckling etc.
- Vaccinate the cows- against important diseases and also guard against insects and pests.
- Every animal should be numbered and particulars pertaining to milk, fat %, feed taken, breeding, drying and calving dates should be recorded.
- Check for mastitis regularly.
- Provide at least two month for dry period.

Care and management of dry and pregnant animal

The good care and management practices given to pregnant animal will give good calf and also high milk yield during the successive lactation.

- Extra concentrate mix of 1.25 to 1.75 kgs should be provided for pregnant animal as pregnancy allowance. Feed good quality of leguminous fodder. The animal should not be not - lean - not fat' condition.
- Provide clean drinking water and protection from thermal stress.
- Do not allow them to mix with other animals that have aborted or that are suffering from or carriers of diseases like brucellosis.
- Allow moderate exercise, which helps in calving normally. Do not tire them by making long distances especially on uneven surfaces.
- Do not allow them to fight with other animals and take care that they are not chased by dogs and other animals.
- Avoid slippery conditions, which causes the animal to fall receiving fractures, dislocation etc.
- If accurate breeding records are available, calculate the expected date of calving. Separate it one or 2 weeks before and shifted to individual parturition pens. These pens are thoroughly cleaned and fresh bedding may be provided.
- Feed one kg extra concentrates during last 8 weeks of gestation. Feed laxative about 3-5 days before and after calving (Wheat bran 3 kgs +0.5 kgs of Groundnut cake + 100 gm. of mineral mixture of salt).
- Symptoms of delivery may be observed i.e. swelling of external genital, swelling of udder; usually majority of animals will deliver without any help. If there is any difficulty, provide veterinary help.
- After parturition external genital, flank should be cleaned arid protect the animal from chill and give warm wafer.
- Placenta will normally leave the cow within 2 - 4 hours after calving. It not takes the help of a veterinarian.
- Take care of the animal before calving from milk fever. Give calcium supplement.
- Some times the udder will be swollen just before calving. Remove the milk partially.
- Take care, of the animal, if at all any abortion.
- Provide always free access to drinking water.

Care and management of bull

The maintenance of breeding bulls in good condition and suitable for breeding is highly essential requirement for the success of breeding programmes. A rising condition is better for reproduction than a falling one. Fat males may produce semen of inferior quality or they may be slow or fail at service. Breeding bull should receive plenty of exercise; will usually produce large ejaculation containing more sperms of higher activity.

A breeding bull should house separately known as "Bull Shed" with sufficient area of floor and proper covering. It is sound practice to provide cool conditions and adequate drinking water. A balanced ration should be fed containing adequate energy, proteins, minerals, and vitamins. Green fodder must be available both before and during breeding season. Most of the bulls are ferocious and so control them properly using nose rings etc. It is of great importance that males should be, fed regularly and not too much at one time, and too little at another. For bulls two mating a day has been found to be openings. Moderate exercise should be provided to keep the breeding bull in active and non fatty conditions. Regular grooming of the breeding bull is practiced. In buffalo bulls regular shaving may be practiced.

Estrous cycle and sign of heat

The combination of physiological events which take place starting with estrous and ending with the next estrous is termed as estrous cycle. The length of oestrous is 21 days. The estrous cycle is divided into 4 phases.i.e. Proestrous , Oestrous, Metoestrus, and Dioestrous. Proestrous makes the animal coming to heat. During this phase the graffian follicle will grow containing hormone oestradiol. This hormone is observed in blood and passed to oviduct and causes the growth of cells lining the oviduct, which facilitates transport of ova. Estrous or heat is the period in which female shows desire to be mated by the male. This period lasts for 24 - 36 hours. During this period the animals shows all most all heat symptoms. At the end of estrous period ovulation will occur.

Metoestrus is the period just after the ovulation during which period level of oestrogen and leutixizing hormones fall and corpusluteum begins to functions. Diaoestrus period predominated by the influence of the progestorone from the corpus leuteum. If the fertilization takes place the pregnancy, with the accompanying high level of progestorone will stop the formation and development of new graffion follicles in the ovaries.

Sign of heat

The various indication of heat are as under-

- The animal will be excited condition. The animal will be in restlessness

and nervousness.

- The animal will be bellow frequency.
- The animal will reduce the intake of feed.
- Peculiar movement of lumbosacral region will b observed.
- The animals which are in heat will lick other animals and smelling other animals.
- The animals will try to mount other animals
- The animals will standstill when other animal try to mount.. This period is known as standing heat. This extends 14-16 hours.
- Frequent micturition (urination) will be observed.
- Clear mucous discharge will be seen from the vulva, sometimes it will be string like the mucous will be seen stick to the near the pasts of valua.
- Swelling of the valva will be seen.
- Congestion and hyperarmia of membrane.
- The tail will be in raised position.
- Milk production will be slightly decreased.
- On Palpation uterus will be turgid and the cervix will be opened.

Artificial Insemination

Merits and demerits

Artificial insemination (A.I.) is deposition of semen into the female genital tract by means of instruments.

Merits of A.I.

There are several advantages by artificial insemination over natural mating or servicing.

- There is no need of maintenance of breeding bull for a herd; hence the cost of maintenance of breeding bull is saved.
- It prevents the spread of certain diseases and sterility due to genital diseases. E.g. contageous abortion, vibriosis.
- By regular examination of semen after collection and frequent checking on fertility make early detection of interior males and better breeding

efficiency is ensured.

- The progeny testing can be done at an early age.
- The semen of a desired size can be used even after the death of that particular sire.
- The semen collected can be taken to the urban areas or rural areas for insemination.
- It makes possible the mating of animals with great differences in size without injury to either of the animal.
- It is helpful to inseminate the animals that are refuse to stands or accept the male at the time of oestrum.
- It helps in maintaining the accurate breeding and cawing records
- It increases the rate of conception.
- It helps in better record keeping.
- Old, heavy and injured sires can be used.

Demerits of A.I.

- Requires well-trained operations and special equipment.
- Requires more time than natural services.
- Necessitates the knowledge of the structure and function of reproduction on the part of operator.
- Improper cleaning of instruments and in sanitary conditions may lead to lower fertility.
- If the bull is not properly tested, the spreading of genital diseases will be increased.
- Market for bulls will be reduced, while that for superior bull is increased.

Pregnancy diagnosis

Pregnancy can be diagnosed by

1. By observing the signs of pregnancy.

(a) The animal will not come to heat again

(b) The animal tends to become sluggish in temperament and tractable

(c) The animal has tendency to grow

(d) Increase in the volume of abdomen

(e) Increase in body weight

(f) The mammary gland become firm, enlarged and tats takes waxy appearance.

2. Examination of uterus spectrum

(a) Ovaries contains corpus luteum throughout the gestation period: C.L. is firm, rounded at the top and slightly elevated from the surface of the ovary.

(b) From the beginning of 2"d month middle of 2"d month one horn of the Uterus is enraged. It enlarged horn is allowed to slide between the thumb and the first two fingers, a sphere can be felt, which has characteristic slippery feeling at the end of 2nd month sensation of foetal membranes can be felt.

By third month uterus detention can be seen. If distended horn is tapped with fingers reveals like a piece of wood floating in fluid beneath. Early in 4th month cotyledons can be felt. By the d of 4tn month uterine artery starts enlarging if the artery is slightly compressed between the fingers and thumb, continuous vibrating fermitus (uterine thrill) can detected which latter changes to pulsation. From 5th month onwards uterus sinks below the pelvic cavity until middle of 6th month. After wards presence of calf can be easily detected.

(b) Barium chloride test: 4 to 5 drops of barium chloride when added to urine of the same volume from a pregnant cow does not effect the colour, but it becomes turbid with non pregnant.

(c) Specific gravity test: 0.25ml of cerical mucoses is placed in copper sulphate with 1.008 specific gravity. If it sinks —> pregnant Floats 4 non pregnant

(d) Sodium hydroxide test: 0.25ml of ceuical mucosa is added to 5ml of 10% sodium hydroxide solution and heat over flame till boiling. Development of orange colour indicates positive and pale colour indicates negative.

(e) Oxidation reduction test: To 3ml of urine 0.6ml of sodium benzoate indicator is added in a test tube. Tubes are inverted and allowed to stand for 30 minutes green colour is developed. If the colour is permanent it is the evidence positive for pregnancy, where as with non pregnant green colour returns to original colour.

Precautions during calving

Parturition is the expulsion of the foetus and its membranes from the uterus through the birth canal. This process is also known as calvning in cattle. In cows the gestation period will be 275 - 285 days and in buffaloes 300 - 310 days. The pregnant animals kept under careful observation when they are approaching parturition and after seeing the symptom of parturition i.e developed udder, enlargement of vulva etc., the pregnant animal is kept in a calving pen with sufficient bedding of' soft straws. The concentrate feeding should be given separately to the animals near parturition with extra rations developing the previous production of the animal .The calved animal is kept warm. The cow should be given plenty of luke warm water to drink.. The cow is washed with clean 'water and sufficient quantity of green grass and food is supplied. After parturition due to straining some times prolapsed of uterus or vagina is seen, in such cases it should be immediately referred to a qualified veterinarian. Some times heavy milking cows will show the symptoms of "Milk fever" which has to be attended by a veterinarian to give 300-500ml of Calcium Borogluconate Intra Venous route, after parturition or delivery of the placenta is not expelled (Retained placenta), in such cases it is removed manually after 24 hours.

General feeding regimes for dairy cows

(a) For non- pregnant dairy cows- It is enough to provide them a maintenance ration of 2.5 Kg per day and about 15-20 Kg. green dodders.

(b) For dry pregnant cows- May be fed 3.5 concentrate feed plus 20-25 Kg. green fodders per day.

(c) For recent calvers- May be fed on the basis of their milk production, for every 2.5 Kg milk we may provide 1 Kg concentrate feed and additional 1 K for their body maintenance plus 30-35 Kg green fodders.

Maximize conception rate in dairy animals

Reproductive efficiency of a female dairy animal is measured mostly in terms of early attainment of sexual maturity and number of calf crops during her entire life span. Under ideal conditions of management and feeding the animal should reach sexual maturity at an early age (2-1/2- 3 years). The first calf should be produced early and subsequently at every 12 to 14 months interval. The productive and reproductive performances of the rural bovines are much lower than the desired optimal level. This may be due to poor quality of germplasm, poor nutrition and management and lack of knowledge of profitable animal husbandry practices.

The most of animal in rural areas reared under grazing, supplementation with bhusa or paddy straw, some times green fodder and occasionally concentrate feeds. These animals are considered to be local or non-descriptive or mixed breeds. The main problems in these rural bovines are as-

- Late sexual maturity
- Long estrous intervals, service period and calving intervals
- Anoestrous due to inactive ovaries and infantile quintal organ or due to sub active ovaries
- Repeat breeding
- Low conception rate
- Improper feeding, breeding and management practices lowers reproductive efficiency
- Improper diagnosis of heat resulting lower conception rate and inexperienced inseminator etc.

Considering the factors affecting the reproductive efficiency in dairy animals in rural areas, which may be attributed to poor quality germplasm, poor nutrition and management and to some extent lack of proper animal husbandry practices and traditional misconception, could play an important role in improving the reproductive efficiency of animals of rural areas. The villagers are not much aware as to when their animals do not get proper attention and are raised on dry fodder and grazing. Thus the age at puberty should have been attended at 2-2.5 years of age is attended as late as 4.5 or 6 years. In this way livestock owner miss at least one crop or two calf crop in their life time. Calving interval is an important reproductive trait, quite high in rural bovines. Longer service period is also a major factor lowering the lower reproductive performance in dairy animals. In post partum animals, involution of uterus is completed within 40-50 days and they are expected to oestrous after this period. However involution of uterus is completed at a considerable adequate time; they do not show sign of heat for quite a long period.

In the light of above discussion, this could be concluded that the reproductive efficiency of rural dairy animals, apart from the animal factors, is affected by human factors. To increase the reproductive efficiency of cattle and buffaloes raise by rural environment under traditional system the farmers are advised to take care of the following aspects-

- All the female calves should be fed properly from the beginning of the birth so that they attain desired body weight and maturity at an early age. The traditional feeding practice should be modified by providing mineral mixture, concentrate and green fodders
- Livestock owners should get educated and trained in modern profitable animal husbandry practices, especially feeding management and care of

growing calves and heifers.

- After calving if animal has not come into heat up to 2.5-3 months the veterinarian should be called for check up of the dairy animals.
- History of the previous calving such as retained placenta, dystocia, metritis etc. should be properly maintained and provided to the veterinary doctor at the time of examination of animals.
- No heat should be missed and insemination should be given by trained person and timely to achieve optimum pregnancy results.
- Post insemination confirmation of pregnancy at 60 days.

Studies on problems of reproduction are of immense economic importance for the development of livestock industry in our country where majorities of bovine kept as a source of rural livelihood. The rural livestock suffer from major reproductive problems leading to their low productive and reproductive performance. This results in an increased economic burden on the already limited resource of their income. Understanding the reproductive problems and the consequential impact on rural economy, by the scientist and clinician engaged in rural development activities. Several studies have been undertaken on bovine reproduction in the organized farm condition, but still little is known about reproduction of bovines in rural situation which constitute a major portion of our animal's wealth. Reproduction manifests their impact on production as well as reproduction potentials. Attempts have been made to show the impact of factors like farmers land holding and season on the lifetime reproductive performance of these bovines. While the numbers of reproductive problems are less in rural bovines the following suggestions are to be taken care of-

- Maintenance of general health of the animals from birth by proper de-worming
- Maintenance of proper record of animals heat symptoms for better breeding
- Proper attention at the time of calving
- Proper post partum feeding during lactation
- Proper and timely insemination of heifers and cows.

Sanitation, hygiene and control of disease on dairy farm

Proper cleaning and sanitation removes most of the germs and parasites along with dirt, thereby remaining germs are few in numbers and weak and they are harmless. The main points for this are as-

- It prevents the spread of diseases
- It helps the efficiency of animals
- It lowers the mortality rate

Sanitation programme including, adequate ventilation, proper lighting, adequate drainage, proper cleaning and proper disinfection with potassium permagnate,bleaching powder, boric acid and lime solution etc.

Control of disease outbreaks: If any disease occurs it is advised to farmers that they should follow the instructions such as-

- Prompt segregation of sick animals
- Vaccination of healthy animals
- Proper disposal of bedding materials
- Cleaning and disinfection of premises and diseased animal
- Consult qualified veterinary consultant
- Report to district livestock authorities
- Make regular use of foot bath

Month wise vaccination and treatment

1. January - De-worming, vaccination for Rinderpest disease
2. February - Spray of insecticides for ticks mites and vaccination for FMD disease
3. March - De-worming and Anthrax vaccination
4. April - Spray of insecticides for ticks mites and vaccination for Hemorrhagic disease
5. May - Spraying of insecticides against ticks and mites
6. June - De- worming and testing of Tuberculosis
7. July - Adoption of hygienic measures in the farm
8. August - De-worming, vaccination for Rinderpest disease
9. September - Spray of insecticides for ticks mites and vaccination for FMD disease
10. October - Scraping and cleaning of animals shed properly

11. November - De-worming
12. December - Check Brucellosis in heifers and adult cows, intrauterine tablets soon after calving and check for regular mastitis.

Conclusion

Scientific Management of dairy animals helps the farmer to channelize his limited resources to maximum returns from his/her dairy farms. The cost of feed, breed and milk prices are generally outside his control. But the maximum production of milk largely depends on best management on scientific lines.

References

Cox J, Varpama S 2000. 'The 'Livestock Revolution' - Development or Destruction? 'Compassion in World Farming's Report into Factory Farming in 'Developing' Countries. June 2000. www.ciwf-livestock-revolution.co.uk/report % 20cover.htm

Durning AB, Brough HB 1991. Taking stock: animal farming and the environment. Worldwatch Paper No. 103. Washington D.C. Worldwatch Institute.

Deore PA 1990. Manual of crossbred cow, First edition, BAIF Div. Res. Foundation, Pune Jacson, W. Livestock farming 1953. USAF Institute, Madison, Wisconsin, USA.

LID 1999. Livestock in Poverty-Focussed Development,Crewkerne, Somerset: Livestock in Development.

World Bank. 2009 Minding the stock: bringing public policy to bear on livestock sector development. Report no. 44010-GLB. Washington, DC.

http://www.fao.org/wairdoes/ilri/x5462eoi/x5462eoi.html

http://www.ndri.res.in/ndri/Design/dairyextension.html

www.dairyfarmguide.com/dairy-animal-breeding-policy-0127.html dated 3.5.12

www.nddb.org/statistics/milkproduction.html dated 25.05.12

16

Handling and Processing of Animal By-products for Economic Returns

Vikas Pathak and V.P. Singh

To sustain the meat industry in profitable manner, efficient utilization of meat by-products is an important issue. In today's competitive world the realization of old saying "Only the last cry of the animal is wasted" has become essential. The utilization of by-products not only ensures additional revenue but also reduce the cost of various mandatory disposal treatments. The live animals are becoming costlier over the dressed carcass which compels the utilization of by-products. Moreover, unused by-products is also a great threat to environment and public-health apart from loosing nutritional resources. According to FAO (2004) estimates 106.75 million livestock and more than 650 million poultry birds were slaughtered annually in India leading to production of 6.3 million MT meat and different by-products of appreciable value. Now a day's various techniques are available for the utilization of by-products for different purposes still significant portion of the by-products are unutilized and poses huge economic losses.

Animal By-products are the portions of animal body which are not included in dressed carcass. Animal by-products can be estimated by subtracting the dressing percentage from 100. Dressing percentage is inversely proportional to the byproducts means lower the dressing percentage, higher will be the yield of by-products and vice versa. In normal cases dressing percentage usually ranges between 50-70 percent and by-products yield in between 30-50 percent. However, the actual percentage may be higher because it excludes the weight of bones in muscles which is a part of dressed carcass.

The utilization of animal by-products is not new concept; it was well established prior to the recorded history. The contribution of by-products in the total income of slaughter house is around 7-15% (Bowater and Gustafson 1988). The contribution of byproducts in different animals may vary as it is about 11.4% of the gross income in beef and 7.5% of the income in pork. These by-products may be utilized in various food products and in the past, by-products were a favorite food in Asia. However, consumers are now a day's more conscious on the health aspects so by-products are now limited to the non-food uses, such as pet foods and pharmaceuticals. The utilization of by-products primarily depends upon available technology, economics and customs.

There are various types of by-products may be obtained from the animal body which may vary in their percentage and weight with respect to the live weight of the animals. The yield of some important by-products according to the live weight basis has been given in Table 1 (Ockerman 1975, Romans *et al.* 1985)

Table 1: Yield of Important by-products based on live weight

By-product	% of Live Weight		
	Beef	Pig	Lamb
Blood	2.4-6	2-6	4-9
Brain	0.08-0.12	0.08-0.1	0.26
Feet	1.9-2.1	1.5-2.2	2.0
Heart	0.3-0.5	0.15-0.35	0.3-1.1'
Kidney	0.07-0.24	0.2-0.4	0.3-0.6
Liver	1.0-4.5	1.1-2.4	0.9-2.2
Lungs	0.4-0.8	0.4-0.85	0.7-2.2
Skirt	0.15-0.33	0.4-0.5	0.5
Spleen	0.1-0.27	0.1-0.16	0.1-0.4
Pancreas	0.06	0.1	0.2

Benefits derived from utilization of slaughter house by – products

By-products constitute about 50 per cent of the live weight of the animals hence their proper utilization will improve returns to the producers. Their mode of handling and utilization further opens the avenues in revenue generation from slaughter houses. There are some common benefits which may be derived from the utilization of by-products.

1. Improvement in environmental sanitation: by the utilization of by-products we can improve the sanitary status of the environment all around the slaughter houses. The by-products like blood, meat trimmings, fleshings, condemned organs and unutilized offals attracts flies, dogs, vultures and jackals causing public nuisance. Further, they get decomposed leading to bad odors and may be one of the cause in spreading of diseases. Moreover, the unutilized by-products may provide the breeding ground

for meat spoilage organisms. So by proper and timely utilization of by-products we can improve the environmental sanitation can decrease the environmental pollution and spread of disease.

2. Improvement in health and productivity of livestock: by the proper use of certain by-products we can enhance the fertility of soil and the health and nutritional status of livestock. In this series bone meal is an excellent example for enhancing the soil fertility. Further fertility of soil may be enhanced by the use of fertilizers based on slaughter house by-products. For the improvement in productivity and better growth of livestock, the feeding of blood meal, bone and meat meal and low grade fats to animals may be the key factors.

3. By the proper utilization of by-products we can set up the secondary rural industries which may also generate employment for rural youth.

4. Price structure of meat may be decided on the basis of income earned through by-products. The extra income through by-products may reduce the cost of production of meat. It not only provides higher returns to the meat industry personnel's but may be helpful in earning of higher returns from the livestock.

In nutshell, the advantages of byproducts utilization may be emphasized as income generation, achievement in sanitary cleanliness of slaughter houses and its environment and availability of creating valuable source of stock feed.

Classification of animal by-products

The animal by-products are generally divided into two classes i.e. edible and inedible. This classification of by-products is mainly based on the purchasing power of consumers, food habits, religion and customs. The byproducts are also categorized as primary and secondary by-products depending upon their way of utilization. Some of the important by-products along with respective secondary by-products are mentioned in Table 2.

Table 2: Primary and secondary by-products of importance

Primary by-product	Secondary by-product
Hide / skin	Leather, collagen sheets
Hide / skin trimmings	Glue
Bones	Bone meal, gelatin, bone morphogenic protein
Bristles	Brushes
Wool	Warm garments, lanolin
Feathers	Feather meal, decorative items, shuttle cock
Blood	Blood meal, albumen, haemoglobin, serum, plasma, fibrin
Intestine	Sausage casings, instrument strings, surgical sutures, tennis racket guts
Horns	Keratin, artifacts
Hooves	Keratin, hoof meal, neat's foot oil
Intestinal contents	Manure
Condemned offals, meat, carcasses	Meat cum bone meal, offal meal, technical fat
Pancreas	Insulin
Lung	Heparin
Liver	Liver extract
Edible offals (lung, liver, spleen etc)	Pet foods
Stomach	Tripe
Calf stomach	Rennin
Adrenal gland	Adrenalin
Brain	Cerebrocides, cholesterol
Fat (tallow, lard)	Soap industry, textile industry

Nutritive value of by-products

Among the by-products, edible meat by-products have high nutritive value and their protein level is almost close to that of lean meat tissue. However amount of collagen is quite high in by-products as compared to the lean meat. The proteins found in animal by-products contain a larger amount of proline, hydroxyproline and glycine and a lower level of tryptophan and tyrosine because of high connective tissue associated with them. The vitamin content of organ meats is usually higher than that of lean meat. Kidney and liver contain the largest amount of riboflavin (1.697-3.630 mg/100g), and have 5-10 times more than lean meat. Liver is the best source of niacin, vitamin B_{12}, B_6, folacin, ascorbic acid and vitamin A. Kidney is also a good source of vitamin B_6, B_{12} and folacin. A 100g serving of liver from pork or beef contributes 450% - 1100% of the RDA of vitamin A, 65% of the RDA of vitamin B6, 3700% of the RDA of vitamin B_{12} and 37% of the RDA of ascorbic acid. Liver, lungs, and spleen are an excellent source of iron, as well as vitamins. The copper content is highest in the livers of beef, lamb and veal. They contribute 90-350% of the RDA of copper (2 mg/day). Livers also contain the highest amount of manganese (0.128-0.344 mg/100g). However, the highest level of phosphorus (393-558 mg/100g) and potassium (360-433 mg/100g) in meat by-products is found in the thymus and sweetbreads. However the use of by-products is limited because of health awareness and preference for muscle food. Before use of such by-products their quality and safety standards are required to be justified.

Blood

Blood accounts about 4-5% of the live weight and has a high level of protein and heme iron. It contains about 20% solids and 90-95% of these solids are high quality food proteins. Blood is utilized in various parts of the world particularly in Europe as a value added products like blood sausage, puddings, pancakes etc. The offensive red colour and less palatability of blood limit its use in meat industry and plasma is the portion of blood that is of greatest interest, because of its functional properties and lack of color. Besides, these edible uses blood can also be utilized for inedible purposes such as blood meal etc. Utilization of blood in form of blood meal is the most common use of blood in India. It is mostly used for the protein supplement in animal feed. Though, the blood is not properly utilized in the developing countries and creating lot of environmental pollution because it has highest biological oxygen demand for its decomposition. The industrial uses of blood may be summarized in Table 3.

Table 3: Uses of blood in different industries

Food	Emulsifier
Feed	Lysine supplement, vitamin stabilizer, milk substitute, nutritional component
Fertilizer	Seed coating, soil pH stabilizer, mineral component
Laboratory	Tissue culture media, haemin, blood agar, peptone, glycerophosphates, albumin, globulin, sphingomyelin, catalase
Medicine	Agglutinin test, immunoglobulin fractionation techniques, blood clotting factors, sutures, fibrinogen, fibrinolysin, fibrin products, serotonin, plasminogen, plasma extenders
Industry	Adhesive, resin extender, finishes for leather and textiles, insecticide spray adjuvant, egg albumin substitute, foam fire extinguisher, porous concrete, ceramic and plastic manufacturer, plastic and cosmetic base formulations.

Hides and skins

Animal hides have been used for shelters, clothing and as food containers by human beings since prehistoric times. The hides accounts 4-11% of the weight of the live animal and are generally one of the most valuable by-products from animals and accounts about 59% of total by-products value (Boyle 1995). Hides and skins are processed into various finished products like leather, shoes and bags, belts, athletic equipment, reformed sausage casing and cosmetic products, sausage skins, edible gelatin and glue. It is a animal by-product of international repute but its quality is mainly determined on the basis of disease status of livestock and harvesting finish of hide.

The nomenclature of hide is quite variable and misleading. As a thumb rule skin from a fully grown large animal is called hide and that from a small animal (sheep, goats and pigs) is called skin. The term skin is also applied to that of calves. The slunk is the term applied to the skin of an unborn calf, which is often used for parchment, light suede or drum beets. The skin from the older calf which has not yet reached maturity is called kip.

Skins and hides are classified into two classes based on the source from which they are obtained, that is slaughtered animals and fallen animals. In case of hides, the contribution of slaughtered animals is to the tune of 25% while fallen animals contribute 75% of the raw materials. In contrast, the trend in small animals is almost reverse, that is 80% of the skins are accounted from slaughtered stock and only 20% from fallen stock. As such hides and skins obtained from slaughtered animals are superior to those obtained from fallen animals. To obtain good quality raw material, hides and skins are to be flayed immediately after death/ slaughter. Undue delay leads to decomposition resulting in inferiority of hides and skins. In Indian conditions they are classified as *Oxen*, which include hides from cow, bullock, bulls, calves etc. Whereas, *Buffs* is the hide obtained from buffaloes of any type.

Flaying and preservation of hide and skin

India is the largest producer as well as exporter of skins in the world. Hide is one of the most valuable by-products which can be processed into leather by tanning while trimmings are used for making glue, artifacts or feed supplement. Flaying refers to the scientific removal of hide/skin from animal after slaughter or death. Defects in skin such as cuts, tear, puncture, holes, insect damage reduce the value. Hence, the job is done by skilled butchers by following specific technique. The quality of skin also depends on feeding and management of animals. Much care needs to be bestowed in the production of hides and skins from slaughtered and fallen animals. The flaying operation should be completed soon after the death of the animal. Some of the following factors will help in the production of good quality raw materials:

1. To facilitate the easy flaying of hides and skins from the subcutaneous tissue of the carcass animals awaiting slaughter should be withheld feed and offered a good supply of potable water. This practice reduces the chances of flay cuts on the hide.
2. Cold water shower just before slaughter to the animals minimizes the blood flow to the peripheral tissue and reduces the amount of remaining blood in the subcutaneous tissue. Thereby, keeping quality of hides and skins may be improved and 'veiny leather' condition and putrefactive changes may also be minimized.

3. Modern scientific flaying techniques by the trained personnel improve the hide quality. The primary requirement for uniformly good shape of hides and skins is opening lines or ripping lines and the unwanted components such as ears, snouts, dew claws and other appendages which are not of any value in leather processing must be trimmed off.
4. The hygienic conditions of the slaughter house must be achieved because dragging of animals violently on the rough floor under unhygienic slaughter floor damages the grain surface of hides and skins. Further, it is also necessary that hides and skins after flaying must be avoided to through on dung and urine contaminated floor. Trampling of hides and skins over the gastro intestinal contents to increase their weight must be avoided.
5. The hide of fallen animals must be flayed immediately after the death of animals and should be preserved immediately to avoid its deterioration.
6. Dung and other adhering contaminants must be washed off immediately after flaying with cold water using a long handled brush. For further use hide either should be sent directly to the tanneries or should be preserved by curing.

Scientific method of flaying

To obtain the good quality hides some of the tips are utmost important:

1. Always use curved knives with rounded and blunt tip provided with thumb rest to facilitate proper hold. It is important to avoid the undesirable cuts on the hides and skins.
2. Flaying must be done by the trained butcher's.
3. Use of flaying cradle facilitates the easy handling and quality hides.
4. To ensure the good patterns and shapes of hides following techniques should be adopted:
   a) Ripping lines
   b) Case on flaying
   c) Blowing air.

Care of hides and skins

Cold water showers or passing of animals through cold water is necessary because it removes the adhering dirt, dung and other particulate matters. Veiny leather condition can also be avoided by the above mentioned technique. Immediately after flaying the skins should be opened and spread out on a concrete floor to allow it to cool down. After washing, loose subcutaneous tissue, meat and fat left on the flesh side should be removed prior to curing. This process is known as fleshing. All sharp irregular flaps on the edge of the skins should be removed. The tail should be cut off. If needed all pieces below knee and hock joint should be removed.

Preservation techniques

Hides and skins roughly contain about 62% moisture and 38% dry matter. The presence of high moisture content and pertinacious material favours the growth of putrefactive bacteria leads to decay and putrefaction of hides and skins. The spoilage in hides or skin may be observed in form of loosing of hair in patches or all over the skin, bad odour, dark patches on the flesh side of the skin. To avoid all these conditions and to prevent physical damages such as wrinkles, insect bites hides or skins should be preserved. The most prevalent method of preservation of hide is curing and sun drying. There are various methods of curing are generally adopted for hides and skins: wet salting or green salting, dry salting or brine curing.

Sun drying

Sun drying is a preferred method of preservation of fallen hides. In this method hides or skins are stretched and tied to wooden or metal frame and dried in open air preferably not under direct sun The place of sun drying should be well ventilated to ensure good air circulation but exposure of direct sun light over hide or skins is always avoided to prevent the damage and reduction of hide or skin strength.

Dry salting

Dry salting is a good method for weight reduction hides and skins for transport and to enhance the keeping quality. It is a recommended method for small hide producers or village flayers or curers far off from main hide market. This is a convenient method for prevention from putrefaction because dry salt prevents the growth of saprophytes.

In dry salting procedures hides or skins are washed and trimmed, then dry salt is rubbed on the flesh side and the skin is stacked or stored in layers overnight. After salting hides or skins are sun dried on a long bamboo poles. Drying may be achieved with in a 2-3 days. In summers when temperature exceeds above 46^0C drying is done under shades whereas, in winter drying may be done in direct sun light. The temperature of drying has great influence on the quality of hides or skins so always avoid high temperature drying. Sun drying may further be classified into ground drying (plinth drying) or air drying. Ground drying should always be avoided because it damages the grain surface by the action of saprophytes which ultimately leads to reduce the value of the skin and hides to half. In General practice, hide or skins are subjected to drying after salting by fixing or tying the skins to hoops made of bamboo and rope frame. Sometimes the hides or skins are stuffed with straw and dried in air. Four sided wooden frames are generally used for drying of hides in dry curing or dry salting. This method facilitates easy washing and fixing of hides length wise and subsequently the sides are evenly suspended. To avoid overstretching and proper drying of hide and skins frames should be placed in slanting position.

Dry salting method has certain advantages over wet salting as it has availability of plenty of air, eliminates the hair slip and putrefaction, it provides better grading possibilities as every cut or bruise, parasitic damage can be observed easily. It also enhances the keeping quality, facilitate cheaper transport as they are lighter and avoided corrosion to the hides and skins.

Wet salting

Wet salting is also known as green salting in which freshly flayed hides and skins are spread on a sloping concave platform to facilitate cooling. Then salt is spread or sprinkled freely on the flesh side and left overnight. In next step, previously sprinkled salt is totally removed and a fresh layer of salt is applied in such a way that the thicker portion of the hide receives more salt than the thinner sides. In further preservation, next hide is spread on the first hide with hair or grain side down and same procedure is repeated. The hides are then left in pile till complete curing. Piles should be tilted and overhauled periodically to prevent development of heat inside because excess heat generation inside the piles leads to deterioration of the hide quality.

Brine curing

Brine curing is generally carried out by immersing the clean hides in a salt solution for a maximum period of 1-2 weeks. For brine curing pots, vats or pits of 4 feet deep may be used. To maintain the saturation of salt periodically more and more salt is added which favours the absorption of salt in hides and loose the moisture from hides to the extent that they are unfavorable to the enzymes present in the hides and subsequent bacterial action. Brine cured hides or skins are superior to hides cured by conventional wet salting or other methods and also preserved the hide for longer period of time. For further enhancement of the keeping quality of brined hides or skins about 2-3% sodium carbonate and 1% naphthalene of the salt weight may be added in the brine solution in the vats. The periodically turning of hides from top to bottom after overnight soaking is required and hides must be cured for about 1-2 weeks. Hides are then drained and packed in the layers and needs to be piled properly before they are lifted to the tanneries.

Brine curing has certain beneficial effects on the hides or skins over other hide preservative techniques. This method is more rapid and uniform, gives greater protection when exposed to adverse conditions during storage and transportation, eliminates salt strains during long storage. It do not require any washing prior to soaking back in the tannery and soaking period is also reduced considerably as compared to wet salted hides. Brine curing produces quality hides with more fullness and brighter grain of leather. The flesh and hair sides are bright and exceptionally free from contamination by dirt without any adhering salt.

The defects most commonly observed in skins and hides are conveniently categorized in to ante mortem defects and post mortem defects. They may be summarized as depicted in Table 4.

Table 4: Defects of hides and skins

Type of defects	Defects	Reasons
Ante mortem defects	Poor substrate of the skin	Genetic nature
	Warble fly infection	*Hypoderma bovis, Hypoderma lineatum, Hypoderma crossi.*
	Pox	Cow pox sheep pox etc.
	Mange	Sarcoptic, Psoroptic and Demodectic
	Grain damage	Bacterial infection, thorn wire injuries, brand marks, contusions (bruises), abrasions, abscesses,
	tick damage	
Post mortem defects	Bad shape, flay cuts, curing defects eg. Salt burn and improper curing.	Improper or careless handling of skins and hides during slaughter and flaying.

Processing of hides and skins into leather

Hide or skin may be converted into leather by chemical treatment and some other processing's which commonly known as tanning. Leather is quite stable and is not affected by decomposition or bacterial decay, even in wet condition and it has typical quality parameters like tensile strength, flexibility, and resilience and abrasion resistance.

Processing steps

1. Washing and soaking: Washing with water helps in removal of salt adhered during curing along with water soluble proteins sticked with blood and lymph, dung and other matter. The soaking of water helps the fibers to absorb water and restores its normal shape and dimensions.
2. Fleshing: it is a process of removal of flesh adhered to the inner side of the skin by scraping with serrated knife. Fleshing facilitate better penetration of the chemicals during further processing.
3. Liming: liming is a process of immersion of hides in vats containing saturated lime solution. It helps in separation of keratinous tissues made up of hair and epidermal cells. It also removes the loosed hair follicles by serrated knife.

4. Washing and deliming: washing at this stage helps in removal of lime and hair. For removal of residual lime soaking is done in weak acid.
5. Bating: bating is a treatment of delimed hides with proteolytic enzymes to make them soft and pliable.
6. Pickling: pickling is a process of soaking the hide or skin in a solution of sulphuric acid and salt for conditioning in tanning operation. It helps in achieving the uniform degree of acidity and cleanliness.

Tanning

Tanning is a process of conversion of hide or skin in to insoluble and non-putrisible leather without destruction of original structure. It is done by treating the conditioned hides and skin with organic, inorganic or synthetic tanning agents. Among these tanning agents, vegetable tanning material and basic chromium sulphate tanning procedures are being commonly used.

Vegetable tanning

In vegetable tanning, extracts of infusion derived from tannin bearing plants or barks are used. In India the most common plants for this purpose are avaram (*Cassia auriculata*) Babul (*Acacia Arabica*), Myrabalan (*Terminalia chebula*) etc. In this process de-limed hide or skin are immersed in progressively stronger tan liquors containing in pits. These pelts are moved at a fixed interval from one pit to another which helps in diffusion of tannin particles into the pelt and also imparts uniform colour. For better spreading of tannin the resolving drum and strong tanning liquor can be used. Vegetable tanning is a common method adopted for the manufacturing of sole belting, harness and saddle and other heavy leather.

Chrome tanning

This is most popular technique of tanning in which chromic sulphate ($Cr(OH)SO_4$) is used to yields soft, supple and strong leather. Chromium treated leather is permeable to air. The main problem in this technique is disposal of chrome because it is toxic in nature.

Post tanning operations

To provide the finish to the leather there are various operations like removal of water and tan liquor, splitting and shaving, neutralizing and dyeing, fat liquoring and stuffing, setting out, shammoying, staking and buffing are required. The furnishing the sole leather oil on grain surface is applied and then it is rolled under a heavy roller. However, chrome leather required glazing after seasoning and drying. It is done by a glass cylinder clamped to the end of a moving arm and the heat generated by the friction softens the wax. This process provides finish and a continuous plastic coat of high luster on the grain surface. After glazing and staking, smooth plating of leathers on the grain side is done in specially designed press. For the production of pleasing grain boarding platforms is generally used. For the production of decorative pattern on vegetable tanned leathers embossing with hydraulic press can be used.

Table 5: Physical properties of leather (shoe upper leather)

Property	Value
Tensile strength	15.3-37.5
Elongation at break (%)	29.5-73.0
Stitch tear strength (N/cm)	1280-2275
Thickness (mm)	1.5-2.4
Bursting strength (kn/cm)	1.1-24.5
Apparent density (g/cm^2)	0.6-0.9
Real density (g/cm^2)	1.4-1.6
Heat resistance	Shrinks depending on moisture

Glandular by-products

Glandular by-products are of great importance for medical and pharmaceutical industry and accounts about 0.28% of the animals' live weight. People are using animal glandular by-products in medical and pharmaceutical industry since time immemorial. Yet there is great scope for developing and processing of these glandular by-products into the valuable products from abattoirs. The important by-products for the medical and pharmaceutical industry have been tabulated (Table 6). There are certain constraints in our meat industry such as lack of abattoir facilities for proper collection, storage and transport; lack of awareness and lack of economically viable available processing technologies which are responsible for the underutilization of these valuable by-products. These by-products are very delicate and fragile and needs immediate preservation. the basic motto glandular by-products preservation is the stoppage of tissue breakdown from bacterial growth that can be easily achieved by rapid freezing. Certain primary handling procedures must be adopted before going to freeze these glands i.e cleaning and trimming of surrounding fat and connective tissue. These glands are then placed onto waxed paper and kept at -18°C or less. On arrival of glands in pharmaceutical plant they must be inspected, chopped and mixed with different solutions for extraction, or placed in a vacuum drier. If dried glands are having too much fat in it, then extra fat may be extracted with the help of solutions such as gasoline, light petroleum, ethylene or acetone. In final step of its preparation defatted glands or extracts are milled into a powder which may be transformed into capsules or can be used in a liquid form. Before going to the sale these glandular products must be checked for its safety and potency.

Table 6: Important glands and organs of medical and pharmaceutical value

Organ/Gland	Use
Liver	Liver extract
Pancreas	Insulin, Pancreatin, TrypsinChymotrypsin, Carboxypeptidase Lipase etc
Pituitary - anterior lobe	ACTH, Prolactin, Growth hormone
Pituitary - posterior lobe	Oxytocin, Vasopressin
Pituitary - middle lobe	MSH
Thyroid	Thyroxin
Adrenals	Adrenaline and cortisones
Lungs	Heparin
Suckling calf stomach	Rennet and pepsin
Gall bladder	Bile, Cholic acid, Cortisones,Bile salts, Cholesterol etc.
Brain	Cholesterol
Blood	Serum, Plasma, Fibrin, Hemoglobin
Pineal gland	Melatonine
Testis	Testosterone
Ovaries	Estradiol, Progesterone, Relaxin

Sausage casings

Sausage casings may be obtained from the various parts of the digestive and excretory tracts like esophagus, large and small intestines, caecum, rectum and bladders and may be processed into natural casings that can serve as containers or packaging materials for comminuted meat products like sausages, salami etc. Natural casings are high value by-products of animal and accounts about 25% of the cost of live animals. In the developing countries like India, casings are manufactured by unorganized cottage industry without any considerable control on quality. So their quality aspects must be checked before use in the meat products as natural coverings. Natural casing are having enough strength and elasticity required for sausages and salami's.

Bones

Bone is specialized form of connective tissue and is left over as a byproduct of boneless meat used for the production of certain emulsion type meat products like sausages, patties etc. Bone contains both organic and inorganic constituents and can be transformed into various types of valuable products like glue, gelatin etc. Various types of bone based industries may be established in rural areas i.e. handicrafts, bone crushing units, mineral based industries, proteins based products etc.

Edible tallow and lard

Edible animal fats like lard and tallow are the important animal fats which are obtained as by-products of slaughter houses and meat packaging industry. Among them Lard is a fat rendered from the clean tissues of healthy pigs whereas, Tallow is hard fat rendered from the fatty tissues of cattle or sheep. Lard and edible tallow are the main by-products obtained during dry or wet rendering. In wet rendering process, the fatty tissues are heated in the presence of water, generally at a low temperature and produces good quality lard or tallow as compared to dry rendering process. These fats can be used as an edible fat without any further processing but they require bleaching and deodorizing treatment before being used in food. These products can be used in deep frying, margarine and shortenings, sausages or emulsified products.

The utilization of animal by-products is a key for the sustainability of meat industry. That is due to high cost of live animals and machinery involved in this industry. The poor sanitary conditions of Indian slaughter houses further poses the problems in its use in various products. So the rendering of by-products is a good option for the optimal uses of these value products. Though, animal by-products have been processed since long into different value added edible and inedible products. Keeping in view economy and increased concerns over health there various technologies have been developed for efficient utilization of these by-products. For better returns and quality products production it is utmost importance that animal by-products be efficient utilized. The utilization of by-products in most effective manner is also necessary for the meat industry to remain viable in future.

References

Bowater FJ, Gustafson MA 1988. Plant layouts, collection and selling meat by-products. Advances in Meat Research 5:381.

Boyle JP 1995. Meat and Poultry facts. American Meat Institute, Washington, 222-26.

Ockerman HW 1975. Chemistry of meat tissue, 8th Ed. Ohio State University, Columbus. pp 786-91.

Romans JR, Jones KW, Costello WJ, Carlson CW, Zeigler PT 1985. The meat we eat, 12th Ed. Interstate, Danville, Illinois, 1015-1056.

17

Vermibiotechnology: A Micro Enterprise for Rural Development and Sustainable Agriculture

Ran Vir Singh

Introduction

After introduction of Green Revolution and White Revolution in our country, India had high growth in food and milk production but farmers of our country had applying more chemical fertilizers, insecticides pesticides, weedicides and various medicines in injudicious way to increase grain, vegetable, fruit, flower, milk, meat and egg production . Due to this reason soil fertilely and soil health is deteriorating continuously. The soil of food bowl states of our country viz. Punjab, Haryana, Western Uttar Pradesh and plain area of Uttarakhand deficient in organic matter, macro and micro and trace elements. Now soil became non living due to death of earthworms and beneficial micro organism. Farmers are using mainly nitrogen, phosphorus and potash in crop production. However there are 15 essential plant nutrition's for growing of crops. Due to more application of chemical fertilizers in crop production these days' farmers are not properly using organic, green and compost manures in crop husbandry. Soil fertility and ground water level has been declined very fast by adopting intensive farming. If soil health of any country has been depleted the productivity will automatically decline. They can not enhance the food productivity even agricultural scientists have developed high yield crops varieties domestic animal and poultry breeds, latest cutting age technologies and biotechnologies. All these innovative efforts are useless for enhancing food and nutrients security. Keeping in view of the above cited facts and to enhance profitability of dairy farming the IVRI, Izatnagar had conceived a research project before 7 years in area of recycling of animal and farm waste management and application of their value added organic products in sustainable crops and animal production. In these projects, our institute had developed following transferable technologies.

- Developed a new Indian earthworm species strain of Jai Gopal for garbage processing.
- Developed improved package of practices for production of vermiculture and vermibiomanure.
- Enriching vermiculture with beneficial microorganism for soil and decomposition of solid waste.
- Developed a optimum design for construction of vermibeds and vermi sheds.
- Developed a vermibiomanure sieving machine.
- Developed a process to produce nutriwash.
- Established garbage processing Earthworm Species Seed Bank at IVRI, Izatnagar.
- In collaboration between IVRI and IIT, Roorkee developed Rotary Drum Composting Machine System.
- In collaboration between IVRI and IARI, New Delhi standardized dozes of value added organic products in various crops.

Concept of organic farming

Organic farming is a natural method of farming system which primarily aims at cultivating the land and raising crops in such a way as to keep the soil alive and in good health by use of organic wastes (crop, animal and farm wastes, aquatic wastes) and other biological materials along with beneficial microbes (biofertilizers) to release nutrients to crops for increased sustainable production in an eco-friendly pollution free environment.

As per the definition of FAO suggested that "Organic agriculture is a unique production management system which promotes and enhances agro-ecosystem health, including biodiversity, biological cycles and soil biological activity, and this is accomplished by using on farm agronomic, biological and mechanical methods in exclusion of all synthetic of farm inputs. The organic agriculture has grown almost 29 fold during the last 5 years. By March 2010 India has brought more than 4.48 million ha area under organic certification process. India has traditionally been a country of organic agriculture, but the growth of modern scientific, input intensive agriculture has pushed it to well. But with the increasing awareness about the safety and quality of foods, long term sustainability of the system and accumulating evidences of being equally productive, the organic farming has emerged as an alternative system of farming ,which not only address the quality and sustainability concerns, but also ensures a debt free, profitable livelihood option.

The international federation for agriculture movements (IFOAM) defined organic agriculture using the principle of health, ecology, fairness and care. The organic agriculture should sustain and enhance the health of soil, plant, animal, human and planet as one and indivisible. Healthy soils produce healthy crops that foster the health of animals and people. Health is the wholeness and integrity of living systems. Organic agriculture should be based on living ecological system and cycles, work with them emulate them and help sustain them. This principle roots organic agriculture with in living ecological systems. It states that production is to be based on ecological systems. Nourishment and well being are achieved through the ecology of the specific production environment. Organic agriculture should build on relationships that ensure fairness with regard to the common environment and life opportunities. Fairness is characterized by equity, respect, justice and steward of the shared world, both among people and in their relations to other living beings. Organic agriculture should provide every one involved with a good quality of life, and contribute to food sovereignty and reduction of poverty. Fairness requires systems of production, distribution and trade that are open and equitable and account for real environmental and social cost. Organic agriculture should be managed in a precautionary and responsible manner to protect the health and well being of current and future generations and the environment. Organic agriculture is a living and dynamic system that responds to internal and external demands and conditions. Practitioners of organic agriculture can enhance, efficiency and increase productivity but this should not be at the risk of jeopardizing health and well being. Consequently, new technologies need to be assessed and existing methods reviewed. This principle states that precaution and responsibility are the key concerns in management, development and technology choices in organic agriculture. Science is necessary to ensure that organic agriculture is healthy, safe and ecologically sound. However, scientific knowledge alone is not sufficient. Practical experience, accumulated wisdom and traditional and indigenous knowledge offer valid solutions tested by time. In totality organic agriculture aims at a sustainable production system based on natural processes. Key characteristics of organic agriculture are

- Relies primarily on local, renewable resources.

- Makes efficient use of solar energy and the production potential of biological systems.
- Maximizes recycling of plant nutrients and organic matter.
- Does not use organism or substances foreign to nature (e.g. GMOs, chemical fertilizers).
- Maintains diversity in production system as well the agricultural landscape.
- Gives farm animal life conditions that correspond to their ecological role and allow them a natural behavior organic agriculture is also a sustainable and environmentally friendly production method, which has particular advantages for small scale farmers. Available evidence indicates that appropriateness of organic agriculture for small farmers in developing countries like India. Organic agriculture contributes to poverty alleviation and food security by a combination of many features, such as :
- Increasing yields in low input areas.
- Conserving biodiversity and nature resources on the farm and in the surrounding area.
- Increasing income or reducing cost of cultivation.
- Producing safe and varied food.
- Being sustainable in the long term.

Animal and farm waste production in India

Livestock and crop production activities generate huge amount of biodegradable waste higher economic growth and rapid urbanization of population in general are usually accompanied with generation of enormous amount of biosolid. India generates about 1677 million tones animal waste, 22.45 million tones poultry dropping waste 500 million tones agro waste and 25-39 million tones municipal solid waste. In addition 10-15% of 75 million tones/annum of fruits and vegetables produced in the country are also available as by products for recycling in agriculture. Food processing industries also producing approximately five million tones/annum of bio solids per year.

Structural components of crop residues

Plant cell wall material of commonly fed straws and stovers is composed of three important constituents: cellulose, lignin and hemicellulose. Lignin is particularly difficult to biodegrade and reduces the bio-availability of the other cell wall constitutents. Cellulose is a long chain of glucose molecules, linked to one another primarily with glycosidic bonds. The simplicity of the cellulosic structure, using repeated identical bonds, means that only a small number of enzymes are required to degrade this material. Ruminants create an environment in their rumen, which encourages microbial degradation, converting cellulose to volatile fatty acids and microbial biomass, which the ruminant can then digest and use. Hemicelluloses are branched polymers of xylose, arabinose, galactose, mannose and glucose. Hemicelluloses bind bundles of cellulose fibrils to form microfibrils, which enhance the stability of the cell wall. They also cross link with lignin creating a complex web of bonds, which provide structural strength but also challenge microbial degradation (Lynch 1992). Lignin is a complex polymer of phenyl propane units, which are cross-linked to each other with a variety of different chemical bonds. This complexity has thus far proven as resistant to detailed biochemical characterization as it is to microbial degradation, which greatly impedes our understanding of its effects. Nonetheless, some organisms, particularly fungi, have developed the necessary enzymes to break lignin apart. The initial reactions are mediated by extra cellular lignin and manganese peroxidases, primarily produced by white-rot fungi. Actinomycetes can also decompose lignin, but typically degrade less than 20 percent of the total lignin present (Crawford 1986). Lignin degradation is primarly an aerobic process and in an anaerobic environment lignin can persist for very long periods (Van Soest 1994). Lignin is the most intractable component of the plant cell wall, and a direct negative relationship between the proportions of lignine and the bioavailability of the substrate exists. The effect of lignin on the bioavailability of other cell wall components is through to be largely a physical restriction, with lignin molecules reducing the surface area available to enzymatic penetration and activity (Haug 1993).

Vermibiotechnology

It is a simple biotechnological process of composting in which animal and farm waste harbour species of earthworms and micro organisms which are used to enhance and accelerate the process of waste conversion into value added organic products viz. vermicompost, nutriwash and high quality of animal protein. The multiplication of earthworms and favourable micro organisms (using earthworm as bioreactors) in organic waste is called vermiculture. The liquid extract from vermicompost and wash of earthworms are termed as nutriwash.

Principle of production of vermibiomanure and value added organic products

1. Air circulation in the beds of animal and agro-waste is increased due to movement of earthworms which enhance the process of aerobic fermentation.

2. Earthworms eat biodegradable waste. These worms grind big particles into small particles by gizzard. This process increase surface area of decomposed feed. Earthworm's gut microorganism and enzymes help in processing of biodegradable feed.

3. Earthowrms act as versatile natural bioreactor to increase the population of beneficial microorganism of soil fertility and garbage processing. The gut of earthworm is having optimum temperature and moisture to increase the population of microorganism. It is appropriate to mention that earthworms promote microbial population either by virtue of their own intestinal mechanism or by their casts, serving as 'Culture Plate for development and multiplication of several strains of becteria. These bacteria are beneficial for decomposition process as well as converting nutriets into available form of plants nutrients.

4. Mucoprotien in gut is mixed in digested substrate through biochemical process. This process help in breakdown of plants nutrients from bridgeable substances into available form for plants.

5. Earthworms eat garbage more than their body weight in one day and excrete about 33% in form of vermicast to consumed amount of feed. Animal and agro-waste harbour microorganism eat in amount of thousands time more than their body weight in one day. Earthworms and microorganism are used to enhance and accelerate the process of waste conversion in value added organic products viz. vermibiomanure, nutriwash, growth promoters and proteinaceous worm biomass. This is principle of nutrition that the animal which are more in smaller body weight but they eat many times more than higher body weight animals.

Earthworm species which harbour on animal and farm waste

There are about 509 species of earthworms in India with various food and burrowing habits.

IVRI, Izatnagar has established a seed bank of earthworm species which harbour on animal and farm waste in 2004. Species of earthworm like *Eudrilus eugeniaeEisenia foetida, Perionyx excavatus* and *Perionyx sansibarious* have been identified efficient earthworm to harbour on animal and farm waste and produce vermicompost (Kale, 2006). There are three types of earthworms like Epegeic, Anageic and Endogeic. It is better to rear all these earthworms which are beneficial both as garbage processor as well as soil processor. Different species of earthworm inhabit different part of soil and have distinct feeding strategies. They can be separated into three ecological groups based on their feeding and burrowing habits.

- Soil surface dwelling or compost preferring species are called *Epigeic species.*
- Top soil dwelling species are called Endogeic.
- Subsoil dwelling species/deep burrowing species are called *Anecic species.*

Studies on suitability of indigenous earthworm species were carried out at IVRI, Izatnagar on recycling of animal and farm waste through vermibiotechnology from 2006 to 2008. In the initial studies various indigenous earthworm species were examined to identity the suitable species on the basis of following characteristics:

- The worm should able to thrive on cow dung and crop residues with high organic content.
- It should be able to adopt variable temperature (0 to 44°C).
- It should have high prolificacy.
- It should have high feed conversion efficiency.
- It should able to multiply beneficial soil and fermentation microorganism.

The studies indicated that *Perionyx ceylensis* designated as "Jai Gopal" was the most suitable earthworm species till reported in the world.

This earthworm species is very much suitable for vermi composting and garbage processing. This is the first report in India. The unique characteristics of this species is given below

- Very good capability of inhabiting on animal and farm waste.
- Very high prolificacy.
- Voracious feeder of animal and farm waste.
- Heat and cold tolerance ability (0 to 44°C).
- Act as bioreactor for beneficial soil and fermentation micro organism.
- Superior quality of vermicast and vermiwash.
- Smallest period of interval from hatching to maturity.
- Long life span.
- Very rich in protein.

IVRI, Izatnagar is maintaining four earthworm species viz. *Eisenia foetida, Eudrilus eugenia, Lampito mauritii* and *Perionyx ceylansis* in seed bank. Principles of selection and mating plan are followed to develop high prolific, voracious feeder heat and cold tolerance character's and improves capability of inhabiting on animal and farm waste. An indigenous earthworm species was collected from 24, South Pargana district of Sunderban region of West Bengal in 2006. This earthworm species was genetically improved through intense selection and mating plan. The sample of earthworm was sent to Zoological Survey of India for identification. They identified this earthworm species on indigenous and its scientific name is *Perionyx ceylansis.*

This is an endemic earthworm which has not any threat for biodiversity of Indian earhtoworm in our country based on experiment conducted at IVRI, Izatnagar *Perionyx ceylansis* is recommended for vermicomposting in India. It is better from two exotic earthworm *Eisenea foetida and Eudrilus eugeneae.*

Earthworms alone are responsible for the decomposing process in the soil is inappropriate. Microbial decomposition involves a variety of organism and a complicated network of exchanges and feed back which are yet to be fully under stood. It is appropriate to mention that earthworms promote microbial population, either by virtue of their own intestinal mechanism or by their casts, serving as 'Culture Plate' for development and multiplication of several strains of bacteria. These bacteria are beneficial for the decomposition process as well as converting nutrients in to available forms for plants.

Benefits of vermiculture and vermibiomanure production

- Generation of self employment for rural youth, women and farmers.
- Cheap and eco-friendly technology and base for sustainable agriculture.
- Application of vermibiomanure and other value added organic products will improve the nutritive quality of food products.
- Enhancement in disease resistance in human and animals by consuming organic food and feed.
- Helpful in maintaining clean environment of village and cities.
- Helpful in reducing production of green house gases.
- Helpful in conservation of soil, water forest, animal and human.
- Availability of raw material in each village and city for production of vermibiomanure and organic manure.
- Vermibiomanure and other value added organic products can be manufacture in each village.
- Availability of all plant nutrients and beneficial soil organism in vermibiomanure.

- Vermibiomanure can be applied in any crop in different combination based on availability.
- Vermibiotechnology is a very easy ecofriendly technology in learning.

Selection of project site for establishment of vermiculture hatcheries and vermibiomanure production unit

Farmers and entrepreneurs should consider following criteria in selection of project site.

- Availability of biodegradable raw material for recycling.
- The project site should be located at highest place and proper drainage system should also be existed.
- There should be a market for sale of vermibiomanure and other value added organic products.

Selection of earthworm species for recycling

The farmers should select garbage processing newly developed Indian earthworm species *Perionyx ceyelensis* whose popular name is designated as "Jai Gopal". This earthworm species is better than exotic earthworm species viz. *Eisenea foetida* and *Eudrilus engeniae* in characteristics of fecundity, growth, heat tolerance and inhabiting habit of animal and farm waste. This species is well adopted in tropical agro climatic condition. The vermiculture of this earthworm species is available @ 250/kg at IVRI, Izatnagar.

Essential requirements for manufacturing of vermibiomanure

Construction of vermi tank and vermi sheds

Farmers and entrepreneurs should construct vermi tank and vermi shed as per design recommended by IVRI, Izatnagar. There are four methods for production of vermibiomanure viz. tank, pit, heap and ring method. Out of these vermi tank is one of the best method for recycling of animal and farm waste because it provide anaerobic environment for microorganism and earthworm species. There is also minimum loss of plant nutrients in this method.

The dimension of vermi tank is strictly followed as 30cm height, 90 cm width and length as per requirement of farmers. If height is increased more than 30 cm, there will an anaerobic fermentation which decrease the growth of earthworm and aerobic decomposition. If width of wall of vermi tank is increased more than 90 cm farmers/workers will feel inconvenient in turning of solid waste and harvesting of finished product. The 60 cm or as per width of wheel barrow space should be kept vacant between two vermi tanks for movement and various operations. The well of vermi tank should construct in half brick in 1:7 ratio of cement and sand. The floor of tank should constructed by brick on face side (37 bricks in one square meter). The gap between brick should be filled by plaster of cement and sand in ratio of 1:12. Vermi sheds should be established on 15^0 clock wise direction of latitude from equatorial line so that these sheds will remain cooler in summer and warmer in winter season. Throughout year foliage trees should be planted at distance of 5 feet from wall of shed in western and southern side. The height of wall of sheds should constructed 3 feet and above this wall 4 feet height iron woven net should be fixed so that birds can not enter inside the sheds. Asbestos sheet or thatch bamboo can be used in ceiling of vermi sheds. The height in centre of shed is kept 10 to 11 feet with 3 to 4 feet slope towards wall. The gate can be fabricated of iron wire and rode with size of 7 feet height and 4 feet wide. There should be proper ventilation in side vermi sheds.

1. Availability of optimum feed for earthworms

Earthworms do not eat green leaves and fresh dung. They eat only decomposed biodegradable solid waste. Firstly green crops residues and leaves of plants should be dried and followed by shredded in 2 to 4 cm size. Afterward dried biodegradable waste should be mixed with bioinnoculam fresh cow dung properly. Windrows up to one meter height are made by mixed waste and allow for decomposition for 10 to 15 days in summer and 5 to 7 days in winter seasons. In first week the temperature will rise up to 60 to 65^0C. After wards the temperature of heap will decline up to 30 to 35^0C in summer. Turning of heap after one week in desirable and followed by sprinkling of water over the earthworms. Forty percent moisture is maintained in heap. The feed is not prepared in deep pit to avoid anaerobic ferment action because the worms do not like to eat it. Proper ventilation are maintained in heap. Due to optimum moisture (40%) high temperature and aerobic micro organism will enhance fermentation of lignin, cellulose and hemicelluloses of crop residues and cow dung, after pre fermentation up to 12 to 15 days in summer and 5 to 7 days in winter. The feed is ready for feeding of worms. The optimum temperature for vermiculture production is 20 to 25^0C. During winter season when temperature decline below 15^0C in this situation to increase temperature of vermi bed 3 to 4 days old fermented biodegradable waste can be feed to earthworm.

List of suitable biodegradable waste for production of vermiculture and vermibiomanure production:

Sl.No.	Source feed ingredient	Suitable biodegradable solid waste for products of vermibiomanure
1.	Animal dung	Cow, buffalo, goat, sheep, horse, ass, wild animal's dung biogas slurry mixed with dry leaves. Poultry dropping and pig feasces can be mixed @ 10 % in cow dung for feeding of worms.
2.	Agriculture waste	Plants stubble, paddy straw, weeds dry leaves fruit rind, vegetable and fruit waste.
3.	Urban Solid Waste	Kitchen waste from household and restaurants waste from market yards and places of worship.
4.	Agro industries	Peel, rind and unused pulp of fruits and vegetables from food processing unit Pressmud and seed husk from vegetable oil refineries fine baggase and boiler ash from sugar factories spent wash, barley waste yeast sludge from breweries and distillery unit, coir of fruits, paper and date expired seeds from seed production units, stem, leaves and flowers after extraction of oil from aromatic oil extraction industry and coir pith from coir industry.

2. Optimum ventilation

Optimum ventilation is essential for supply of oxygen to earthworms and aerobic microorganism. Earthworms respire from his whole body in presence of optimum moisture. The oxygen is very essential for microbial and biochemical process in the gut. The vermibed should not filled by feed with pressure. The 2/3 portion of vermibed should filled by new feed and 1/3 portion should be kept for finished product of vermibiomanure. Before harvesting finished product. It is kept in 1/3 portion of bed up to 10 to 12 days more so that extra water can be evaporated and all earthworms will move towards new feed and all cocoon will hatch so that hatchling will also move toward new feed. The height of vermi tank should not kept more than one feet to check anaerobic fermentation. Proper and optimum ventilation in vermi bed is only possible when height of tank is kept up to one feet.

3. Optimum moisture

The whole body/skin of earthworm acts as respiratory system to exchange gases, therefore, it is very much essential to moist the whole body of earthworm. For more activity and more enhancement of recycling ability of earthworms, it is also essential to maintain 40 to 50 percent moisture in feed. There should be proper drainage of excess water from vermibeds and flooring of bed is constructed through brick on face. The gap between two bricks should be filled by 1:12 ratio of cement and sand so that extra water cum easily seepage in ground. Farmer can test excess water in feed by making shape of ball when ball is pressed by hands than water is seepage. It means excess moisture, when ball is not properly formed by pressing with hands. It mean below moisture level than 40% in feed. If ball is properly formed, it means feed contain optimum moisture 40 to 50% level. Vermi tank should not filled by water. To maintain optimum level of moisture in feed sprinkling is followed by putting sprinkler on pipe. Always avoid sprinkling on finished product. Never use saline water for sprinkling on vermi beds.

4. Optimum temperature

The 20 to 30°C ambient temperature is always good and optimum for vermiculture and vermibiomanure production. The *Eudrilus eugeniae* will not survive more than 35°C and below than 10°C ambient temperature. However *Eisenia foetida* will survive between 50 to 40°C temperature. The newly developed Indian earthworm species "Jai Gopal" is having ability to survive between 2 to 42°C ambient temperature for management of optimum temperature in vermi beds. We should allow pre fermentation of biodegradable material in one meter height windrows up to 10 to 15 days during summer in sheds. The moisture level in feed is maintained between 40 to 50° percent only if it increase more than 50 percent in vermi bed or in heap. The dung fly will lay eggs followed by maggots in feed emerged. We should follow prevention to create this situation in summer. In northern state of our country 15th November on wards the ambient temperature starts decline. Therefore we can feed earthworms directly after 3 to 4 days pre fermentation of feed so that temperature of vermibed will be increased. It enhances the process of recycling. In situation of less ventilation and high moisture in vermitank, temperature of feed in bed is increased which is harmful for recycling process. Turning of feed is desirable to maintain optimum moisture and temperature in the bed, when it is required.

5. Optimum pH and removal of toxic material from feed

Seven pH of earthworm feed is optimum for faster growth when ever pH levels increase or decrease in the feed; it has significant effect on recycling processing. Saline water is not suitable for vermiculture production. Salt and toxic feed materials have also harmful effect on growth of earthworms and microorganisms. Always neutral PH feed and water should be used.

6. Mulching on vermibeds

There is wide variation in climatic parameters in different agro climatic zone of our country. When temperature of bed starts declining below 15^{0}C temperature, in that season, mulching using old jute begs, paddy straw and dry leaves is desirable on structure of beds. When temperature starts increase more than 15^{0}C in vermibeds under closed shed, mulching should be removed from surface of beds. Mulching can also be done. In open area to protect worms from high temperature, check evaporation and protect worms from birds.

7. Protection of earthworms from their predator

Due to 67% protein in earthworms becomes best food for birds, mole, frog, and mongoose and mountain rat. Surroundings of vermi sheds should be properly cleaned so that these predators can not hide. Iron wire net should be used in shed to enter birds inside sheds. Frog should be catch with hand and through away from side. Mountain rat can be killed using rat kill biscuit in shed. Mole can be trapped by keeping a tub in shed with dimension of 30 to 40 cm depth filled with indigenous cow urine. As per behavior, mole attracts toward dirty smelly things in house holds. Indigenous cow urine is having foully smelled. Therefore, mole is attracted towards a cow urine tub where it will fall in the cow urine filled and trapped. A lime of ash around vermished can protect to enter red ants inside the beds. The vermiculture production site is made free from termites. An anti termite treatment is performed at site of the project. The feed of earthworms should not carry any termite. It should be checked properly before feeding to earthworms.

Operation of vermiculture hatcheries and vermibiomanure production unit

For successful operation of these units a young rural youth is trained from Technology Incubator in area of Vermibiotechnology from IVRI, Izatnagar or Krishi Vighyan Kendra of any State Agricultural University. So that he can learn to all improved package of practices of vermiculture and vermibiomanure production by their hands. He can also learn different stage of life cycle of earthworm and identification and breeding of garbage processing earthworms. After training any nationalized bank can sanction loan to establish these units. NABARD is providing 25% subsidy on establishment of vermiculture hatcheries. Approximate Expenditure and Income

1. Fixed cost on capital

Sl. No.	Items	Numbers	Cost (Rs.)
1.	Land for project	60x 40 feet = 2400^2 feet	Own land of enterpreneure
2.	Cost of levelling of land, fencing etc.	-	20,000
3.	Construction of vermished (50 x 32 feet)	1	1,00,000
4.	Vermibeds (20x 3 x 1 feet)	12	30,000
5.	Godown cum labour house (20 x 8 feet)	1	1,00,000
6.	Vermibiomanure sieving machine	1	10,000
7.	Bag sewing machine	1	5000
8.	Electronic digital weighting machine (50 kg capacity)	1	4000
9.	Dunlop water supply pipe	50 meter	6,000
10.	Wheel barrow trolly	2	6000
11.	India marka hand pump motor driven pump	1	30,000
12.	Spade, belcha, tub, plumbing and fitting tools etc	-	4000
13.	Electric fitting	-	6000
14.	Training and consultancy fee	Per candidate	5000
	Total		3,26,000

2. Non recurring cost

1.	Construction cost of vermisheds, vermibeds and fencing depreciation cost (20% per year)	:	Rs. 30,000
2.	Equipment depreciation cost (10% per year)	:	Rs. 17,100
3.	Fixed cost + interest (up to 5 years) per year	:	Rs. 6926 month x 12 = Rs. 83,112
	Total		Rs. 1, 30,212

3. Recurring cost

Sl. No.	Items	Numbers	Cost (Rs.)
1.	Cost of cow dung and solid waste (Rs. 16 per quintal)	1500 quintal	24,000
2.	Salary of one skilled labour per month (Rs. 5000 per month)	12	60,000
3.	Cost of HDPE / Taiwan cloth packing bags	1500	22,500
4.	Vermiculture (Rs. 250 per kg)	200 kg	50,000
5.	Electric city charges	Rs. 500 per month	6000
	Total		1,62,500

Total Expenditure

Total Non recurring cost	:	Rs. 3, 26,000
Total recurring cost	:	Rs. 1, 62,500
Grand Total	:	Rs. 4, 88,500

4. Details of expenditure and income from project Expenditure

Sl. No	Details	First year(Rs)	Second year (Rs)	Third Year (Rs)	Fourth Year (Rs)	Fifth Year (Rs)	Total (Rs)
1.	Non recurring	1,30,212	1,30,212	1,30,212	1,30,212	1,30,212	6,51,060
2.	Recurring cost	1,62,500	1,12,500	1,12,500	1,12,500	1,12,500	6,12,500
3.	Total Expenditure	2,92,712	2,42,712	2,42,712	2,42,712	2,42,712	12,63,560
Income							
4.	Vermi 600 Kg biomanureX Rs 500	3,00,000	3,00,000	3,00,000	3,00,000	3,00,000	15,00,000
5.	Vermi culture 2000kg X Rs 200	4,00,000	4,00,000	4,00,000	4,00,000	4,00,000	20,00,000
	Total income	7,00,000	7,00,000	7,00,000	7,00,000	7,00,000	35,00,000
6.	Net income per years	4,07,288	4,07,288	4,07,288	4,07,288	4,07,288	23,36,440
7.	Net income per months	33,940	38,107	38,107	38,107	38,107	186,360

References

Lynch JM 1992. Substrate availability in the production of composts, Proceedings of the International Composting Research Symposium, H.A.J. Hoitink and H. Keener (Editors), Pp 24-35

Crawford DL 1986. The role of actinomycetes in the decomposition of lignocellulose. FEMS Symposium 34: 715-728

Van Soest PJ 1994. The Nutritional Ecology of the Ruminant, 2nd edition, Cornell University Press Ithaca, N Y. 476 pp

Haug RT 1993. The Practical Handbook of Compost Engineering. Lewis Publishers, Boca Raton, F1. 717 pages.

18

Vaccines: A Boon for the Animals

Ram Niwas, D P Singh and Abed M Al-Bial

In its original concept, vaccination aims to mimic the development of naturally acquired immunity by inoculation of nonpathogenic but still immunogenic components of the pathogen in question, or closely related organisms. The term "vaccine" (from the Latin term "vacca," meaning cow) was first coined by Edward Jenner to describe the inoculation of humans with the cowpox virus to confer protection against the related human smallpox virus and illustrates the close relationship between human and animal infectious disease sciences. The famous proverb saying 'Prevention is better than cure' implies perfectly to the vaccines. Vaccines have proven to be boon for animals which can be regard a potential weapons against infectious diseases caused mainly by bacteria and viruses. The major aims of veterinary vaccines are to improve the health and welfare of companion animals, increase production of livestock in a cost-effective manner, and prevent animal-to-human transmission from both domestic animals and wildlife. Vaccines have long been used to combat infectious disease; however the last decade has witnessed a revolution in the approach to vaccine design and development. Apart from improving animal health and productivity, veterinary vaccines have a significant impact on public health through reductions in the use of veterinary pharmaceuticals and hormones and their residues in the human food chain. No longer is there a need to rely on the laborious classical methods such as attenuation or killing the pathogen. Now sophisticated technologies such as genomics, proteomics, functional genomics, and synthetic chemistry may be used for the rational identification of antigens, where as the synthesis of complex glycans, the generation of engineered carrier proteins, and much more also available for the identification of these antigens. There is a shortage of published work in this field therefore the topic which we have taken is more exciting.

What is a vaccine

A vaccine is pathogenic microorganism which is weakened or killed or its extract that capable to stimulate body immune system causing production of immunoglobulin which is able to clear the pathogen from the body and gives power to recognize it as foreign and finally destroy it.

Pathway of vaccine development

The first era of vaccine was the use of primitive vaccines or crude extracts of the pathogens although these were not safer option. Subsequently development of cell culture technique in 1950_s led to cell culture derived vaccines which can be regarded as the second era of vaccines. These were attenuated and inactivated vaccines and are still the 'Work Horse' in veterinary practices. With the development of recumbent DNA technology there came the third era of vaccines which are used only for human purpose. There are some differences in vaccination practices in human and animals. In case of human quality control strictly regulated by WHO (World Health Organization) and other side there is more relaxation for manufacture of vaccines of veterinary use.

The Objectives of vaccination

There are three basic objectives in vaccination:-

- To provide immunity to the animal or group of animals (active immunity)
- To provide immunity to the offspring of an animal via vaccination of the dam (passive immunity)
- Or to provide immunity to the animal or group of animals and their offspring (active and passive immunity)

Different kinds of vaccine

At present whole virus vaccines, either live or killed, constitute the vast majority of vaccines in the use. However, recent advances in molecular biology had provided alternative methods for producing vaccines. Listed below are the possibilities;-

1. Live whole virus vaccines
2. Killed whole virus vaccines
3. Subunit vaccines;- purified or recombinant viral antigen

4. Recombinant virus vaccines
5. Anti-idiotype antibodies
6. DNA vaccines

1. Live vaccines (attenuated vaccines)

Live virus vaccines are prepared from attenuated strains that are almost or completely devoid of pathogenicity but are capable of inducing a protective immune response. They multiply in the human host and provide continuous antigenic stimulation over a period of time; Primary vaccine failures are uncommon and are usually the result of inadequate storage or administration. Another possibility is interference by related viruses as is suspected in the case of oral polio vaccine in developing countries. Several methods have been used to attenuate viruses for vaccine production.

Use of a related virus from another animal - the earliest example was the use of cowpox to prevent smallpox. The origin of the vaccinia viruses used for production is uncertain.

Administration of pathogenic or partially attenuated virus by an unnatural route - the virulence of the virus is often reduced when administered by an unnatural route. This principle is used in the immunization of military recruits against adult respiratory distress syndrome using enterically coated live adenovirus type 4, 7 and (21).

Passage of the virus in an "unnatural host" or host cell - the major vaccines used in man and animals has all been derived this way. After repeated passages, the virus is administered to the natural host. The initial passages are made in healthy animals or in primary cell cultures. There are several examples of this approach: the 17D strain of yellow fever was developed by passage in mice and then in chick embryos. Polioviruses were passage in monkey kidney cells and measles in chick embryo fibroblasts. Human diploid cells are now widely used such as the WI-38 and MRC-5. The molecular basis for host range mutation is now beginning to be understood.

Development of temperature sensitive mutants - this method may be used in conjunction with the above method.

2. Inactivated whole virus vaccines

These were the easiest preparations to use. The preparation was simply inactivated. The outer virion coat should be left intact but the replicative function should be destroyed. To be effective, non-replicating virus vaccines must contain much more antigen than live vaccines that are able to replicate in the host.

Preparation of killed vaccines may take the route of heat or chemicals. The chemicals used include formaldehyde or beta- propiolactone. The traditional agent for inactivation of the virus is formalin. Excessive treatment can destroy immunogenicity whereas insufficient treatment can leave infectious virus capable of causing disease. Soon after the introduction of inactivated polio vaccine, there was an outbreak of paralytic poliomyelitis in the USA use to the distribution of inadequately inactivated polio vaccine. This incident led to a review of the formalin inactivation procedure and other inactivating agents are now available, such as Beta-propiolactone. Another problem was that SV40 was occasionally found as a contaminant and there were fears of the potential oncogenic nature of the virus.

3. Subunit vaccines

Originally, non-replicating vaccines were derived from crude preparations of virus from animal tissues. As the technology for growing viruses to high titres in cell cultures advanced, it became practicable to purify virus and viral antigens. It is now possible to identify the peptide sites encompassing the major antigenic sites of viral antigens, from which highly purified subunit vaccines can be produced. Increasing purification may lead to loss of immunogenicity, and this may necessitate coupling to an immunogenic carrier protein or adjuvant, such as an aluminum salt. Examples of purified subunit vaccines include the HA vaccines for influenza A and B, and HBsAg derived from the plasma of carriers.

4. Recombinant viral proteins

Virus proteins have been expressed in bacteria, yeast, mammalian cells, and viruses. E. Coli cells were first to be used for this purpose but the expressed proteins were not glycosylated, which was a major drawback since many of the immunogenic proteins of viruses such as the envelope glycoproteins, were glycosylated. Nevertheless, in many instances, it was demonstrated that the non-glycosylated protein backbone was just as immunogenic. Recombinant hepatitis B vaccine is the only recombinant vaccine licensed at present.

An alternative application of recombinant DNA technology is the production of hybrid virus vaccines. The best known example is vaccinia; the DNA sequence coding for the foreign gene is inserted into the plasmid vector along with a vaccinia virus promoter and vaccinia thymidine kinase sequences. The resultant recombination vector is then introduced into cells infected with vaccinia virus to generate a virus that expresses the foreign gene. The recombinant virus vaccine can then multiply in infected cells and produce the antigens of a wide range of viruses. The genes of several viruses can be inserted, so the potential exists for producing polyvalent live vaccines. HBsAg, rabies, HSV and other viruses have been expressed in vaccinia.

Hybrid virus vaccines are stable and stimulate both cellular and humoral immunity. They are relatively cheap and simple to produce. Being live vaccines, smaller quantities are required for immunization. As yet, there are no accepted laboratory markers of attenuation or virulence of vaccinia virus for man. Alterations in the genome of vaccinia virus during the selection of recombinant may alter the virulence of the virus. The use of vaccinia also carries the risk of adverse reactions associated with the vaccine and the virus may spread to susceptible contacts. At present, efforts are being made to attenuate vaccinia virus further and the possibility of using other recombinant vectors is being explored, such as attenuated poliovirus and adenovirus.

5. Synthetic peptides

The development of synthetic peptides that might be useful as vaccines depends on the identification of immunogenic sites. Several methods have been used. The best known example is foot and mouth disease, where protection was achieved by immunizing animals with a linear sequence of 20 amino acids. Synthetic peptide vaccines would have many advantages. Their antigens are precisely defined and free from unnecessary components which may be associated with side effects. They are stable and relatively cheap to manufacture. Furthermore, less quality assurance is required. Changes due to natural variation of the virus can be readily accommodated, which would be a great advantage for unstable viruses such as influenza.

Synthetic peptides do not readily stimulate T cells. It was generally assumed that, because of their small size, peptides would behave like haptens and would therefore require coupling to a protein carrier which is recognized by T-cells. It is now known that synthetic peptides can be highly immunogenic in their free form provided they contain, in addition to the B cell epitope, T- cell epitopes recognized by T-helper cells. Such T-cell epitopes can be provided by carrier protein molecules, foreign antigens or within the synthetic peptide molecule itself.

Synthetic peptides are not applicable to all viruses. This approach did not work in the case of polioviruses because the important antigenic sites were made up of 2 or more different viral capsid proteins so that it was in a concise 3-D conformation.

Advantages of defined viral antigens or peptides include

1. Production and quality control simpler
2. No NA or other viral or external proteins, therefore less toxic.
3. Safer in cases where viruses are oncogenic or establish a persistent infection

4. Feasible even if virus cannot be cultivated

Disadvantages

1. May be less immunogenic than conventional inactivated whole-virus vaccines
2. Requires adjuvant
3. Requires primary course of injections followed by boosters
4. Fails to elicit CMI.

6. Anti-idiotype antibodies

The ability of anti-idiotype antibodies to mimic foreign antigens has led to their development as vaccines to induce immunity against viruses, bacteria and protozoa in experimental animals. Anti-idiotypes have many potential uses as viral vaccines, particularly when the antigen is difficult to grow or hazardous. They have been used to induce immunity against a wide range of viruses, including HBV, rabies, Newcastle disease virus and FeLV, reoviruses and polioviruses.

7. DNA vaccines

Recently, encouraging results were reported for DNA vaccines whereby DNA coding for the foreign antigen is directly injected into the animal so that the foreign antigen is directly produced by the host cells. In theory these vaccines would be extremely safe and devoid of side effects since the foreign antigens would be directly produced by the host animal. In addition, DNA is relatively inexpensive and easier to produce than conventional vaccines and thus this technology may one day increase the availability of vaccines to developing countries. Moreover, the time for development is relatively short which may enable timely immunization against emerging infectious diseases. In addition, DNA vaccines can theoretically result in more long-term production of an antigenic protein when introduced into a relatively nondividing tissue, such as muscle.

Indeed some observers have already dubbed the new technology the "third revolution" in vaccine development—on par with Pasteur's ground-breaking work with whole organisms and the development of subunit vaccines. The first clinical trials using injections of DNA to stimulate an immune response against a foreign protein began for HIV in 1995. Four other clinical trials using DNA vaccines against influenza, herpes simplex virus, T-cell lymphoma, and an additional trial for HIV were started in 1996.

The technique that is being tested in humans involves the direct injection of plasmids - loops of DNA that contain genes for proteins produced by the organism being targeted for immunity. Once injected into the host's muscle tissue, the DNA is taken up by host cells, which then start expressing the foreign protein. The protein serves as an antigen that stimulates an immune responses and protective immunological memory.

The differences/advantages/limitation of different kind of vaccines are summarized in table

Parameters	Attenuated vaccines (live vaccines)	Inactivated vaccines (killed vaccines)	DNA vaccines
Route of administration	I/M, S/C, oral, nasal	I/M, S/C	S/S, I/M
Required dose	Low	High	Nil
Booster dose	No	Multiple	Multiple
Adjuvant in formulation	Not required	Required	Under Research
Duration of Immunity	Longer	Relatively shorter	Longer
Antibody Responses	IgG, IgA	IgG	IgG
Cell Mediated Immune (CMI) response	Yes	No	No
Cold chain requirement	Yes	No	No

Because live vaccines replicate inside host cells, bits of virus antigen are presented to the cell surface and recognized by cytotoxic cells.

Potential safety problems

Live vaccines

1. Underattenuation
2. Mutation leading to reversion to virulence
3. Preparation instability
4. Contaminating viruses in cultured cells
5. Heat lability
6. Should not be given to immunocompromized or pregnant patients

Killed vaccines

1. Incomplete inactivation
2. Increased risk of allergic reactions due to large amounts of antigen involved

Methods of administration

The routes of vaccine administration will depend on various factors including the easiest method of introducing the antigen to a particular species, the labour involvement, the type of immunity required (i.e. circulating or local [mucosal]), whether the vaccine is live or dead, any adjuvants present that will alter the duration of exposure. The ease of administration is a very important practical consideration of their usefulness. At one time most inactivated vaccines were given by injection. However it is now possible to administer some of them by mouth via the food or water, as a spray or through the skin.

Injections

Most injected vaccines are given either subcutaneously (sc) or intramuscularly (im). The former will usually remain at site longer than the latter. They are the usual form of application in ruminants and produce mainly IgG. Intraperitoneal (ip) injections are the most common routes of vaccination in fish which are sedated or anaesthetised to aid the procedure.

Intra-dermal injections

This method of application is unusual and is mainly used in contagious pustular dermatitis (orf) in sheep. The application probably results in cell-mediated immunity. A modification of this technique for fowlpox is the wing web puncture or feather follicles.

Oral

A common route of application and it is used in particular with parasitic infections such as lungworm in cattle and coccidiosis in poultry. Many live viral poultry vaccines are placed in the water.

Aerosol (nebulisation or fogging)

Mainly used for mass poultry vaccination of young birds. The vaccines tend to be live and viral. They are quickly absorbed and produce both local and circulating immunity. In poultry, coarse sprayer cabinets are used in hatcheries as well as a range of coarse sprayers on farms. Finer, more uniform, sprays are also used to produce aerosol or controlled droplet administration (CDA).

Intranasal

This is undertaken for respiratory diseases in cattle. This produces a relatively high IgA level. The vaccines are live viral. Some can be used to vaccinate animals against viruses in the face of disease. This is because of the rapid formation of non-specific interferon which assists in viral destruction and then the production of local IgA within a few days.

Eye drop

Sometimes used on young birds with a live viral vaccine, they are rapidly absorbed and allow both local and systemic immunity to develop.

Water application

This provides vaccination for fish using their environment to distribute the vaccine. This is the equivalent to aerosol vaccination in poultry. However drinking water application is commonly used in poultry where the target organ is the gut or, less frequently, the respiratory system.

Feed application

Occasionally vaccines are used in the feed of pigs (not currently in the UK), poultry and sometimes fish. The main problems are ensuring adequate stability of the vaccine and its even distribution and uptake by the animals. In poultry administration is by spray onto the feed and is usually applied by a mechanical sprayer.

Egg (In-ova) injection

Some vaccines are injected into the egg on the transfer into the hatchery to allow the chick shortly after birth to have some specific immunity to the infection.

Vaccination schedule for various kinds of livestock species are summarized in the table

Factors affecting the vaccine efficiency

Usually modern vaccines, properly administered, will achieve a high level of protection\regardless of the husbandry system in which the animal is kept.

There are a few animal diseases, usually inherited, in which the animal is unable to make a proper immune response. These are very uncommon and are usually self-limiting because, if the animal is born alive, it will soon die on exposure to infection. However in animals of all ages there may be reasons for the ability of the animal to make immunity (immunocompetence) being reduced (immunosuppressed) or taking longer to develop (immunocompromised) or absent (immunodeficient).

Besides the above, vaccination will be unlikely to work if the natural infection has entered the animal before the vaccine. Equally vaccination may not stop the infection or disease when challenge is very heavy or the animal's health is compromised in other ways. Reasons for producing a poor immunity to an infection can include:-

Vaccination schedule for various kinds of livestock species are summarized in the table

S N	Disease	Animal	Vaccine	Dose	Immunity	Time of vaccination
1.	Foot & MouthDiseases (FMD)	All clovenfooted animals	Polyvalent FMD vaccine	3 ml. S/C	1 Year	February & December
2.	HemorrhagicSepticemia (HS)	Cattle, Buffalo	HS Vaccine	5 ml S/C	6 month & 1Year	May-June
3.	Black Quarter(BQ)	Cattle, Buffalo	BQ Vaccine	5 ml S/C	6 month & 1Year	May-June
4.	Anthrax	All species ofanimals	Anthrax SporeVaccine	1 ml S/C	1 Year	May-June
5.	Enterotoxemia(ET)	Sheep & Goat	ET Vaccine	5 ml S/C	1 Year	May-June
6.	ContagiousCaprine Pleuro Pneumonia (CCPP)	Sheep & Goat	IVRI Vaccine	0.2 ml S/C	1 Year	-
7.	Peste Des PettisRuminants (PPR)	Sheep & Goat	PPR Vaccine	1 ml S/C	3 Year	-
8.	Brucella	Female cattle &buffalo Calf age4-8 monthsonly	BrucellaVaccine	2 ml S/C	1 Year	-
9.	Theileriosis	Cattle & calvesabove 2months of age	TheileriaVaccine	3 ml S/C	1 Year	-
10.	Rabies	All species ofanimals	Rabies PostBite Vaccine	1 ml S/C	1 Year 0, 3,7,14,28 & 90days	-

Note - Before any vaccination deworming should be compulsory to get better results.

- **The infection itself** – a number of infections suppress immunity and these particularly include some viruses and mycoplasma.
- **Stress** – this comes in many different forms for an animal
- **Nutrition** – an inadequate overall diet may suppress immunity
- **Deficiency in individual feed ingredients** – especially proteins, vitamin A and E, selenium, etc.
- **The production status of the animal** – immune response is often less in the mother close to birth
- **Age of animal** – very young or old is often imunosuppressed
- **Other diseases** – especially cancer, metabolic diseases (e.g. pregnancy toxaemia in sheep, fatty liver in cattle)
- **Vaccination process** – insufficient vaccine given, out-of-date, stored incorrectly, not administered by correct route, period too long from making up vaccine to its administration.

Handling vaccines and their storage

Vaccines, particularly live ones, need to be treated with respect when handling, storing and using them. It is important to read and follow all instructions on the SPC and product literature to ensure that the vaccines will produce their maximum activity. Many live vaccines need to be kept under a specific storage environment that may involve keeping them away from light and in cool conditions. It is important to regularly check that a refrigerator is working at the required temperature. Often once the vaccine is made up it will only be viable for a short period and again still needs to be kept in the conditions stated on the SPC. Only sufficient vaccine should be made up as can be used in the stated viability period. Other vaccine that is required to be made up later must still be kept under the required storage conditions to ensure its efficacy is maintained. Dead vaccines usually require two doses and these should be administered at the interval indicated by the manufacturer. When live or dead vaccines need boosting then this should be done at the intervals stated by the manufacturers.

Benefits of vaccination

Vaccines and vaccination strategies should form an integral part of all farm/ herd health planning. Some diseases that are prevalent or a constant threat should be considered in detail in such a plan as how best to control or prevent their occurrence. In such instances it would be helpful to show the cost-benefit of such control measures to that particular farm or herd/flock. Other diseases that are less common or are of less severity should also be included in the plans in case these occur in the area or on the farm. The criteria for controlling (including vaccination strategies) the entry and/or spread of disease on that farm should be detailed. Ideally the risk assessment and a rough cost benefit for the strategy should be included and agreed in the proposals.

In all cases where disease is likely to occur, vaccination will produce benefit to healthy animals. Usually this is in the form of preventing signs of illness when disease occurs on the farm or by reducing the severity or losses associated with the disease. Any lack of disease is taken for granted, and as such it is often hard to impress on producers the cost benefit of disease control and vaccination. Whilst the effects of disease may be all too apparent when there has been an outbreak which is then controlled or removed by vaccination, memories tend to be short. Then the use of vaccination in the control strategy may be challenged or might be discontinued.

Where disease has not apparently occurred, the financial advantages are much less easy to determine on the individual farm or to convince the farmer about, unless a detailed examination is carried out of the on-farm benefits and they are compared with the costs of disease when it occurs. Thus to most farmers the costs of calf pneumonia outbreaks appear only to be due to the result of veterinary involvement and the treatments provided. However it can be shown that the costs of such inputs and involvement in a bovine respiratory disease outbreak in suckler calves as well as those born in the dairy herd and reared for beef or heifer replacements are only about 40 % of the total costs of a disease outbreak. The other costs can include increase in management inputs, mortality, labour, growth reductions and delayed deaths. They are hidden or not costed but on all farms they provide the majority of costs.

In sheep, the costs of outbreaks of pasteurellosis have been shown to result in a cost benefit in favour of vaccination of over 7:1. The problem with this disease is that outbreaks are sporadic and variable in extent. Thus often one or a few dead sheep are accepted as part of the normal farming process. This means that in individual farm problems, because of the wide variation in incidence, mortality and related disease problems in each group of sheep, the benefit over the cost of clostridial/pasteurellosis vaccination varied from 2:1 to about 40:1. Often the relationships between one disease and others not necessarily related are not recognised by those dealing with the problem.

Perhaps one of the clearest advantages of vaccination was seen in the 1990s following the introduction of pig enzootic pneumonia (EP) vaccination into herds where the disease was present. In these herds, once the vaccination regime was introduced, there was often a reduction in age of animals reaching slaughter weight of about a fortnight. Conversely removal of EP vaccination resulted in the time to slaughter lengthening by an almost equal amount. In salmon farming, the introduction of the oil-based bacterial vaccines around the early 1990s dramatically reduced the amount of antibiotics to a fraction of that previously used.

Demerits/Side effects from vaccination

Considering the many millions of doses of vaccine sold annually and used in farm animals, adverse side effects are very rare. The types of reaction will vary according to the vaccine used. However it is usually live vaccines that can make the animal show a temperature rise with associated reduction or cessation of feeding, dullness and reduced milk production, if lactating. With dead vaccines the most common side effect is local and is a reaction around the site of vaccination especially if there is an adjuvant present to delay the antigen absorption. However reactions can occur in vaccines without the presence of adjuvants. It is for this reason that most vaccines which are injected will indicate the site for their administration. Usually it will be advised that the vaccine, particularly if subcutaneous, should be introduced into an area of the animal not used for human consumption such as behind the animal's ear or in the area of the chest wall behind the elbow. Then if there is any residual vaccine left or any reaction to it, there will be no involvement of an edible part of the carcase. Occasionally a relatively large volume of a cold vaccine into a small or young animal has resulted in cold shock.

Conclusion

Vaccinology has become a recognized science thatcombines disciplines of immunology, microbiology, protein chemistry, and molecular biology with practical considerations of production costs, regulatory affairs, and commercial returns. The ultimate aim of any new vaccine is to provide a product that will be used to protect animals and humans against disease. More recently, vaccines have also found applications in animal production andreproduction processes. Veterinary vaccines have already made enormous impacts not only on animal health, welfare, and production but also on human health. A continuous interchange between animal and human disease control agencies and scientists will be essential to be prepared for the ever-present threat of new, emerging diseases.

References

Bagley Clell V 2001. Vaccination program for beef calves. Animal Health Factsheet Beef 40, Utah State University Extension.

Beard CW, Brugh M 1975. Immunity to Newcastle Disease. Am J Vet Res, 36(4):509-512.

Beard CW, Easterday BC 1967. The influence of the route of administration of Newcastle disease virus on host response. J Infect Dis, 117:55-61.

Carter PB 2003. The current state of veterinary vaccines: is there hope for the future? J Vet Med Ed 30:152-154.

Davis BS, Chang GJ, Cropp B, Roehrig JT, Martin DA, Mitchell CJ, Bowen R, Day MJ 2006. Vaccine side effects: fact and fiction. Vet. Microbiol. 117:51-58.[PubMed].

Delves PJ, Lund T, Roitt IM 2002. Antifertility vaccines Trends Immunol. 23:213-219. [PubMed]

Lincoln SD 2005. Cattle vaccines and their use. Beef Cattle Handbook, from Cattle Producers Library. BCH-3015: 1-4.

Bunning ML 2001. West Nile virus recombinant DNA vaccine protects mouse and horse from virus challenge and expresses in vitro a noninfectious recombinant antigen that can be used in enzyme-linked immunosorbent assays. J Virol 75:4040-4047. [PMC free article] [PubMed]

Mason HS, Haq TA, Clements JD, Arntzen CJ, Vaccine 1998. 16 1336–1343. Hein MB, Yeo TC, Wang F, Sturtevant A, Ann NY Acad Sci, 1996, 792: 51–56.

http://www.ncbi.nlm.nih.gov/pmc/articles/PMC1932753

University of Nebraska LincolnExtensionPublications.2007.http://www.ianrpubs.unl.edu/pages/publicationD.jsp?publicationId=437

19

Ethnoveterinary Management of Livestock Diseases by Inhabitants of Central Himalaya

Lalit Tiwari, Nitin Rai, Rajeev Kumar Sharma and P C Pande

Introduction

Harshberger first used the term ethnobotany in 1895-96, for the plants used by the local tribes and people, for their basic needs and requirements. They also used plant, plant parts and plant products for animal healthcare and other practices. This indigenous or traditional animal health care practice is called ethnoveterinary medicine. Ethnoveterinary medicine is defined as a holistic comprehension of the indigenous systems of animal health, their interpretation through western medicine and the development of effective and appropriate technologies (McCorkle 1986). McCorkle *et al.* (1989) define Ethnoveterinary medicine (EVM) as dealing with the folk beliefs, knowledge, skills, methods and practices pertaining to the health care of animals. McCorkle *et al.* (1996) give a description of ethnoveterinary medicinal research as the holistic interdisciplinary study of the local knowledge and the socio-cultural structures and environment associated with animal health care and husbandry.

EVM consists of local people's knowledge pertaining to animal health and production. According to Mathias (1996 & 2002), it includes:

- **Information:** Recognizing sick animals, knowing which plants have medicinal properties and how to collect and prepare them.
- **Practices:** Plants, plant parts, plant products, animal parts, animal products, minerals and metals are used against animal diseases.

- **Tools and technologies:** Bamboo syringes, acupuncture needles, storage of animal feed.
- **Beliefs:** Feeding blessed food, protecting animals through amulets, magico religious practices.
- **Experimentation:** Trying different medicines, cross-breeding.
- **Human resources:** Specialists such as healers and bonesetters.
- **Breeds:** Numerous local breeds, which are the result of centuries of selection.
- **Materials:** Stones used as tourniquets to stop bleeding.

Indian History of Veterinary Medicines

The earliest information on the art of caring for animals in India is provided by the sacred texts of the Vedic religion. These texts, written between 1500 and 600BC were also transmitted orally from generation to generation, over the centuries, before being written down. The oldest Vedic literature consists of collections of hymns, liturgical chants and sacrificial or magical formulae, mainly in verses, which constitute the *Veda* proper (1500 to 1000BC) (Mazars 1998). Towards the end the Vedic period, Indian medicine began to adopt observation and rational procedure, which developed into a coherent system known as *Aryurveda*. This knowledge served as a model for veterinary medicine. The oldest existing veterinary text from India is a treatise entitled *Asvayurvedasiddhanta* (complete *Ayurvedic* system for horses; probably earlier than the 10th century AD) attributed to a certain Salihotra. The same author apparently wrote two books on horses: *Asvalakanasastra* (Treatise on the Marks of Horses) and *Asvaprasamsa* (in Praise of Horses). Subsequently, various treatises on horses and diseases of horses were composed. Among the best-known works of this specialist literature are *Asvacikitsita* (Therapeutics of horses), written by Nakula around the year 1000AD and *Asvavaidyaka* (Medicine of horses) by Jayadatta, which probably originated in the 13th century (Mazars 1998).

Methodology

For the documentation of ethnoveterinary knowledge, personal interview method was adopted. Females are keener observers of veterinary diseases, due to their association with milking and have knowledge about problems related to lactation, milk letdown, milk quality, etc. During the time of interview, traditional healers were hesitant to disclose the information about local treatments. There might be two possible reasons for this: (1) the knowledge is widely valued and considered confidential by traditional healers, (2) the healers thought that this information could be used for my own commercial benefits. Extensive field survey of Uttarakhand Himalaya was made during the 2004 to 2008 to collect the desired information. Published available literature related to ethno-veterinary medicines were also consulted (Gaur *et al.* 1992, Samal *et al.* 2002 & 2003, Tiwari and Pande 2004, Bisht *et al.* 2004, Tiwari and Pande 2005, pande *et al.* 2006, Tiwari and Pande 2006, Tiwari and Pande 2006a, Tiwari and Pande 2006b, Shah *et al.* 2007, Tiwari *et al.* 2007, Pande *et al.* 2007, Shah *et al.* 2008, Tiwari and Pande 2009, Tiwari and Pande 2010, Tiwari and Pande 2011 and Tiwari *et al* 2011). During the survey data were gathered from the knowledgeable person who practices and had experience about animal husbandry and veterinary medicines. The information was further verified by cross checking from different aged men and women.

Animal Husbandry in Central Himalaya

The Central Himalayan Region covers the new state of Uttarakhand (embodying the Kumaon and Garhwal), which came into existence on November 9, 2000 as the 27th state of India is bounded by China (Tibet) on the north, Nepal on the east, Uttar Pradesh on the south and Himanchal Pradesh on the north-west and lies between 28° 53' 24" and 31° 27' 50" N latitude and between 77° 34' 27" and 81° 02' 22" E longitudes. It has an area of about 53, 483 sq. km. supports a human population of 70.51 lakhs About 71 per cent of this population lives in rural areas in 15,117 villages (*Uttaranchal Forest Statistical* 200).

Farmers in Uttarakhand Himalaya generally face paucity of resource and women largely manage agricultural operations this is why livestock play a crucial role in hilly economy. More than 70% population of the area depends totally on animals for their basic needs, requirements and economic sources. The livestock composition is consists of 48.17% cattle, 29.70% buffaloes and 21.93% of goats. Cattle population is higher because of large numbers of indigenous bull (37% of cattle) reared for cultivation purposes (Singh *et al.* 2004).

In Uttarakhand Himalaya, livestock feeding depends on pastures and grasslands. The grassland/pasture lands are intermixed with big or small patches of open forest. These lands are used for grazing of cattle throughout the year. The common fodders used by the villagers are wild grasses, tree leaves, leguminous plants, nuts, rice, wheat straw and other agricultural by-products. Villagers use traditional methods for the preservation of these fodder plants. Fodder grasses are allowed to grow in areas protected from grazing during the rainy season (from mid June to mid September). These protected grasses are cut, spread and sun-dried. Thereafter small bundles are made and stored on roofs, open fields, wastelands, caves, cowshed and trees etc. in the form of *luta*. Though the value of indigenous knowledge of breeding practices has been largely ignored, the people of higher altitude still believe in it. The most remarkable breeding policy that in used by the Bhotiya tribes is the breeding of yaks with cows; the hybrid product is called *jubbu* (male) and *jumma* (female). Another technique adopted is the breeding between horses and donkey for production of mule, which, seems to be very useful for transportation purposes (Tiwari and Pande 2006a).

Even after the lapse of 11 year of the formation of the state of Uttarkhand, sufficient veterinary infrastructure has not been developed. The area comprises of approximately 82 blocks and 15366 villages but has only 248 veterinary hospitals and 506 veterinary dispensaries. The total population in the year 2003 of cattle was 1936163, of goats were 1092418 and of poultry were 922937 cocks and hens (Annual Report 2006-2007, Department of Animal Husbandry, Dairying & Fisheries, Ministry of Agriculture, Government of India). These figures indicate that only 745 veterinary hospitals and dispensaries in all are available for 3451518 domestic animals. The most of the hospitals are situated in district headquarters or in blocks and facilities are only utilized by roadside villages. Veterinary hospitals or dispensaries are far away from remote villages. This shows that the government veterinary infrastructure is insufficient to provide the required service. This is one of the reasons for the scope of EVM in Uttarakhand being very wide. Due to lack of Government veterinary infrastructure, people use their own traditional treatments and formulations to treat animal diseases. Here another cause of the popularity of EVMs is that, the modern allopathic medicines are costly and poor animal owners cannot afford them (Tiwari and Pande, 2011).

Himalayan EVM

Diseases are basic problems for the people and animals alike. They have always fought with diversified types of disease since prehistoric periods. Veterinary healthcare infrastructure is very poor in Uttarakhand; people use their own traditional treatments against these animal diseases. They use in the vicinity herbs, minerals, animal products and other natural resources as well as magico-religious treatments (Table 1). That is why a high level of traditional animal health care knowledge is distributed unevenly through Himalayan societies. Villagers, especially women are traditionally considered to take a subsidiary role in the care of animals. They are responsible for milking the animals and they are often the first to notice signs of ill health in animals. They thus have a greater knowledge about animal nature, behavior, demeanour, diseases, disorders and their treatments. In some remote areas of Uttarakhand Himalaya many traditional healers are present who have great-undocumented traditional knowledge about animal diseases, herbal treatments, herbal formulations, etc. The persons who prescribe these medicines are (Tiwari and Pande 2010).

1. The housewife of the animal owner who prescribes the household remedies
2. The old person of the village or traditional healers (*vaidyas*)

Table1: Ethnoveterinary plants of Central Himalaya.

S.No.	Vernacular Name	Botanical	Family	Part/Product used	Diseases and disorders
1.	Aalu	*Solanumtuberosum*	Solanaceae	Tuber	Burn
2.	Aaru	*Prunus persica*	Rosaceae	Leaf	Skin disease, tympany, external parasites, FMD
3.	Adrak	*Zingiber officinale*	Zingiberaceae	Rhizome	Hoof diseases, constipation, tympany
4.	Ajwain	*Trachyspermumammi*	Apiaceae	Seed	Remove sterility, diarrhoea, galactagogue, food poisoning
5.	Akhrot	*Juglans regia*	Juglandaceae	Fruit	Eye diseases, broken horns, snout blisters, bone fracture
6.	Atis	*Aconitumheterophyllum*	Ranunculaceae	Root	Stomachache, internal parasites
7.	Banj	*Quercus leucotrichophora*	Fagaceae	Leaf	Glactagogue
8.	Bel	*Aegle marmelos*	Rutaceae	Leaf	External parasites
9.	Bethuwa	*Chenopodium album*	Chenopodiaceae	Leaf	FMD, haematuria, cuts, wounds
10.	Bhang	*Cannabis sativa*	Cannabaceae	Resin	Stomachache, internal injury, dysentery
11.	Bhatt	*Glycine max*	Fabaceae	Seed	Skin disease, diarrhoea, glactagogue
12.	Bhimal	*Grewia optiva*	Tiliaceae	Leaf	Galactagogue, constipation, skin diseases
13.	Boja	*Acorus calamus*	Araceae	Rhizome	Hoof infection, external parasites
14.	Chana	*Cicer arietinum*	Fabaceae	Seed	Skin disease, glactagogue
15.	Chir	*Pinus roxburghii*	Pinaceae	Resin	Bone fracture, external parasites, FMD
16.	Choru	*Angelica glauca*	Apiaceae	Root	Dysentery, stomachache, constipation, tympany
17.	Deodar	*Cedrus deodara*	Pinaceae	Oil	Eczema, hoof disease
18.	Dhan	*Oryza sativa*	Poaceae	Husk	Remove sterility, retention of placenta
19.	Dhania	*Coriandrum sativum*	Apiaceae	Seed	Tympany, diarrhoea, food poisoning
20.	Dub	*Cynodon dactylon*	Poaceae	Whole plant	Skin disease, internal injury, food poisoning
21.	Ganna	*Saccharum officinarum*	Poaceae	Jaggery	Fever, pimple, haemorrhagic septicaemia
22.	Gehun	*Triticum aestivum*	Poaceae	Grain	Strength, diarrhoea, skin diseases, food poisoning, glactagogue
23.	Ghingharu	*Pyracantha crenulata*	Rosaceae	Leaf	Snout blisters, burn
24.	Gurja	*Tinospora cordifolia*	Menispermaceae	Stem	Diarrhoea, strength, skin disease, haematuria, mastitis, galactagogue
25.	Hajari	*Tagetes erecta*	Asteraceae	Leaf	Wounds, mastitis, broken horn

Contd.

S.No.	Vernacular Name	Botanical	Family	Part/Product used	Diseases and disorders
26.	Haldi	*Curcuma domestica*	Zingiberaceae	Rhizome	Mouth blisters, sprains, internal parasites, FMD
27.	Heeng	*Ferula asafetida*	Apiaceae	Resin	Neck sore, tympany
28.	Jau	*Hordeum vulgare*	Poaceae	Flour	Haematuria, remove sterility, post- calving care
29.	Kela	*Musa paradisiaca*	Musaceae	Plant juice	Glactagogue, tympany, diarrhea, haematuria
30.	Keruwa	*Asparagus racemosus*	Liliaceae	Tuber	Glactagogue, wounds, haemorrhagic septicaemia
31.	Khumani	*Prunus armeniaca*	Rosaceae	Fruit	Gastric troubles, strength
32.	Kilmora	*Berberis aristata*	Berberidaceae	Root	Eye diseases, wounds
33.	Lahsan	*Allium sativum*	Alliaceae	Clave	Food poisoning, tympany, rheumatism
34.	Madira	*Echinochloa frumentacea*	Poaceae	Seed	Glactagogue, diarrhoea
35.	Makka	*Zea mays*	Poaceae	Flour	Food poisoning
36.	Marsha	*Amaranthus caudatus*	Amaranthaceae	Seed	Dysentery, haemachuria, glactagogue
37.	Masoor	*Lens culinaris*	Fabaceae	Seed	Bone fracture, haematuria
38.	Mehal	*Pyrus pashia*	Rosaceae	Fruit	Eye diseases
39.	Mircha	*Capsicum annuum*	Solanaceae	Fruit	Hoof infection, internal parasites, skin disease, dog bite
40.	Mooli	*Raphanus sativus*	Brassicaceae	Root	Food poisoning, tympany, dysentery
41.	Moong	*Vigna radiata*	Fabaceae	Seed	Post-calving care, internal injury
42.	Neem	*Azadirachta indica*	Meliaceae	Leaf	Burn, tympany, broken horns,FMD
43.	Pati	*Artemisia nilagirica*	Asteraceae	Leaf	Diarrhea, sun strokes
44.	Paudina	*Mentha arvensis*	Lamiaceae	Leaf	Tympany, constipation, fever, diarrhoea
45.	Piyaj	*Allium cepa*	Alliaceae	Bulb	Food poisoning, skin diseases, hoof diseases
46.	Ram-bansh	*Agave americana*	Agavaceae	Leaf	Broken horns, cough
47.	Sanuf	*Foeniculum vulgare*	Apiaceae	Seed	Diarrhoea, stomachache
48.	Sarson	*Brassica campestris*	Brassicaceae	Seed	Constipation, dog bite, food poisoning
49.	Sisona	*Urtica dioica*	Urticaceae	Twig	Bone fracture, burn, internal injury
50.	Tulasi	*Ocimum tenuiflorum*	Lamiaceae	Leaf	Constipation, wounds
51.	Ugal	*Fagopyrum*	Polygonaceae	Seed	Glactagogue *esculentum*

The mainly common animal diseases and disorders in Uttarakhand are haemorrhagic septicaemia (*galghontu, panjebadh, dodhjeebh, panjebhi*), mastitis (*thanela, thand, ninod, dudaro*), foot and mouth diseases (*kurani, kurpaka, khuriya, kuriyal*), haematuria (*lu gaud, khoon gaud, sitlagu, loyetaro*), contagious exthyma (*mauni, muyal, minaki, fisana, atela*), ecto- and endo-parasitic infection, eczema (*kuras*), indigestion (*apach, pisyaan*), tympanitis (*afaara, tilaksi, vatana, baghi, pet foolana*), diarrhoea (*chheru lagana, chheruwa, chirani, chiraka*), dysentery (*cheru lagana, cheruk*), constipation (*kabj, pet sukhana, ajeer*), bone fracture, broken horn, wounds, etc. Villagers treat these diseases and disorders thorough the traditional treatments. For the treatment, they use natural resources like plants and plants products, minerals, metals, chemicals, animal products, etc. However, this important veterinary knowledge is in danger of extinction due to rapid modernization.

Himalayan Medicinal Wealth

Nature has bestowed Uttarakhand Himalaya with rich forests that are made up of thousands of species of tress, shrubs, herbs and climbers. These combine to form forest type which varies from the sub-tropical forests in the lower hills to the dense temperate forests and the dry vegetation on the upper Darma, Niti valley of Uttarakhand Himalaya.

The people of the Uttarakhand Himalayas are a racial mixture of various tribes. Garhwalies and Kumaonies are the principal communities of the Uttarakhand. Whereas, the Bhotias, the Rajis, the Buxas and the Jaunsaries are the important tribal communities that inhabit the Uttarakhand Himalayan region. Among them Bhotias, Rajis, Buxas and Jaunsaries are prominent but a few others are also of the similar category. The people of Gangi village (Buderas), the people of Rawain area (Rawanltas) of Uttarakshi district and Gujars are also tribal.

The Indian Himalayan region alone supports about 18,440 species of plants (Angiosperms: 8000 spp., Gymnosperms: 44 spp., Pteridophytes: 600 spp., Bryophytes: 1736 spp., Lichens: 1159 spp. and Fungi: 6900 spp.) of which about 45% have medicinal properties (Singh and Hazra 1997). According to Samant *et al.* (1998) out of the total species of vascular plants, 1748 spp. has medicinal properties. Pande *et al.* (2006) documented total 1338 ethnomedicinal plants and 364 EVM plants from Uttarakhand Himalayan region.

Conclusion

Present chapter clearly shows that Indian Himalaya has great wealth of ethnoveterinary knowledge. But in the process of modernization this undocumented knowledge is vanishing very rapidly. Actually this information survived by being passed from one generation to next but now a days young generation is not taking interest in animal husbandry practices. Due to this apathy the valuable knowledge about EVM is disappearing. The Government should recognize EVM and set up enabling legal frameworks that facilitate the use and integration of EVM and healers into animal health care provision and consider the joint delivery of human and animal health care in pastoral areas. Documentation of plants and there uses will be very useful to inventing new pharmaceutical research in the field of veterinary medicines.

References

Annual Report 2006-2007. Department of Animal Husbandry, Dairying and Fisheries, Ministry of Agriculture, Government of India.

Bisht N, Pande PC, Tiwari L 2004. Ethno-veterinary practices of Kapkot block of Bageshwar district, Uttaranchal. Asian Agri-History 8(4): 209-314.

Gaur RD, Bhatt KC, Tiwari JK 1992. An ethnobotanical study of Uttar Pradesh Himalaya in relation to veterinary medicines. J Indian bot Soc. 72:139-144.

Mathias Evelyn 1996. How can ethnoveterinary medicine be used in field projects? Indigenous Knowledge and Development Monitor. *http://www.nuffic.nl/ciran/ikdm/4-2/articles/mathias.html*

Mathias Evelyn 2001. Introducing ethnoveterinary medicines. *http://www.ethnovetweb.com*

Mazars G 1998. Veterinary medicines in ancient and medieval India. Studies in History of Medicine and Science 16(1-2): 27-36.

McCorkle CM 1986. An introduction to ethnoveterinary research and development. Journal of Ethnobiology 6(1): 129-149.

McCorkle CM 1989. Veterinary Anthropology. Human Organization, pp: 156-162.

McCorkle CM, Mathias-Mundy E, Schillhorn Van Veen TW 1996. Ethnoveterinary Research and Development. IT Publications, South Africa, pp: 338.

Pande PC, Tiwari L, Pande HC 2006. Folk-Medicine and Aromatic Plants of Uttaranchal. Bishen Singh and Mahendra Pal Singh, Dehra Dun.

Pande PC, Tiwari L, Pande HC 2006. Folk Medicine and Aromatic Plants of Uttaranchal Himalaya. Bishen Singh Mahendra Pal Singh, Dehradun, India.

Pande PC, Tiwari L, Pande HC 2007. Ethnoveterinary plants of Uttaranchal-a review. Indian J Traditional Knowledge 6(3): 444-458.

Samal PK, Shah A, Tewari SC, Agarwal DK 2002. Indiginous animal health care practices and their relevance to bioresources conservation in Indian Central Himalaya. International J Sustain Dev World Ecol 8: 167-178.

Samal PK, Shah A, Tiwari S, Mehara B, Agrawal DK 2003. Indigenous animal health care practices in India Central Himalaya: empirical evidences. Indian J Traditional Knowledge 2(1): 40-50.

Samant SS, Dhar U, Palni LMS 1998. Medicinal Plants of Indian Himalaya: Diversity, Distribution, Potential Values. Gyanodaya Prakashan, Nainital.

Shah R, Pande PC, Tiwari L 2007. Folklore biomedicine for some veterinary diseases and disorders in Western part of Almora District. In: Traditional Knowledge Systems and Archaeology: with Special Reference to Uttarakhand. Agrawal DP, Shah M, Jamal S (eds). Aryan Books International, New Delhi, pp 34-45.

Shah R, Pande PC, Tiwari L 2008. Traditional veterinary herbal medicines of western part of Almora district, Uttarakhand Himalaya. Indian J Traditional Knowledge 7(2): 355-359.

Singh DK, Hajra PK 1997. Floristic diversity. In: Biodiversity Status in the Himalaya. Gujral (ed.). British Council, New Delhi, pp. 23-38.

Singh PR, Singh M, Jaiswal RS, Raikwar RK 2004. Feeding of existing concentrate mixture to dairy animals in the Kumaon hills of Uttaranchal. Indian J Animal Research 38(2):147-149.

Tiwari L, Pande PC 2004. Traditional veterinary practices in south-eastern part of Chamoli district, Uttaranchal. Indian J Traditional Knowledge 3(4): 397-406.

Tiwari L, Pande PC 2005. Traditional Veterinary Medicinal Plants of Bhilangana Valley of Tehari district, Uttaranchal Himalaya. Asian Agri History 9(3): 253-262.

Tiwari L, Pande PC 2006. Ethno-veterinary medicines of holy Doonagiri hills of Uttaranchal. J Econ Taxon Bot 30: 151-162.

Tiwari L, Pande PC 2006a. Indigenous veterinary practices of Darma valley of Pithoragarh district, Uttaranchal Himalaya. Indian J Traditional Knowledge 5(2): 201-206.

Tiwari L, Pande PC 2006b. Ethnoveterinary plants of Uttarkashi district, Uttaranchal, India. Ethnobotany 18(1&2): 139-144.

Tiwari L, Pande PC 2009. Ethnoveterinary Plants of Johar Valley of Pithoragarh district, Uttaranchal Himalaya. VEGETOS 22(1): 55-62.

Tiwari L, Pande PC 2010. Ethnoveterinary Medicines in Indian Perspective: Reference to Uttarakhand Himalaya. Indian J Traditional Knowledge 9(3): 611-617.

Tiwari L, Pande PC 2011. Traditional animal husbandry practices of Central Himalaya. Asian Agri-History 15(4): 263-282.

Tiwari L, PC Pande 2010. Himalayan medicine system and its Ayurvedic perspectives. Asian Agri-History 14(3): 261-272.

Tiwari L, Sharma RK, Amminuddin, Pande PC 2011. Notes on therapeutic potential of Unani medicinal plants in ethnoveterinary medicine. Hippocratic J Unani medicine 6(4):131-144.

Tiwari D, Pande PC, Tiwari L 2007. Ethno-veterinary herbal medicines of Dwarahaat area of Uttaranchal Himalaya. Indian Forester 133: 379-390.

20

Quality Standards of Medicinal Plants with Special Reference to Regulatory Aspects

Nitin Rai, Lalit Tiwari and R K Sharma

The quality of medicinal plants is paramount in respect of its therapeutic value. The therapeutic value of a plant solely depends on its genuineness in respect of its identity and optimum quantity of therapeutically active phyto-constituents. The medicinal plants are resourced from wild and cultivated sources. The major sources of medicinal plants are wild sources viz. forest areas, open land, non-cultivated sources etc. The medicinal plants collected from wild sources remain questionable for their quality especially when they have been procured from trade channels owing to fair chances of adulteration, substitution and inappropriate storage condition which lead to deterioration in quality. All the medicinal plants which are available in dried form in the trade are termed as crude drugs and are always subject to quality check in a laboratory on the basis of pharmacognostical, physico-chemical, phyto-chemical, microbiological and other analytical specifications. These specifications are available in pharmacopoeial monographs under regulatory frame work. The pharmacopoeial standards are experimentally established and are also approved standards incorporated in a pharmacopoeia. A pharmacopoeia is a book of standards applicable to drug substance and their dosage forms of common usage in a country. Pharmacopoeia is prepared by recognized authority appointed by the Government of a particular country. It is an essential requirement of a regulatory framework in the sphere of Drugs and Cosmetics Act 1940 and Rules there under (Anonymous 1940). Pharmacopoeial standards are experimentally established and approved standards.

This chapter provides details on Pharmacopeial standards on crude drugs/ medicinal plants used in Ayurvedic, Siddha and Unani systems of medicine. All the Pharmacopoeial standards published in respect of Ayurvedic, Unani and Siddha drugs so far, are elaborated in foregoing text for their brief historical background and contents.

Pharmacopoel publication on quality standards

Ayurvedic system of medicine

The crude drugs incorporated in Ayurvedic Pharmacopeia are ingredients of classical formulations incorporated in Ayurvedic Formulary of India, Table 1 (Anonymous, 1978 and 2000). The published volumes of Ayurvedic Pharmacopeia of India comprises monographs on 519 single plant drugs, 21 single metal and minerals drugs, 15 aqueous extracts, 15 hydro alcoholic extracts and 152 classical formulations. There are also two companion volumes on TLC atlas and macroscopic and microscopy atlas for the monographs of Ayurvedic Pharmacopoeia of India, Pt. I, Vol. I and V respectively. The details of published monographs in different volumes of Ayurvedic Pharmacopeia are given in Table 2.

Table 1: Single drugs used in preparation of classical formulations included in ayurvedic formulary of India, part I & II

Sl. No.	Origin	Nos. of drugs in part-I	Nos. of drugs in part-II
1.	Plant Origin Single Drugs	351	271

Table 2. Ayurvedic pharmacopoeial publications

Sl. No.	Pharma copoeia cation	Year of Publi	Speci fications	No. of Monographs
1.	The Ayurvedic Pharmacopeia of India, Pt. I, Vol. I	1986	Single plant origin drugs	80
2.	The Ayurvedic Pharmacopeia of India, Pt. I, Vol. II	1999 b	Single plant origin drugs	78
3.	The Ayurvedic Pharmacopeia of India, Pt. I, Vol. III	2001 b	Single plant origin drugs	100
4.	The Ayurvedic Pharmacopeia of India, Pt. I, Vol. IV	2004	Single plant origin drugs	68
5.	The Ayurvedic Pharmacopeia of India, Pt. I, Vol. V	2006 c	Single plant origin drugs	92
6.	The AyurvedicPharmacopeia of India, Pt. I, Vol. VI	2008 e	Single plant origin drugs	101
7.	The AyurvedicPharmacopeia of India,Pt. I, Vol. VIII (Aqueous & Hydro alcoholic extracts)	2008 g	Aqueous & hydroalcohalic extracts drugs	60
8.	TLC Atlas of Ayurvedic Pharmacopoeial drugs Pt. I, Vol. I	2009 b	TLC profile of single plant drugs	80
9.	Macroscopy and Microscopy Atlas of API Drugs Pt.- Vol. V	2009 c	Macroscopy and Microscopic profile of single plant drugs	92

Unani System of Medicine

Plant drugs used as ingredients in the formulations described in these formularies are subjected for pharmacopoeial standardisation (Table 3). Six volumes of single Unani Pharmacopoeia of India Part-I, comprising 298 monographs on single plant drugs have been published (Anonymous 1998a, 2007a, b, c, 2008 j, 2009d, e). Unani Pharmacopoeia of India, Part II, comprises 100 monographs on classical Unani formulations (Anonymous 2009e, 2010d) (Table 4).

Table 3: Single drugs used in preparation of classical formulations included in national formulary of unani medicine, part I to V

Sl. No.	Origin	Nos. of Drugs in Part-I	Nos. of Drugs in Part-II	Nos. of Drugs in Part-III	Nos. of Drugs in Part-IV	Nos. of Drugs in Part-IV
1.	Plant Origin Single Drugs	342	251	181	339	342

Table 4: Unani pharmacopoeial publications

Sl. No	Pharmacopoeia	Year of Publication	Specifications	No. of Monograohs
1.	The Unani Pharm acopeia of India, Pt. I, Vol.I	1998 a	Single origin plant origin drugs	45
2.	The Unani Pharm acopeia of India, Pt. I, Vol.II	2007 a	Single plant origin drugs	50
3.	The Unani Pharmacopeia of India, Pt. I, Vol.III	2007 b	Single plant origin drugs	53
4.	The Unani Pharm acopeia of India, Pt. I, Vol.IV	2007 c	Single plant origin drugs	50
5.	The Unani Pharmacopeia of India, Pt. I, Vol.V	2008 j	Single plant origin drugs	52
6.	The Unani Pharmacopeia of India, Pt. I, Vol.VI	2009 d	Single plant origin drugs	48

Siddha system of medicine

Siddha system of medicine has two volumes of pharmacopoeia providing 139 monographs on single plant drugs (Table 6). Siddha formularies have two parts covering 402 classical compound formulations of system. The Siddha Formulary of India, Part-I is available in Tamil (Anonymous 1984) and English version (Anonymous 1992) whereas Part – II is published in Tamil version only (Anonymous 2011). The format of the monographs on single drugs (plant origin) is identical to The Ayurvedic Pharmacopoeia of India and The Unani Pharmacopeia of India. The selection of single drugs is made from the two published Siddha Formularies (Table 5), which comprises siddha classical formulations (Anonymous 1992, 2011).

Table 5: Single drugs used in preparation of classical formulations included in siddha formulary of India, part I & II

Sl. No.	Origin	Nos. of drugs in part-I	Nos. of drugs in part-II
1.	Plant Origin Single Drugs	311	287

Table 6: Siddha pharmacopoeial publications

Sl. No.	Pharmacopoeia	Year	Specifications of Publication	No. of Monogrphs
1.	The Siddha Pharmacopeia of India, Pt. I, Vol.I	2008 i	Single plant origin drugs	73
2.	The Siddha Pharmacopeia of India, Pt. I, Vol.II	2010 c	Single plant origin drugs	66

Besides these Ayurvedic, Unani and Siddha pharmacopoeia, Indian Pharmacopoeia and Homoeopathic Pharmacopoeia of India also comprises monographs used in modern and homoeopathic medicines. All the medicinal plants/crude drugs having quality specification in the pharmacopoeias of Ayurveda, Unani and Siddha are enumerated in (Table 9).

Table 7: Medicinal plants specification in the Ayurvedic, Unani and Siddha Pharmacopoeia

S. No.	Medicinal plant species	Morphological part as crude drug	APIRef.	UPIRef.	SPIRef.
1.	*Abies webbiana* Lindl.	Dried leaf	Vol. IV		
2.	*Abrus precatorius* Linn.	Root	Vol. II	Vol. IV	
3.	*Abrus precatorius* Linn.	Seed	Vol. I	Vol. I	
4.	*Abutilon indicum* (Linn.) Sw.	Root	Vol. I		
5.	*Acacia catechu* Willd.	Heart wood	Vol. I		
6.	*Acacia leucophloea* Willd.	Stem bark	Vol. II	Vol. VI	
7.	*Acacia nilotica* (Linn.) Willd. ex Del. Sp. *indica* (Benth.)	Stem bark	Vol. I		
8.	*Acacia nilotica* (Linn.) Willd. ex Del. Sp. *indica* (Benth.)	Gum	Vol. VI		
9.	*Acacia pennata* (L.) Willd.	Leaf	Vol. VI		
10.	*Acacia suma* Buch. -Ham	Heart wood	Vol. V		
11.	*Acalypha fruticosa* Forsk.	Root	Vol. VI	Vol. II	
12.	*Acalypha indica* L.	Whole Plant	Vol. VI	Vol. II	
13.	*Achyranthes aspera* Linn.	Root	Vol. III	Vol. IV	
14.	*Achyranthes aspera* Linn.	Whole plant	Vol. II	Vol. I	
15.	*Aconitum chasmanthum* Stapf	Root	Vol. II	Vol. IV	
16.	*Aconitum heterophyllum* Wall. ex Royle	Root	Vol. I	Vol. I	Vol. I
17.	*Acorus calamus* Linn.	Rhizome	Vol. II	Vol. V	
18.	*Adhatoda vasica* Nees		Leaf	Vol. I	Vol. VI
19.	*Adhatoda zeylanica* Medic.	Root	Vol. IV		
20.	*Adiantum cappillus- veneris* L	Whole Plant	Vol. VI		

Contd.

S. No.	Medicinal plant species	Morphological part as crude drug	APIRef.	UPIRef.	SPIRef.
21.	*Adiantum lunulatum* Burm.	Whole Plant	Vol. III		
22.	*Aegle marmelos* Corr.ex Roxb.	Fruit pulp	Vol. I	Vol. I	
23.	*Aegle marmelos* Corr.ex Roxb.	Root	Vol. III		Vol. I
24.	*Aegle marmelos* Corr.ex Roxb.	Stem bark	Vol. IV		
25.	*Aerva lanata* Juss. ex Schult	Whole Plant	Vol. V		Vol. I
26.	*Agaricus alba* Linn.	Fruit		Vol. VI	
27.	*Ailanthus excelsa* Roxb.	Stem bark	Vol. III		
28.	*Alangium salviifolium* (Linn.f) Wang.	Leaf	Vol. V	Vol. V	
29.	*Albizia lebbek* Benth.	Stem bark	Vol. III		
30.	*Alhagi pseudalhagi* (Bieb.) Desv.	Whole Plant	Vol. II	Vol. VI	
31.	*Allium cepa* Linn.	Fresh bulb		Vol. III	
32.	*Allium cepa* Linn.	Fruit		Vol. III	
33.	*Allium sativum* Linn.	Bulb	Vol. III	Vol. V	Vol. II
34.	*Aloe barbadensis* Mill.	Leaf	Vol. I	Vol. I	
35.	*Alpinia calcarata* Rosc.	Rhizome	Vol. VI	Vol. II	
36.	*Alpinia galanga* Willd.	Rhizome	Vol. V	Vol. II	Vol. I
37.	*Alstonia scholaris* R.Br.	Stem bark	Vol. I		
38.	*Alternanthera sessilis* DC.	Whole Plant	Vol. II		Vol. I
39.	*Althaea officinalis* Linn.	Root	Vol. V	Vol. V	
40.	*Althaea officinalis* Linn.	Seed	Vol. V	Vol. V	
41.	*Amaranthus tricolor* Linn.	Whole Plant	Vol. III		
42.	*Amomum subulatum* Roxb.	Seed	Vol. II	Vol. IV	
43.	*Amomum subulatum* Roxb.	Fruit	Vol. VI		
44.	*Amorphophallus campanulatus* Blume ex	Decne	Corm	Vol. III	
45.	*Anacyclus pyrethrum* DC	Root	Vol. II	Vol. II	Vol. II
46.	*Andrographis paniculata* Ness.	Stem & Leaf			
47.	*Anethum gravelones* Linn.	Fruit		Vol.V	
48.	*Anethum sowa* Kurz	Fruit	Vol. II		
49.	*Anethum sowa* Kurz	Seed		Vol. VI	
50.	*Angelica archangelica* Linn. var. *himalaica* (C.B. Clarke) Krishna Badhawar	Root	Vol. V		
51.	*Angelica glauca* Edgew	Root	Vol. V		
52.	*Anisomeles malabarica* R.Br.	Whole Plant	Vol. VI		
53.	*Anogeissus latifolia* Wall ex. Bedd.	Stem bark	Vol. VI		
54.	*Anogeissus latifolia* Wall.	Fruit	Vol. VI		
55.	*Anthocephalus cadamba* Miq.	Stem bark	Vol. II		
56.	*Apium graveolens* L.	Whole plant	Vol. VI		
57.	*Apium graveolens* L.	Seed		Vol. II	
58.	*Apium leptophyllum* (Pers.) F.V.M. ex Benth.	Fruit	Vol. I		
59.	*Aquilaria agallocha* Roxb.	Heart wood	Vol. IV	Vol. VI	
60.	*Aquilaria malaccensis* (Lam.)	Heart wood			Vol. II
61.	*Areca catechu* Linn.	Seed	Vol. I	Vol. I	
62.	*Argyreia nervosa* (Burm.f.) Boj. syn. *Argyreia speciosa* Sweet	Root	Vol. V		
63.	*Aristolochia bracteata* Retz	Leaf	Vol. VI		Vol. II

Contd.

S. No.	Medicinal plant species	Morphological part as crude drug	APIRef.	UPIRef.	SPIRef.
64.	*Aristolochia indica* Linn.	Root	Vol. III	Vol. V	
65.	*Artemisia absinthium* Linn.	Whole plant	Vol. VI		
66.	*Artemisia absinthium* Linn.	Stem		Vol. II	
67.	*Artocarpus heterophyllus* Lamk	Root bark	Vol. VI		
68.	*Asarum europaeum* Linn.	Rhizome	Vol. VI	Vol. V, VI	
69.	*Asparagus officinalis* Linn.	Seed		Vol. VI	
70.	*Asparagus officinalis* Linn.	Root	Vol. VI		
71.	*Asparagus raecemosus* Willd.	Root	Vol. IV	Vol. VI	Vol. II
72.	*Asphodelus tenuifolius* Cav.	Leaf		Vol. III	
73.	*Asphodelus tenuifolius* Cav.	Seed		Vol. VI	
74.	*Asteracantha longifolia* Nees	Root	Vol. II		
75.	*Asteracantha longifolia* Nees	Seed	Vol. II	Vol. III	
76.	*Asteracantha longifolia* Nees	Whole plant	Vol. II		
77.	*Azadirachta indica* A.Juss.	Flower	Vol. V	Vol. V	Vol. I
78.	*Azadirachta indica* A.Juss.	Fruit	Vol. V	Vol. V	Vol. I
79.	*Azadirachta indica* A.Juss.	Leaf	Vol. II	Vol. II., IV	Vol. II
80.	*Azadirachta indica* A.Juss.	Root bark	Vol. V	Vol. V	
81.	*Azadirachta indica* A.Juss.	Seed		Vol. III	
82.	*Azadirachta indica* A.Juss	Bark		Vol. III	
83.	*Azadirachta indica* A.Juss.	Stem bark	Vol. II	Vol. IV	Vol. I
84.	*Bacopa monnieri* (Linn.) Penn.	Whole plant	Vol. II	Vol. IV	Vol. I
85.	*Baliospermum montanum* Muell.-Arg.	Root	Vol. III		
86.	*Balsamodendron caudata* Mauch	Leaf	Vol. VI	Vol. II	
87.	*Barleria prionitis* Linn.	Whole plant	Vol. III		
88.	*Barleria strigosa* Willd.	Root	Vol. V		
89.	*Barringtonia acutangula* Gaertn.	Fruit	Vol. III	Vol. VI	
90.	*Bauhinia racemosa* Lamk.	Flower	Vol. VI		
91.	*Bauhinia racemosa* Lamk.	Stem bark		Vol. II	
92.	*Bauhinia variegata* Linn.	Stem bark	Vol. I		
93.	*Benincasa hispida* (Thunb.) Cogn.	Fruit	Vol. IV		Vol. II
94.	*Benincasa hispida* (Thunb.) Cogn.	Seed		Vol. III	
95.	*Berberis aristata* DC	Stem	Vol. II	Vol. IV	Vol. I
96.	*Berberis aristata* DC.	Fruit	Vol. VI		
97.	*Berberis aristata* DC.	Root extract		Vol. II	
98.	*Bergenia ciliata* Sternb.	Rhizome	Vol. I		
99.	*Betula utilis* D.Don.	Stem bark	Vol. V	Vol. V	
100.	*Blepharis persica* (Burm.f.) O. Kuntze.	Seed	Vol. IV		
101.	*Boerhaavia diffusa* Linn.	Root	Vol. III		
102.	*Boerhaavia diffusa* Linn.	Whole plant	Vol. I		Vol. I
103.	*Boerhaavia verticillata* Poiret	Root	Vol. V		
104.	*Bombax ceiba* Linn.	Stem bark	Vol. III	Vol. V	
105.	*Borago officinalis* Linn.	Leaf		Vol. II	
106.	*Borago officinalis* Linn.	Flower		Vol. VI	
107.	*Borassus flabellifer* Linn.	inflorescence	Vol. III		
108.	*Boswellia serrata* Roxb.	Exude	Vol. IV		
109.	*Brassica campestris* Linn.	Seed	Vol. III	Vol. V	

Contd.

S. No.	Medicinal plant species	Morphological part as crude drug	APIRef.	UPIRef.	SPIRef.
110.	*Brassica campestris* Linn.	Oil	Vol. VI		
111.	*Brassica rapa* Linn.	Seed & Tuber		Vol. III	
112.	*Buchanania lanzan* Spreng.	Seed	Vol. II	Vol. IV	
113.	*Buchanania lanzan* Spreng.	Stem bark	Vol. IV		
114.	*Butea monosperma* (Lam.) Kuntze	Stem bark	Vol. II	Vol. V	
115.	*Butea monosperma* (Lam.) Kuntze.	Flower	Vol. IV	Vol. V	
116.	*Butea monosperma* (Lam.) Kuntze.	Gum	Vol. IV	Vol. VI	
117.	*Butea monosperma* (Lam.) Kuntze.	Seed	Vol. IV	Vol.VI	Vol. II
118.	*Butea monosperma* Taub. syn. B. forndosa	Seed	Vol. V	Vol. II	
119.	*Butea monosperma* Taub. syn. B. forndosa	Flower	Vol. V	Vol. V	Vol. II
120.	*Caesalpinia bonducella* Flem.	Seed	Vol. V		
121.	*Caesalpinia bonduc* (Linn.) Roxb.	Seed		Vol. V	
122.	*Caesalpinia sappan* Linn.	Heart wood	Vol. IV		
123.	*Cajanus cajan* (Linn.) Millsp.	Root	Vol. III		
124.	*Cajanus cajan* (Linn.) Millsp.	Seed	Vol. IV		
125.	*Calamus rotang L.*	Rhizome	Vol. VI		Vol. II
126.	*Calamus thwaitesii* Becc.	Rhizome	Vol. VI		
127.	*Callicarpa macrophylla* Vahl.	Fruit	Vol. IV		
128.	*Callicarpa macrophylla* Vahl.	inflorescence	Vol. II		
129.	*Calotropis procera* (Ait.) R. Br.	Leaf	Vol. I	Vol. I	
130.	*Calotropis procera* (Ait.) R. Br.	Root	Vol. I		
131.	*Calotropis procera* (Ait.) R. Br.	Flower		Vol. II	
132.	*Calotropis procera* (Ait.) R. Br.	Root bark		Vol. IV	
133.	*Calotropis procera* (Ait.) R.Br.	Stem bark	Vol. III		
134.	*Calycopteris floribunda* Lam.	Leaf	Vol. V		
135.	*Calycopteris floribunda* Lam.	Root	Vol. V		
136.	*Calycopteris floribunda* Lam.	Stem	Vol. V		
137.	*Cannabis sativa* Linn.	Leaf	Vol. I	Vol. I	Vol. I
138.	*Capparis sepiaria* Linn.	Fruit	Vol. V		
139.	*Capparis spinosa* Linn.	Root	Vol. V	Vol. V	
140.	*Cardiospermum halicacabum* Linn.	Root	Vol. V		
141.	*Cardiospermum halicacabum* Linn.	Seed	Vol. V	Vol. V	
142.	*Careya arborea* Roxb.	Seed	Vol. V	Vol. V	
143.	*Carica papaya* Linn.	Fruit	Vol. VI		
144.	*Carica papaya* Linn.	Root	Vol. VI		
145.	*Carissa carandas* Linn.	Root	Vol. III		
146.	*Carissa carandas* Linn.	Stem bark	Vol. II		
147.	*Carthamus tinctorius* L.	Flower	Vol. VI	Vol. III	
148.	*Carthamus tinctorius* Linn.	Fruit	Vol. VI		
149.	*Carthamus tinctorius* Linn.	Leaf	Vol. VI		
150.	*Carum carvi* Linn.	Fruit	Vol. I	Vol. I	
151.	*Cassia absus* Linn.	Seed		Vol. II	
152.	*Cassia angustifolia* Vahl.	Leaf	Vol. I	Vol. I	
153.	*Cassia angustifolia* Vahl.	Pod			
154.	*Cassia fistula* Linn.	Root	Vol. I		
155.	*Cassia fistula* Linn.	Fruit pulp		Vol. I	

Contd.

S. No.	Medicinal plant species	Morphological part as crude drug	APIRef.	UPIRef.	SPIRef.
156.	*Cassia seena* Linn.	Leaf			Vol. II
157.	*Cassia fistula* Linn.	Stem bark	Vol. V		Vol. II
158.	*Cassia fistula* Linn.	Fruit pulp			Vol. I
159.	*Cassia tora* Linn.	Seed	Vol. III	Vol. II	
160.	*Cedrelia toona* Roxb.	Stem bark	Vol. V		
161.	*Cedrus deodara* (Roxb.) Loud.	Heart wood	Vol. IV	Vol. VI	Vol. II
162.	*Celastrus paniculatus* Willd.	Seed	Vol. II	Vol. IV	Vol. I
163.	*Celosia argentea* Linn.	Seed	Vol. VI		
164.	*Centaurea behan* Linn.	Root		Vol. III	
165.	*Centella asiatica* (Linn.) Urban	Whole plant	Vol. IV		
166.	*Centipeda minima* Linn.	Whole plant		Vol. III	
167.	*Centratherum anthelminticum* Kuntze		Fruit	Vol. V	
168.	*Ceratonia siliqua* Linn.	Fruit		Vol. VI	
169.	*Chlorophytum arundinaceum* Baker	Root		Vol. II	
170.	*Chrysanthemum indicum* L.	Leaf	Vol. VI		
171.	*Cicca acida* (Linn.) Merrillsyn. *Phyllanthus distichus* Muell-Arg.	Fruit	Vol. V		
172.	*Cicer arietinum* Linn.	Whole plant	Vol. VI		
173.	*Cinchona officinalis* Linn.	Bark		Vol. III	
174.	*Cichorium intybus* Linn.	Seed		Vol. VI	
175.	*Cinnamomum cassia* Blume.	Leaf		Vol. III	
176.	*Cinnamomum cassia* Blume.	Stem bark		Vol. III	
177.	*Cinnamomum tamala* Nees & Eberm.	Leaf	Vol. I	Vol. I	Vol. I
178.	*Cinnamomum tamala* Nees & Eberm.	Stem bark		Vol. III	
179.	*Cinnamomum zeylanicum* Blume	Oil	Vol. VI		
180.	*Cinnamomum zeylanicum* Blume	Stem bark		Vol. I	
181.	*Cinnamomum zeylanicum* Breyn	Bark	Vol. I		
182.	*Cissampelos pareira* Linn.	Root	Vol. I		
183.	*Cissus quadrangularis* Linn.	Aerial Part	Vol. VI		Vol. II
184.	*Cissus quadrangularis* Linn.	Stem	Vol. III		
185.	*Citrullus colocynthis* Schrad.	Fruit	Vol. III		Vol. I
186.	*Citrullus colocynthis* Schrad.	Fruit pulp		Vol. VI	
187.	*Citrullus colocynthis* Schrad.	Leaf	Vol. II		
188.	*Citrullus colocynthis* Schrad.	Root	Vol. II	Vol. II, IV	
189.	*Citrus limon* (Linn.) Burm.f.	Fruit	Vol. IV		
190.	*Citrus medica* Linn.	Fruit	Vol. III		
191.	*Citrus medica* Linn.	Pericarp		Vol. III	
192.	*Citrus reticulate* Blanco.	Rind fruit		Vol. III	
193.	*Cleome gynandra* Linn.	Seed	Vol. I		
194.	*Clerodendrum phlomidis* Linn.	Root	Vol. III	Vol. VI	
195.	*Clerodendrum serratum* (Linn.) Moon	Root	Vol. III		Vol II
196.	*Clitoria ternatea* Linn.	Leaf	Vol. IV		
197.	*Clitoria ternatea* Linn.	Root	Vol. II		Vol. I
198.	*Coccinia grandis* (Linn.) Voigt	Leaf	Vol. VI		Vol. II
199.	*Coccinia grandis* (Linn.) Voigt	Stem	Vol. VI		Vol. II
200.	*Coccinia indica* W. & A.	Whole plant	Vol. III		

Contd.

S. No.	Medicinal plant species	Morphological part as crude drug	APIRef.	UPIRef.	SPIRef.
201.	*Cochlospermum relogiosum* Linn.	Gum		Vol. VI	
202.	*Cocos nucifera* Linn.	Endosperm	Vol. III		
203.	*Coix lacrymajobi* Linn.	Root	Vol. V		
204.	*Coldenia procumbens* Linn.	Whole plant	Vol. VI		Vol. II
205.	*Coleus amboinicus* Lour.	Leaf	Vol. VI		
206.	*Coleus forskohlii* Briq.	Root	Vol. V		
207.	*Coleus forskohlii* Briq.	Extract	Vol VIII		
208.	*Commiphora myrrha* (Nees) Engl.	Gum-resin		Vol. II	
209.	*Commiphora wightii* (Arn.) Bhand.	Exudate	Vol. I	Vol. I	
210.	*Convolvulus pluricaulis* Choisy	Whole plant	Vol. II		
211.	*Corallocarpus epigaeus* Benth.ex Hook. f.	Root	Vol. VI		Vol. II
212.	*Cordia dichotoma* Forst f.	Fruit	Vol. VI		
213.	*Cordia dichotoma* Forst f.	Stem bark	Vol. VI		
214.	*Coriandrum sativum* Linn.	Fruit	Vol. I	Vol. I	Vol. I
215.	*Coscinium fenestratum* (Gaertn.) Colebr.	Root	Vol. V		
216.	*Coscinium fenestratum* (Gaertn.) Colebr.	Stem	Vol. V		
217.	*Costus speciosus* (Koeni) Sm.	Rhizome	Vol. V		
218.	*Corylus avellana* Linn.	Fruit		Vol. III	
219.	*Crateva magna* (Lour.) DC.	Stem bark			Vol. I
220.	*Crateva nurvala* Buch.-Ham.	Stem bark	Vol. I		
221.	*Crocus sativus* Linn.	Style & Stigma	Vol. IV	Vol. VI	Vol. II
222.	*Crotolaria juncea* Linn.	Seed	Vol. III		
223.	*Croton tiglium* Linn.	Seed	Vol. II	Vol. IV	Vol. I
224.	*Cryptolepis buchanani* Roem. & Schult.	Root	Vol. IV		
225.	*Cucumis melo* Linn var. *utilissimus* Duthie & Fuller	Seed	Vol. II	Vol. III, IV	
226.	*Cucumis sativus* Linn.	Seed	Vol. V	Vol. V	Vol. II
227.	*Cucurbita moschata* Duch. ex. Lam.	Kernel		Vol. VI	
228.	*Cuminum cyminum* Linn.	Fruit	Vol. I		Vol. I
229.	*Curculigo orchioides* Gaertn.	Rhizome	Vol. IV		Vol. II
230.	*Curcuma amada* Roxb.	Rhizome	Vol. V	Vol. V	
231.	*Curcuma longa* Linn.	Rhizome	Vol. I	Vol. I	Vol. I
232.	*Curcuma zedoaria* Rosc.	Rhizome	Vol. IV		Vol. II
233.	*Cuscuta reflexa* Linn.	Seed		Vol. II	
234.	*Cuscuta reflexa* Roxb.	Whole plant		Vol. III	
235.	*Cyamopsis tetragonolobus* Linn.	Gum	Vol. VIII		
236.	*Cymbopogon citrates* (DC) Stapf.	Whole plant	Vol. V		
237.	*Cymbopogon martini* (Roxb.) Wats	Whole plant	Vol. V	Vol. V	
238.	*Cynodon dactylon* (Linn.) Pers.	Root	Vol. III	Vol. IV	
239.	*Cynodon dactylon* (Linn.) Pers.	Whole plant	Vol. IV		
240.	*Cydonia oblonga* Mill.	Seed		Vol. II	
241.	*Cyperus rotundus* Linn.	Rhizome	Vol. III	Vol. V	
242.	*Dalbergia sissoo* Roxb.	Heart wood	Vol. III	Vol. V	
243.	*Dalbergia sissoo* Roxb.	Stem bark	Vol. III		
244.	*Datura metel* Linn.	Seed	Vol. III	Vol. II, IV	
245.	*Datura metel* Linn.	Whole plant	Vol. IV		

Contd.

S. No.	Medicinal plant species	Morphological part as crude drug	APIRef.	UPIRef.	SPIRef.
246.	*Daucus carota* Linn.	Seed		Vol. VI	
247.	*Daucus carota* Linn.	Root		Vol. II	
248.	*Delphinium denudatum* Wall.	Root		Vol. VI	
249.	*Dendrophthoe falcata* (Linn.f.) Etting	Aerial root	Vol. V		
250.	*Dendrophthoe falcata* (Linn.f.) Etting	Flower	Vol. V		
251.	*Dendrophthoe falcata* (Linn.f.) Etting	Fruit	Vol. V		
252.	*Dendrophthoe falcata* (Linn.f.) Etting	Leaf	Vol. V		
253.	*Dendrophthoe falcata* (Linn.f.) Etting	Stem	Vol. V		
254.	*Desmodium gangeticum* DC	Root	Vol. III		
255.	*Desmodium gangeticum* DC.	Whole plant	Vol. VI		
256.	*Desmostachya bipinnata* Stapf	Root stock	Vol. III		
257.	*Dioscorea bulbifera* Linn.	Rhizome	Vol. IV		
258.	*Diospyros peregrina* Gurke	Fruit	Vol. VI		
259.	*Diospyros tomentosa* Roxb.	Stem bark	Vol. V		
260.	*Dorema ammoniacum* Don.	Oleogum resin		Vol. III	
261.	*Doronicum hookeri* C.B.Clarke	Rhizome	Vol. VI		
262.	*Dracaena cinnabari* Balf. f.	Exudate	Vol. VI		
263.	*Dryopteris filix-mas* (Linn.) Schott	Rhizome	Vol. VI		
264.	*Eclipta alba* (Linn.) Hassk.	Whole plant	Vol. II	Vol. IV	
265.	*Eclipta prostrata* (Linn.) Hassk.	Whole plant			Vol. II
266.	*Elaeocarpus sphaericus* (Gaertn). K.Schum	Seed	Vol. IV		Vol. II
267.	*Elettaria cardamomum* Maton	Fruit	Vol. I	Vol. I	Vol. II
268.	*Eleusine coracana* Gaertn.	Root	Vol. V		
269.	*Embelia ribes* Burm.f.	Fruit	Vol. I	Vol. I	Vol. I
270.	*Emblica officinalis* Gaertn.	Fruit	Vol. I	Vol. I	
271.	*Emblica officinalis* Gaertn.	Fruit plup	Vol. I		
272.	*Enicostemma axillare* (Lam.) A. Raynal	Whole plant	Vol. VI		Vol. II
273.	*Erythrina indica* Lam.	Stem bark	Vol. II		
274.	*Eucalyptus globulus* Labill	Oil	Vol. VI		
275.	*Eucalyuptus globulus* Labill.	Leaf	Vol. V		
276.	*Euphorbia hirta* Linn.	Whole plant	Vol. VI		
277.	*Euphorbia lanzan* Spreng.	Whole plant	Vol. II		
278.	*Euphorbia neriifolia* Linn.	Stem	Vol. I		
279.	*Euphorbia nivulia* Buch.- Ham.	Latex	Vol. VI		
280.	*Euphorbia prostrata* W. Ait.	Whole plant		Vol. V	
281.	*Euphorbia royleana* Bioss.	Latex		Vol. III	
282.	*Euphorbia royleana* Bioss.	Leaf		Vol. III	
283.	*Euphorbia thymifolia* Linn.	Whole plant	Vol. V		
284.	*Fagonia cretica* Linn.	Whole plant	Vol. V	Vol. V	
285.	*Feronia limonia* (Linn.) Swingle	Fruit plup	Vol. II		
286.	*Ferula foetida* Regel.	Oleo-Gum-Resin	Vol. I	Vol. I	Vol. I
287.	*Ferula jaeschkeana* Vatke	Leaf	Vol. V		
288.	*Ficus arnottiana* Miq.	Rhizome	Vol. V		
289.	*Ficus bengalensis* Linn.	Arial root	Vol. IV	Vol. VI	
290.	*Ficus bengalensis* Linn.	Stem bark	Vol. I		
291.	*Ficus carica* Linn.	Fruit		Vol. II	

Contd.

S. No.	Medicinal plant species	Morphological part as crude drug	APIRef.	UPIRef.	SPIRef.
292.	*Ficus glomerata* Roxb.	Fruit	Vol. III		
293.	*Ficus hispida* Linn.f.	Stem bark		Vol. VI	
294.	*Ficus hispida* Linn.f.	Fruit	Vol. III		
295.	*Ficus hispida* Linn.f.	Root	Vol. III	Vol. IV	
296.	*Ficus lacor* Buch.-Ham	Stem bark	Vol. II	Vol. VI	
297.	*Ficus lacor* Buch.-Ham.	Fruit	Vol. IV		
298.	*Ficus racemosa* Linn.	Bark	Vol. I	Vol. I	Vol. I
299.	*Ficus religiosa* Linn.	Bark	Vol. I		
300.	*Flacourtia indica* (Burm.f.) Merr.	Fruit	Vol. VI		
301.	*Flacourtia indica* Merr.	Leaf	Vol. IV		
302.	*Flacourtia indica* Merr.	Stem bark	Vol. IV		
303.	*Foeniculum vulgare* Mill.	Fruit	Vol. I	Vol. I	Vol. I
304.	*Fritillaria roylei* Hook	Blub	Vol. V		
305.	*Fumaria parviflora* Lam.	Whole plant	Vol. IV	Vol. VI	
306.	*Garcinia pedunculata* Roxb.	Fruit rind	Vol. VI		
307.	*Garcinia pedunculata* Roxb.	Fruit & Extract			
308.	*Gardenia gummifera* Linn. f.	Exudate	Vol. VI		
309.	*Gaultheria fragrantissima* Wall.	Leaf	Vol. VI		
310.	*Gentiana kurroo* Royle.	Rhizome	Vol. VI		
311.	*Gentiana olivieri* Griseb.	Flower		Vol. III	
312.	*Gisekia phamaceoides* Linn.	Leaf	Vol. VI		
313.	*Glinus lotoides* Linn.	Whole plant	Vol. VI		Vol. II
314.	*Gloriosa superba* Linn.	Tuber	Vol. III		
315.	*Glycyrrhiza glabra* Linn.	Stem & Root	Vol. I		Vol. I
316.	*Glycyrrhiza glabra* Linn.	Root		Vol. I	
317.	*Gmelina arborea* Roxb.	Fruit	Vol. II		
318.	*Gmelina arborea* Roxb.	Root bark	Vol. I		
319.	*Gmelina arborea* Roxb.	Stem	Vol. III		
320.	*Gmelina arborea* Roxb.	Stem bark	Vol. IV		
321.	*Gossypium herbaceum* Linn.	Seed	Vol. I	Vol. I	
322.	*Grewia tenax* (Forsk.) Aschers & Schwf.	Stem bark	Vol. II		
323.	*Gymnema sylvestre* R.Br.	Leaf	Vol. V	Vol. II	
324.	*Gymnema sylvestre* R.Br.	Root	Vol. V	Vol. V	Vol. I
325.	*Habenaria intermedia* D.Don	Tuber	Vol. V		
326.	*Hedychium spicatum* Buch.Ham.	Rhizome	Vol. I		
327.	*Helicteres isora* Linn.	Fruit		Vol. III	
328.	*Heliotropium indicum* Linn.	Whole plant	Vol. VI		
329.	*Hemidesmus indicus* R.Br.	Root	Vol. I		Vol. I
330.	*Hibiscus abelmoschus* Linn.	Seed	Vol. IV		
331.	*Hibiscus sabdariffa* Linn.	Root	Vol. III		
332.	*Hiptage benghalensis* Kurz.	Flower	Vol. VI		
333.	*Holarrhena antidysenterica* (Linn.) Wall.	Seed	Vol. III		
334.	*Holarrhena antidysenterica* (Linn.) Wall.	Stem bark	Vol. I		
335.	*Holoptelea integrifolia* Planch	Stem bark	Vol. V		
336.	*Holoptelea integrifolia* Planch.	Fruit	Vol. III		
337.	*Hordeum vulgare* Linn.	Fruit	Vol. II	Vol. VI	

Contd.

S. No.	Medicinal plant species	Morphological part as crude drug	APIRef.	UPIRef.	SPIRef.
338.	*Hordeum vulgare* Linn.	Whole plant	Vol. IV		
339.	*Hydnocarpus pentandra*\ (Buch.-Ham.) Oken	Seed	Vol. VI		Vol. II
340.	*Hyoscyamus niger* Linn.	Seed	Vol. V	Vol. V	Vol. I
341.	*Hyssopus officinalis* Linn.	Flower		Vol. II	
342.	*Illicium verum* Hook F.	Fruit	Vol. VI		Vol. II
343.	*Imperata cylindrica* (Linn.) Beauv	Root	Vol. V		
344.	*Indigofera aspalathoides* Vahl ex DC.	Root & Stem	Vol. VI		
345.	*Indigofera tinctoria* Linn.	Leaf	Vol. II		
346.	*Indigofera tinctoria* Linn.	Root	Vol. II		Vol. I
347.	*Indigofera tinctoria* Linn.	Whole plant	Vol. III		Vol. I
348.	*Inula racemosa* Hook.f.	Root	Vol. IV		
349.	*Ipomoea digitata* Linn.	Root	Vol. V	Vol. V	
350.	*Ipomoea nil* Linn.	Seed		Vol. II	
351.	*Iris ensata* Thunb.	Root		Vol. II	
352.	*Jasminum officinale* Linn.	Leaf	Vol. III	Vol. IV	
353.	*Jatropha glandulifera* Roxb.	Seed	Vol. V		
354.	*Juglans regia* Linn.	Cotyledon	Vol. II		
355.	*Juglans regia* Linn.	Kernel		Vol. IV	
356.	*Juniperus communis* Linn.	Fruit	Vol. III	Vol. IV	
357.	*Lagenaria siceraria* (Mol.) Standl.	Fruit	Vol. III		
358.	*Lallemantia royleana* Benth.	Seed		Vol. VI	
359.	*Lawsonia inermis* Linn.	Leaf	Vol. IV	Vol. II	
360.	*Lens culinaris* Medic.	Seed	Vol. III	Vol. VI	
361.	*Leonotis nepetaefolia* R.Br.	Root	Vol. III		
362.	*Lepidium sativum* Linn.	Seed	Vol. I		
363.	*Leptadenia reticulata* W.& A.	Root	Vol. VI		
364.	*Leucas cephalotes* Spreng.	Whole plant	Vol. II		
365.	*Lilium polyphyllum* D.Don	Tuber	Vol. III		
366.	*Linum usitatissimum* Linn.	Seed	Vol. I	Vol. I	
367.	*Litsea chinensis* Lam.	Stem bark	Vol. V	Vol. V	
368.	*Litsea chinensis* Lam.	Wood	Vol. V		
369.	*Lodoicea maldivica* (Poir.) Pers.	Fruit	Vol. IV		
370.	*Luffa acutangula* (Linn.) Roxb.	Whole plant	Vol. III		
371.	*Lycium barbarum* L.	Aerial Part	Vol. VI		
372.	*Madhuca indica* (Linn.) J.F. Gmel	Flower	Vol. II		
373.	*Malaxis acuminata* D. Don	Tuber	Vol. V		
374.	*Mallotus philippensis* Muell.-Arg.	Fruit	Vol. I	Vol. I	
375.	*Malva sylverstris* Linn.	Fruit		Vol. III	
376.	*Mangifera indica* Linn.	Seed	Vol. III		
377.	*Mangifera indica* Linn.	Stem bark	Vol. III	Vol. IV	
378.	*Maranta arundinacea* Linn.	Rhizome	Vol. VI		
379.	*Marsdenia tenacissima* Wight. & Arn.	Root	Vol. II		
380.	*Martynia annua* Linn.	Seed	Vol. III		
381.	*Matricaria chamomilla* Linn.	Flower		Vol. II	
382.	*Melia azedarach* Linn	Fruit		Vol. III	

Contd.

S. No.	Medicinal plant species	Morphological part as crude drug	APIRef.	UPIRef.	SPIRef.
383.	*Melia azedarach* Linn	Leaf		Vol. III	
384.	*Melia azedarach* Linn.	Stem bark	Vol. IV	Vol. III	
385.	*Mentha viridis* Linn.	Aerial Part	Vol. V	Vol. V	
386.	*Merremia tridentata* (Linn.) Hall.f.	Whole plant	Vol. VI		
387.	*Mesua ferrea* Linn.	Stem	Vol. II	Vol. IV	
388.	*Michelia champaca* Linn.	Flower	Vol. IV		Vol. II
389.	*Mimosa pudica* Linn.	Whole plant	Vol. II		
390.	*Mimusops elengi* Linn.	Ripe fruit		Vol. III	
391.	*Mimusops elengi* Linn.	Flower bud		Vol. III	
392.	*Mimusops kauki* (Linn.) Dub.	Stem bark		Vol. III	
393.	*Mollugo cerviana* Seringe	Whole plant	Vol. VI		Vol. II
394.	*Momordica charantia* Linn.	Fresh fruit	Vol. II	Vol. IV	
395.	*Momordica dioica* Roxb ex willd	Root	Vol. V		
396.	*Monochoria vaginalis* Presl	Rhizome	Vol. VI		Vol. II
397.	*Moringa oleifera* Lam.	Leaf	Vol. II	Vol. V	
398.	*Moringa oleifera* Lam.	Root bark	Vol. IV		
399.	*Moringa oleifera* Lam.	Stem bark			Vol. II
400.	*Moringa oleifera* Lam.	Seed	Vol. IV		
401.	*Moringa oleifera* Lam.	Stem bark	Vol. IV		
402.	*Mucuna pruriens* (Linn.) DC.	Seed			Vol. II
403.	*Mucuna prurita* Hook.	Root	Vol. IV		
404.	*Mucuna prurita* Hook.	Seed	Vol. III	Vol. II, IV	
405.	*Musa paradisiacal* Linn.	Fresh ripe fruit		Vol. III	
406.	*Musa paradisiacal* Linn.	Rhizome		Vol. IV	Vol. II
407.	*Moringa pruriens* (Linn) DC.	Seed		Vol. II	
408.	*Murraya koenigii* (Linn.) Spreng	Leaf	Vol. VI		
409.	*Musa paradisiaca* Linn.	Flower	Vol. IV		
410.	*Musa paradisiaca* Linn.	Fruit	Vol. III		
411.	*Musa paradisiaca* Linn.	Fruit		Vol. III	
412.	*Myrica esculenta* Buch. -Ham. ex D.Don	Fruit	Vol. III	Vol. IV	
413.	*Myrica esculenta* Buch. -Ham. ex D.Don	Stem bark		Vol. III	Vol. II, IV
414.	*Myristica fragrans* Houtt.	Seed	Vol. I	Vol. I	Vol. I
415.	*Myristica fragrans* Houtt.	Arial part		Vol. VI	
416.	*Myroxylon balsamum* Linn.	Trunk			
417.	*Nardostachys jatamansi* DC.	Rhizome	Vol. I		
418.	*Nelumbo nucifera* Gaertn.	Flower	Vol. II		Vol. I
419.	*Nelumbo nucifera* Gaertn.	Rhizome	Vol. III		Vol. I
420.	*Nepeta hindostana* (Roth.) Haines	Seed		Vol. II	
421.	*Nerium indicum* Mill.	Leaf	Vol. I	Vol. I	
422.	*Nerium indicum* Mill.	Root	Vol. III	Vol. IV	
423.	*Nigella sativa* Linn.	Seed	Vol. I	Vol. I	
424.	*Nigella sativa* Linn.	Leaf			Vol. I
425.	*Nymphaea alba* Linn.	Flower	Vol. III		
426.	*Nymphaea stellata* Willd.	Flower	Vol. III		
427.	*Ocimum sanctum* Linn.	Leaf	Vol. II		
428.	*Ocimum sanctum* Linn.	Seed	Vol. IV		

Contd.

S. No.	Medicinal plant species	Morphological part as crude drug	APIRef.	UPIRef.	SPIRef.
429.	*Ocimum trnuiflorum* Linn.	Leaf			Vol. II
430.	*Ocimum sanctum* Linn.	Whole plant	Vol. II	Vol. V	
431.	*Onosma bracteatum* Wall.	Aerial part	Vol. III		
432.	*Onosma bracteatum* Wall.	Leaf		Vol. V	
433.	*Onosma hispidum* Wall. ex. D. Don.	Root		Vol. III	
434.	*Operculina turpethum* (Linn.) Silva Manso	Root	Vol. III	Vol. V	
435.	*Orchis latifolia* Linn.	Tuber		Vol. VI	
436.	*Orchis mascula* Linn.	Tuber		Vol. III	
437.	*Oroxylum indicum* vent.	Root	Vol. III		
438.	*Oryza sativa* Linn.	Fruit	Vol. III		
439.	*Oryza sativa* Linn.	Root	Vol. II		
440.	*Ougeinia oojeinensis* (Roxb.) Hochr	Heart wood	Vol. V		
441.	*Oxalis corniculata* Linn.	Whole plant	Vol. III		Vol. II
442.	*Paederia foetida* Linn.	Whole plant	Vol. II		
443.	*Paeonia emodi* Wall.	Tuber		Vol. III	
444.	*Paganum harmala* Linn.	Seed		Vol. II	
445.	*Pandanus odoratissimus* Linn.	Root	Vol. VI		Vol. II
446.	*Pandanus odoratissimus* Linn. f.	Male flower		Vol. III	
447.	*Pandanus tectorius* Soland ex Parkinson	Root	Vol. I		
448.	*Papaver somniferum* Linn.	Seed	Vol. V	Vol. II	Vol. I
449.	*Parmelia perlata* (Huds.) Ach.	Whole plant	Vol. III	Vol. V	
450.	*Paspalum scrobiculatum* Linn.	Grain	Vol. V		
451.	*Pavetta indica* L.	Root	Vol. VI		
452.	*Pavonia odorata* Willd.	Whole plant	Vol. VI		
453.	*Pennisetum typhoides* (Burm.f.) Stapf & C.E. Hubb.	Leaf base	Vol. VI		
454.	*Pergularia daemia* (Forsk) Chiov.	Whole plant	Vol. VI		
455.	*Peristrophe bicalyculata* Nees	Root	Vol. III		
456.	*Peristrophe bicalyculata* Nees	Seed	Vol. V	Vol. V	
457.	*Phaseolus radiatus* Linn.	Seed	Vol. III		
458.	*Phoenix dactylifera* Linn.	Fruit	Vol. IV	Vol. VI	Vol. II
459.	*Phoenix dactylifera* Linn.	Fresh fruit	Vol. IV		Vol. II
460.	*Phyla nodiflora* (Linn.) Greene	Whole plant	Vol. V		
461.	*Phyllanthus amarus* Schum & Thonn.	Whole plant			Vol. I
462.	*Phyllanthus embelica* Linn.	Fruit			Vol. I
463.	*Phyllanthus embelica* Linn.	Fresh fruit			Vol. I
464.	*Phyllanthus fraternus* Webst.	Root, Leaf & Stem	Vol. I		
465.	*Phyllanthus amarus* Schum. & Thonn.	Whole plant			
466.	*Physalis alkekengi* Linn.	Fruit	Vol. V	Vol. V	
467.	*Picrorhiza kurroa* Royle ex Benth.	Rhizome	Vol. II	Vol. IV	Vol. I
468.	*Pimpinelia anisum* Linn.	Fruit	Vol. V	Vol. II	
469.	*Pinus gerardiana* Wall	Kernel	Vol. VI	Vol. VI	
470.	*Pinus longifolia* Roxb.	Exudate	Vol. V	Vol. V	
471.	*Pinus roxburghii* Sargent.	Heart wood	Vol. III		
472.	*Pinus roxburghii* Sargent.	Root	Vol. III	Vol. V	
473.	*Piper betle* Linn.	Leaf	Vol. III		Vol. II

Contd.

S. No.	Medicinal plant species	Morphological part as crude drug	APIRef.	UPIRef.	SPIRef.
474.	*Piper cubeba* Linn. f.	Fruit	Vol. I	Vol. I	
475.	*Piper longum* Linn.	Fruit	Vol. IV		Vol. I
476.	*Piper longum* Linn.	Stem	Vol. II		
477.	*Piper nigrum* Linn.	Fruit	Vol. III	Vol. IV	Vol. I
478.	*Piper retrofractum* Vahl.	Stem	Vol. II		
479.	*Piper retrofractum* Vahl.	Fruit			
480.	*Pistacia lentiscus* Linn.	Leaf	Vol. I		
481.	*Pistacia lentiscus* Linn.	Resin	Vol. V	Vol. V	
482.	*Pistacia vera* Linn.	Seed		Vol. III	
483.	*Pistia stratiotes* Linn.	Whole plant	Vol. VI		
484.	*Pastinaca secacul* Linn.	Rhizome		Vol. III	
485.	*Plantago lanceolata* L.	Leaf	Vol. VI		
486.	*Plantago major* L.	Seed		Vol. II	
487.	*Plantago ovata* Forssk.		Husk		
488.	*Plantago ovata* Forssk.	Seed		Vol. II	
489.	*Plectranthus amboinicus* Lour.	Leaf			Vol. II
490.	*Pluchea lanceolata* C.B.Clarke	Leaf	Vol. III		
491.	*Plumbago indica* Linn.	Root	Vol. VI		
492.	*Plumbago zeylanica* Linn.	Root	Vol. I	Vol. I	
493.	*Polygonatum cirrhifolium* Royle	Rhizome	Vol. VI		
494.	*Polygonatum cirrhifolium* Royle	Rhizome	Vol. V		
495.	*Polygonum bistorta* Linn.	Rhizome		Vol. III	
496.	*Polypodium vulgare* Linn.	Rhizome		Vol. II	
497.	*Pongamia pinnata* Pierre	Leaf	Vol. II	Vol. IV	
498.	*Pongamia pinnata* Pierre	Root	Vol. II	Vol. IV	
499.	*Pongamia pinnata* Pierre	Root bark	Vol. II	Vol. IV	
500.	*Pongamia pinnata* Pierre	Seed	Vol. I	Vol. I	Vol. I
501.	*Pongamia pinnata* Pierre	Stem bark	Vol. II	Vol. IV	Vol. I
502.	*Pongamia pinnata* Pierre	Leaf		Vol. IV	
503.	*Portulaca oleracea* Linn.	Whole plant	Vol. II	Vol. IV	
504.	*Prosopis cineraria* Druce	Leaf	Vol. VI		Vol. II
505.	*Prunus Amygdalus* Batsch..	Seed		Vol. II	
506.	*Prunus avium* Linn.	Root	Vol. VI		
507.	*Prunus avium* Linn.	Stem bark	Vol. VI		
508.	*Prunus avium* Linn.	Seed	Vol. V		
509.	*Prunus cerasoides* D.Don	Heart wood	Vol. III		
510.	*Psoralea corylifolia* Linn.	Fruit	Vol. I	Vol. I	Vol. I
511.	*Pterocarpus antalinus* Linn.	Heart wood		Vol. V	
512.	*Pterocarpus marsupium* Roxb.	Heart wood	Vol. I		
513.	*Pterocarpus marsupium* Roxb.	Stem bark	Vol. III		
514.	*Pterocarpus santalinus* Linn.f.	Heart wood	Vol. III	Vol. V	Vol. II
515.	*Pueraria tuberosa* DC.	Tuber	Vol. II		
516.	*Pueraria tuberosa* DC.	Tuber	Vol. V		
517.	*Punica granatum* Linn.	Fruit	Vol. IV		Vol. II
518.	*Punica granatum* Linn.	Fruit rind	Vol. IV		Vol. II
519.	*Punica granatum* Linn.	Leaf	Vol. IV	Vol. II	

Contd.

S. No.	Medicinal plant species	Morphological part as crude drug	APIRef.	UPIRef.	SPIRef.
520.	*Punica granatum* Linn.	Seed	Vol. II	Vol. IV, VI	
521.	*Quercus infectoria* Olive.	Gall	Vol. IV	Vol. III	Vol. II
522.	*Raphanus sativus* Linn.	Root	Vol. II	Vol. V	
523.	*Raphanus sativus* Linn.	Seed	Vol. III	Vol. V	
524.	*Raphanus sativus* Linn.	Whole plant	Vol. II		
525.	*Rauvolfia serpentina* Benth.ex Kurz	Root	Vol. V	Vol. V	
526.	*Rheum emodi* Wall.	Root		Vol. II	
527.	*Rhus coriaria* Linn.	Pericarp		Vol. III	
528.	*Rhus parviflora* Roxb.	Aerial Part	Vol. V		
529.	*Ricinus communis* Linn	Leaf	Vol. III		
530.	*Ricinus communis* Linn.	Root	Vol. I		
531.	*Ricinus communis* Linn.	Seed	Vol. III	Vol. IV	
532.	*Ricinus communis* Linn.	Oil			
533.	*Rosa centifolia* Linn.	Flower	Vol. III	Vol. IV	
534.	*Rosa damascene* Mill.	Flower		Vol. III	
535.	*Rubia cordifolia* Linn.	Stem	Vol. III	Vol. IV	
536.	*Rumex maritimus* Linn.	Seed		Vol. II	
537.	*Rumex vesicarius* Linn.	Seed		Vol. III	
538.	*Ruta graveolens* Linn.	Leaf		Vol. VI	
539.	*Saccharum bengalense* Retz.	Root	Vol. III		
540.	*Saccharum officinarum* Linn.	Jaggery	Vol. VI		
541.	*Saccharum officinarum* Linn.	Root stock	Vol. IV		
542.	*Saccharum officinarum* Linn.	Stem	Vol. II		Vol. II
543.	*Saccharum spontaneum* Linn.	Root stock	Vol. III		
544.	*Salix alba* Linn.	Leaf	Vol. VI		
545.	*Salix caprea* Linn.	Flower		Vol. II	
546.	*Salmalia malabarica* (DC.) Schott. & Endl.	Gum		Vol. II	
547.	*Salvadora persica* Linn.	Fruit	Vol. V	Vol. V	
548.	*Salvadora persica* Linn.	Leaf	Vol. V	Vol. V	
549.	*Salvadora persica* Linn.	Root bark	Vol. V	Vol. V	
550.	*Salvia haematodes* Linn.	Root		Vol. III	
551.	*Santalum album* Linn.	Heart wood	Vol. III	Vol. VI	Vol. II
552.	*Saraca asoca* (Roxb.) De Willd.	Stem bark	Vol. I		
553.	*Saussurea lappa* C.B. Clarke	Root	Vol. I	Vol. I	
554.	*Saussurea costus* (Falc.) Lipsch.	Root			Vol. I
555.	*Scindapsus officinalis* Schott.	Fruit	Vol. II		
556.	*Scirpus kysoor* Roxb.	Rhizome	Vol. I		
557.	*Selinum candollei* DC.	Root	Vol. II		
558.	*Selinum vaginatum* C.B. Clarke	Fruit	Vol. VI		
559.	*Selinum vaginatum* C.B. Clarke	Rhizome	Vol. VI		
560.	*Semecarpus anacardium* Linn.f	Fruit	Vol. II	Vol. II, IV	Vol. II
561.	*Sesamum indicum* Linn.	Oil	Vol. VI		
562.	*Sesamum indicum* Linn.	Seed	Vol. IV	Vol. II, IV	Vol. II
563.	*Sesbania bispinosa* W.F.Wight	Root	Vol. V		
564.	*Sesbania bispinosa* W.F.Wight	Stem	Vol. V		
565.	*Sesbania sesban* (Linn.) Merr.	Leaf	Vol. II		Vol. I

Contd.

S. No.	Medicinal plant species	Morphological part as crude drug	APIRef.	UPIRef.	SPIRef.
566.	*Shorea robusta* Gaertn f.	Heart wood	Vol. VI		
567.	*Sida rhombifolia* Linn.	Root	Vol. III		
568.	*Sisvambrium irio* Linn.	Seed	Vol. V	Vol. V	
569.	*Smilax china* Linn.	Tuber	Vol. V	Vol. V	Vol. I
570.	*Solanum anguivi* Lam.	Whole plant	Vol. VI		
571.	*Solanum indicum* Linn.	Root	Vol. II		
572.	*Solanum nigrum* Linn.	Whole plant	Vol. II	Vol. IV	
573.	*Solanum surattense* Burm.f.	Whole plant	Vol. I		Vol. I
574.	*Solanum surattense* Burm.f.	Shoot		Vol. II	
575.	*Sphaeranthus indicus* Linn.	Leaf	Vol. III		
576.	*Sphaeranthus indicus* Linn.	Whole plant	Vol. IV		
577.	*Sphaeranthus indicus* Linn.	Inflorescence		Vol. II	
578.	*Spondias pinnata* (Linn.f.) Kurz.	Stem	Vol. III		
579.	*Spondias pinnata* (Linn.f.) Kurz.	Stem bark	Vol. II		
580.	*Stereospermum chelonoides* (L.F.) DC.	Stem bark	Vol. IV		
581.	*Stereospermum chelonoides* (L.F.) DC.	Root			Vol. I
582.	*Stereospermum suaveolens* DC.	Root	Vol. III		
583.	*Streblus asper* Lour.	Stem bark	Vol. III		
584.	*Strychnos nux-vomica* Linn.	Seed	Vol. IV	Vol. II	Vol. II
585.	*Strychnos potatorum* Linn.	Seed	Vol. IV		Vol. II
586.	*Swertia chirayita* (Roxb.ex Flem) Karst.	Whole plant	Vol. I	Vol. I	
587.	*Symplocos racemosa* Roxb.	Stem bark	Vol. I	Vol. I	
588.	*Synantherias syeatica* Schott Gen.Aocja	Tuber	Vol. VI		
589.	*Syzygium aromaticum* (Linn.) Merr. & L.M. Perry	Bud	Vol. I		Vol. I
590.	*Syzygium aromaticum* Merril & Perry	Oil	Vol. VI		
591.	*Syzygium aromaticum* Merril & Perry	Bud		Vol. I	
592.	*Syzygium cuminii* (Linn.) Skeels	Seed	Vol. II	Vol. IV	
593.	*Syzygium cuminii* (Linn.) Skeels	Stem bark	Vol. II	Vol. IV	Vol. II
594.	*Tamarindus indica* Linn.	Fruit	Vol. IV	Vol. VI	Vol. II
595.	*Tamarix gallica* Linn.	Gall		Vol. VI	
596.	*Taxus baccata* Linn.	Leaf	Vol. III		
597.	*Taxus wallichiana* Zucc.	Leaf			Vol. II
598.	*Tecomella undulata* G. Don.	Stem bark	Vol. VI		
599.	*Tectona grandis* Linn. f.	Heart wood	Vol. III		
600.	*Teramnus labialis* Spreng.	Whole plant	Vol. III		
601.	*Terminalia arjuna* (Roxb) W. & A.	Stem bark	Vol. II	Vol. IV	Vol. I
602.	*Terminalia bellirica* Roxb.	Fruit	Vol. I	Vol. I	Vol. I
603.	*Terminalia chebula* Retz.	Fruit	Vol. I	Vol. I	Vol. I
604.	*Terminalia chebula* Retz.	Fruit extract	Vol VIII		
605.	*Thespesia populnea* Soland ex Correa	Stem bark	Vol. V		Vol. I
606.	*Tinospora cordifolia* (Willd.) Miers ex Hook.f. & Thoms.	Stem	Vol. I	Vol. I	Vol. I
607.	*Toddalia asiatica* (Linn.) Lam.	Whole plant	Vol. VI		
608.	*Trachyspermum ammi* (Linn.) Sprague	Fruit	Vol. I	Vol. VI	Vol. II
609.	*Tragia involucrata* Linn.	Whole plant	Vol. IV		
610.	*Trapa natans* Linn.	Seed	Vol. IV		

Contd.

S. No.	Medicinal plant species	Morphological part as crude drug	APIRef.	UPIRef.	SPIRef.
611.	*Trianthema decandra* Linn.	Whole plant	Vol. VI		
612.	*Trianthema portulacastrum* Linn.	Root	Vol. IV		
613.	*Tribulus terrestris* Linn.	Whole plant	Vol. VI		Vol. II
614.	*Tribulus terrestris* Linn.	Fruit	Vol. I	Vol. I	Vol. I
615.	*Tribulus terrestris* Linn.	Root	Vol. I		Vol. I
616.	*Trichosanthas bracteata* (Lam.) Voigt	Root	Vol. V		
617.	*Trigonella foenum-graecum* Linn.	Seed	Vol. II	Vol. II, IV	Vol. I
618.	*Triticum aestivum* Linn.	Seed		Vol. VI	
619.	*Typha australis* Schum. & Thonn.	Root	Vol. V		
620.	*Typha elephantina* Roxb.	Root	Vol. V		
621.	*Typhonium trilobatum* Schott	Tuber	Vol. VI		
622.	*Uraria picta* Desv.	Whole plant	Vol. IV		
623.	*Valeriana wallichii* DC.	Rhizome	Vol. I	Vol. I	
624.	*Vallaris solanacea* Kuntze	Root	Vol. V		
625.	*Vateria indica* Linn.	Exudate	Vol. IV	Vol. VI	
626.	*Vernonia cinerea* Lees.	Whole plant	Vol. III		
627.	*Vernonia anthelmintica* (Linn.) Willd.	Fruit			Vol. I
628.	*Vetiveria zizanioides* (Linn.) Nash	Root	Vol. III	Vol. IV	Vol. II
629.	*Vicia faba* Linn.	Seed		Vol. II	
630.	*Vigna trilobata* (Linn.) Verdc.	Whole plant	Vol. IV		
631.	*Vigna unguiculata* (Linn.) Walp.	Seed	Vol. I	Vol. I	Vol. I
632.	*Viola odorata* Linn.	Flower		Vol. II	
633.	*Viola pilosa* Blume.	Leaf		Vol. III	
634.	*Vitex negundo* Linn.	Fruit	Vol. V	Vol. III	
635.	*Vitex negundo* Linn.	Leaf	Vol. III	Vol. V	Vol. II
636.	*Vitex negundo* Linn.	Root	Vol. IV		
637.	*Vitex negundo* Linn.	Fruit		Vol. III	
638.	*Vitis vinifera* Linn.	Fruit	Vol. III	Vol. III, VI	Vol. II
639.	*Wedelia calendulacea* Lees.	Whole plant	Vol. VI		
640.	*Wedelia chinensis* Merril.	Aerial part			Vol. II
641.	*Withania somnifera* Dunal	Root	Vol. I	Vol. I	Vol. I
642.	*Withania somnifera* Dunal	Root extract			
643.	*Woodfordia fruticosa* Kurz	Flower	Vol. I		
644.	*Wrightia tinctoria* Syn.*Nerium tinctorium* Roxb.	Seed		Vol. II	
645	*Xeromphis spinosa* Keay	Fruit	Vol. I	Vol. I	
646	*Zanthoxylum armatum* DC.	Fruit	Vol. IV		
647.	*Zanthoxylum armatum* DC.	Stem bark	Vol. II		
648.	*Zingiber officinale* Rosc.	Rhizome	Vol. II	Vol. IV	Vol. I
649.	*Zingiber officinale* Rosc.	Rhizome	Vol. I		
650.	*Ziziphus jujuba* Mill.	Fruit pulp	Vol. III		
651.	*Ziziphus jujuba* Mill.	Stem bark	Vol. III		
652.	*Ziziphus mauritiana* Lam.	Fruit bark		Vol. VI	
653.	*Ziziphus sylopyra* Willd.	Fruit	Vol. V		

Abbreviations: API: Ayurvedic Pharmacopoeia of India; UPI: Unani Pharmacopoeia of India; SPI: Siddha Pharmacopoeia of India

Non-Pharmacopoeial Publication on Quality Standards

Pharmacopoeial publications viz. The Ayurvedic Pharmacopoeia of India (API), Unani Pharmacopoeia of India (UPI) and Siddha Pharmacopoeia of India (SPI) are Government of India publications and published by officially constituted respective Pharmacopoeial committees. These pharmacopoeia are duly accredited in Drugs & Cosmetics Act & Rules (Anonymous 1940) under schedule - I for regulatory purposes. Besides these, following publications were also titled with 'Pharmacopoeia' or 'Pharmacopoeial' or 'Quality Standards' words. These publications do not come under the provisions of regulatory framework, but may serve as good reference for quality control purposes.

1. Pharmacopoeial standards for ayurvedic formulations

This publication was brought out by Central Council for Research in Ayurveda & Siddha (an autonomous organization under Department of AYUSH, Ministry of Health & Family Welfare) New Delhi comprising standards 431 compound formulations of Ayurveda. The formulations for standardisation are based on Ayurvedic Formulary of India, Part- I. This publication is pioneer publication on the standardisation of classical Ayurvedic Formulations (Anonymous 1987).

2.Indian herbal pharmacopoeia

This joint publication of Regional Research Laboratory (Council of Scientific & Industrial Research) Jammu Tawi and Indian Drug Manufacturers' Association, Mumbai is published in two volumes. Volume first provide the quality control standards for 20 medicinal plants and second volume also covers detailed standards on another 20 medicinal and aromatic plants (Anonymous,1998b,1999c). A combined and revised volume comprising 52 monographs were also published by Indian Drug Manufacturers' Association, Mumbai (Anonymous, 2002). This publication is worth in content and is desk reference for a quality control laboratory.

3. Pharmacopoeial standards of herbal plants

This publication is mere compilation of available data and information under a format by (Karnick 1994) in two volumes. First Volume comprises 79 plants and second volume covers 151 medicinal plants.

4. Quality standards of Indian medicinal plants

These series publications (nine volumes published so far) on standards of medicinal plants comprise quality standards on 309 medicinal plants by Indian Council of Medical Research (ICMR). Each monograph consists of details on nomenclature, description (macroscopic, microscopic and powder), TLC identity test, assay /analytical methods, quantitative standards, adulterants/substitutes, pharmacology, therapeutic category, safety aspects, dosage and references. Monographs are followed by relevant appendices providing protocols for Evaluation of crude drugs, phytochemical evaluation, isolation of markers, drying and storage, pesticide, residues and microbial contamination (Anonymous 2003, 2005a, 2005b, 2006b, 2008b, 2008c, 2008d, 2010b, 2011)

5. Quality assessment of selected Indian medicinal plants

The joint publication of National Medicinal Plants Board, New Delhi and National Remedies Pvt. Ltd., Bangalore is outcome of project on development parameters for selected Indian Medicinal plant. Current publication is first volume and comprises analytical specification for ten medicinal plants (Agarwal and Murali 2010).

A quality control laboratory should refer the pharmacopoeia not only for the compliance of regulatory provisions but also to maintain in-house standards too. In a pharmaceutical concern where plant drugs are used for formulations, an inventory of such plant drugs should be prepared and check with different pharmacopoeia so as to ascertain the pharmacopoeial standards thereof. Since Pharmacopoeial standards are specific, consistent and reproducible, should invariably be practiced for quality control purposes.

References

Agarwal A, Balasubramanian M 2010. Quality Assessment of Selected Indian Medicinal Plants. NMBP, New Delhi and Natural Remedies Pvt Ltd, Bangalore.

Anonymous 1940. Drugs & Cosmetics Act and Rules. Govt. of India, Ministry of Health & Family Welfare, New Delhi.

Anonymous 1955. Pharmacopoeia of India. Second ed. Manager of Publications, Govt. of India New Delhi.

Anonymous 1978. The Ayurvedic Formulary of India. Part-I, (English ed.) Govt. of India, Ministry of Health & Family Welfare, New Delhi.

Anonymous 1981. National Formulary of Unani Medicine. Part-I, (English ed.), Govt. of India, Ministry of Health & Family Welfare, New Delhi.

Anonymous 1984. Siddha Formulary of India. Part-I, (Tamil ed.), Govt. of India, Ministry of Health & Family Welfare, New Delhi.

Anonymous 1986. The Ayurvedic Pharmacopoeia of India. Part- I, Volume–I, First edition, Govt. of India, Ministry of Health & Family Welfare, New Delhi.

Anonymous 1987. Pharmacopoeial Standards for Ayurvedic Formulations. Central Council for Research in Ayurveda and Siddha, New Delhi.

Anonymous 1992.Siddha Formulary of India. Part-I, (English ed.), Govt. of India, Ministry of Health & Family Welfare, New Delhi

Anonymous 1998a. The Unani Pharmacopoeia of India. Part-I, Vol.-I, Govt. of India, Ministry of Health & Family Welfare, New Delhi.

Anonymous 1998b. Indian Herbal Pharmacopoeia. Vol.-I, R.R.L., Jammu & I.D.M.A., Mumbai.

Anonymous 1999a. National Formulary of Unani Medicine. Part-II, (English ed.), Govt. of India, Ministry of Health & Family Welfare, New Delhi.

Anonymous 1999b. The Ayurvedic Pharmacopoeia of India. Part- I, Volume–II, First edition, Govt. of India, Ministry of Health & Family Welfare, New Delhi.

Anonymous 1999c. Indian Herbal Pharmacopoeia. Vol.-II, R.R.L., Jammu & I.D.M.A., Mumbai.

Anonymous 2000. The Ayurvedic Formulary of India. Part-II, (English ed.) Govt. of India, Ministry of Health & Family Welfare, New Delhi.

Anonymous 2001a. National Formulary of Unani Medicine. Part-III, (English ed.), Govt. of India, Ministry of Health & Family Welfare, New Delhi.

Anonymous 2001b. The Ayurvedic Pharmacopoeia of India. Part- I, Volume–III, First edition, Govt. of India, Ministry of Health & Family Welfare, New Delhi.

Anonymous 2002. Indian Herbal Pharmacopoeia. Rev.ed. I.D.M.A., Mumbai.

Anonymous 2003. Quality Standards of Indian Medicinal Plants. Vol. I, Indian Council of Medical Research, New Delhi.

Anonymous 2004. The Ayurvedic Pharmacopoeia of India. Part- I, Volume–IV, First edition, Govt. of India, Ministry of Health & Family Welfare, New Delhi.

Anonymous 2005a. Quality Standards of Indian Medicinal Plants. Vol. II, Indian Council of Medical Research, New Delhi.

Anonymous 2005b. Quality Standards of Indian Medicinal Plants. Vol. III, Indian Council of Medical Research, New Delhi.

Anonymous 2006a. National Formulary of Unani Medicine. Part-IV (English ed.), Govt. of India, Ministry of Health & Family Welfare, New Delhi.

Anonymous 2006b. Quality Standards of Indian Medicinal Plants. Vol. IV, Indian Council of Medical Research, New Delhi.

Anonymous 2006c. The Ayurvedic Pharmacopoeia of India. Part- I, Volume–V, First edition, Govt. of India, Ministry of Health & Family Welfare, New Delhi.

Anonymous 2007a. The Unani Pharmacopoeia of India. Part-I, Vol.-II, Govt. of India, Ministry of Health & Family Welfare, New Delhi.

Anonymous 2007b. The Unani Pharmacopoeia of India. Part-I, Vol.-III, Govt. of India, Ministry of Health & Family Welfare, New Delhi.

Anonymous 2007c. The Unani Pharmacopoeia of India. Part-I, Vol.-IV, Govt. of India, Ministry of Health & Family Welfare, New Delhi.

Anonymous 2008a. National Formulary of Unani Medicine. Part-V (English ed.), Govt. of India, Ministry of Health & Family Welfare, New Delhi.

Anonymous 2008b. Quality Standards of Indian Medicinal Plants. Vol. V, Indian Council of Medical Research, New Delhi.

Anonymous 2008c. Quality Standards of Indian Medicinal Plants. Vol. VI, Indian Council of Medical Research, New Delhi.

Anonymous 2008d. Quality Standards of Indian Medicinal Plants. Vol.VII, Indian Council of Medical Research, New Delhi.

Anonymous 2008e. The Ayurvedic Pharmacopoeia of India. Part- I, Volume–VI, First edition, Govt. of India, Ministry of Health & Family Welfare, New Delhi.

Anonymous 2008f. The Ayurvedic Pharmacopoeia of India. Part- I, Volume–VII, First edition, Govt. of India, Ministry of Health & Family Welfare, New Delhi.

Anonymous 2008g. The Ayurvedic Pharmacopoeia of India. Part- I, Volume–VIII, First edition, Govt. of India, Ministry of Health & Family Welfare, New Delhi.

Anonymous 2008h. The Ayurvedic Pharmacopoeia of India. Formulation, Part- II, Volume–I, First edition, Govt. of India, Ministry of Health & Family Welfare, New Delhi.

Anonymous 2008i. The Siddha Pharmacopoeia of India. Part-I, Vol.-I, Govt. of India, Ministry of Health & Family Welfare, New Delhi

Anonymous 2008j. The Unani Pharmacopoeia of India. Part-I, Vol.-V, Govt. of India, Ministry of Health & Family Welfare, New Delhi.

Anonymous 2009a. The Ayurvedic Pharmacopoeia of India. Formulation, Part- II, Vol.–II, First edition, Govt. of India, Ministry of Health & Family Welfare, New Delhi.

Anonymous 2009b. TLC Atlas of Pharmacopoeial Drugs. Part I, Volume–I, First edition, Govt. of India, Ministry of Health & Family Welfare, New Delhi.

Contd.

Anonymous 2009c. Macroscopy and Microscopy of API Drugs. Part-I, Volume–V, First edition, Govt. of India, Ministry of Health & Family Welfare, New Delhi.

Anonymous 2009d. The Unani Pharmacopoeia of India. Part-I, Vol.-VI, Govt. of India, Ministry of Health & Family Welfare, New Delhi.

Anonymous 2009e. The Unani Pharmacopoeia of India. Formulations. Part-II, Vol.-I, Govt. of India, Ministry of Health & Family Welfare, New Delhi.

Anonymous 2010a. Pharmacopoeia of India. Sixth ed. Manager of Publications, Govt. of India New Delhi.

Anonymous 2010a. The Ayurvedic Pharmacopoeia of India. Formulation, Part- II, Volume–III, First edition, Govt. of India, Ministry of Health & Family Welfare, New Delhi.

Anonymous 2010b. Quality Standards of Indian Medicinal Plants. Vol. VIII, Indian Council of Medical Research, New Delhi.

Anonymous 2010c. The Siddha Pharmacopoeia of India. Part-I, Vol.-I, Govt. of India, Ministry of Health & Family Welfare, New Delhi.

Anonymous 2010d. The Unani Pharmacopoeia of India. Formulations. Part-II, Vol.-II, Govt. of India, Ministry of Health & Family Welfare, New Delhi.

Anonymous 2011. Quality Standards of Indian Medicinal Plants. Vol. IX, Indian Council of Medical Research, New Delhi.

Anonymous 2011.The Siddha Formulary of India. Part-II, (Tamil ed.) Govt. of India, Ministry of Health & Family Welfare, New Delhi.

Karnick CR 1994. Pharmacopoeial Standards of Herbal Plants. Vol I & II. Sri Satguru Publications, Delhi.

Singh Harikishan 1994. Pharmacopoeias and Formularies. Vallabh Prakashan, Delhi.

FSC
www.fsc.org
MIX
Papier | Fördert
gute Waldnutzung
FSC® C083411

Zeitfracht Medien GmbH
Ferdinand-Jühlke-Straße 7
99095 Erfurt, Deutschland
produktsicherheit@kolibri360.de